AF615957

MULTIWAVELENGTH COSMOLOGY

ASTROPHYSICS AND SPACE SCIENCE LIBRARY

VOLUME 301

MULTIWAVELENGTH COSMOLOGY

Proceedings of the "Multiwavelength Cosmology" Conference, held on Mykonos Island, Greece, 17–20 June, 2003

Edited by

MANOLIS PLIONIS

Institute of Astronomy & Astrophysics,
National Observatory of Athens, Greece
and
Instituto Nacional de Astrofísica Optica y Electrónica,
Puebla, Mèxico

KLUWER ACADEMIC PUBLISHERS
DORDRECHT / BOSTON / LONDON

A C.I.P. Catalogue record for this book is available from the Library of Congress.

ISBN 1-4020-1971-8 (HB)
ISBN 0-306-48570-2 (e-book)

Published by Kluwer Academic Publishers,
P.O. Box 17, 3300 AA Dordrecht, The Netherlands.

Sold and distributed in North, Central and South America
by Kluwer Academic Publishers,
101 Philip Drive, Norwell, MA 02061, U.S.A.

In all other countries, sold and distributed
by Kluwer Academic Publishers,
P.O. Box 322, 3300 AH Dordrecht, The Netherlands.

Printed on acid-free paper

Printed in the Netherlands.

Contents

PREFACE

Manolis Plionis
Intitute of Astronomy & Astrophysics, National Observatory of Athens, Greece
Instituto Nacional de Astrofísica, Optica y Electrónica, Mèxico
mplionis@astro.noa.gr

Cosmology, the science of dealing with the origin, nature and nowadays the evolution of the Universe, is as old as human civilisation. It is to the human nature the search for an understanding and comprehension of the mysteries of the Universe within which we live and flourish, unfortunately however which we also destroy and kill for vanity and profit.
The recent scientific efforts have brought a revolution on our understanding of the *Cosmos*. Amazing results is the outcome of amazing experiments! The huge scientific, technological & financial effort that has gone into building the 10m class telescopes as well as many space and balloon observatories, essential to observe the multitude of cosmic phenomena in their manifestations at different wavelengths, from γ-rays to the millimetre and the radio, has given and is still giving its fruits of knowledge.
Many of the recent scientific achievements in Observational and Theoretical Cosmology were presented in our "Multiwavelength Cosmology" conference, that took place in the beautiful Mykonos island of the Aegean between the 17th and the 20th of June 2003. More than 180 Cosmologists from all over the world gathered for a four-day intense meeting in which recent results from large ground based surveys (AAT/2-df, SLOAN) and space missions (WMAP, Chandra, XMM, ISO, HST) were presented and debated, providing a huge impetus to our knowledge of the *Cosmos*. The conference was devoted mostly on the constraints on Cosmological models and galaxy formation theories that arise from the study of the high-z Universe, of clusters of galaxies and their evolution, of the cosmic microwave background, the large-scale structure and star-formation history.
This conference would not have been possible to realize if not for the support of the Instituto Nacional de Astrofísica, Optica y Electrónica of Mèxico, of the National Observatory of Athens and the Municipality of Mykonos island and especially of its Mayor, Mr. Christos Veronis.

M. Plionis (ed.), Multiwavelength Cosmology, 1-2.

Many thanks are due to the scientific organizing committee, especially C. Bennet, H. Böhringer, C. Canizares, C. Frenk, A. Fabian, K. Gorski, J. Peebles, P. Rosati and M. Postman, the local organizing committee (T. Akylas, S. Basilakos, A. Georgakakis, I. Georgantopoulos, O.Giannakis, S. Kitsionas, V. Kolokotronis), our segretaries (O. Koumentakou and E. Parigori), the representative of Mykonos Municipal Cultural agency (DEPAM), Ms. Annita Gianopoulou and Mr. Nondas Ntardas for his valuable Mykonos nightlife advise. Also I would like to thank the staff of Kluwer for their kind cooperation and understanding and for their efforts that resulted in the fine appearance of this volume.

I would like to end this brief introduction with some thoughts regarding these troubled times. Although it is long since the human race has risen from the subconscious, most of its actions are still dictated by instincts and greed. We have come a long way only to realize that we still have to go an even longer way towards consciousness, humanity and respect for the different. I often wonder if we are the same race that has had the amazing ability and imagination to build the devices to observe the Universe back in time, as it was just a few hundred thousands years after it was born, with the race that invades, loots and kills for apparently no good reason, except for a malicious drive for ever bigger profits. The year 2003, when our "Multiwavelength Cosmology" meeting took place, will be remembered not only for the scientific advancements of the human race but probably more for the "criminal" behaviour from part of the world's leadership.

A FEW WORDS OF WELCOME

Christos Goudis
Director, Institute of Astronomy & Astrophysics, National Observatory of Athens, Lofos Nymfon, Thesio 11810, Athens, Greece

Whenever I am confronted with an international astronomical conference, symposium or assembly, I always recall scenes from the wandering scholars of the past and some verses of our Greek poet Kostis Palamas, relevant to the occasion:

And the gypsies came, those who know the whereabouts of the planets and all the secrets of the stars, those who talk to the stars, and who by gazing at those foresee lives, loves, densities, deaths.

And the gypsies came, the learned, the thoughtful, the absorbed in the explanation of the unexplained...

It is exactly this "explanation of the unexplained" which occupies the human mind since culture and civilization started characterizing collectively the human beings.

Cosmology, being a Greek word, and I dare say, a Greek concept - since they were Greek Physicists, mainly from Ionia, who firstly thought of principal elements and intrinsic relations underlying and transforming the structure and evolution of the Cosmos- has something from the classical Greek spirit concerning the necessity of pure knowledge.

It was Aristotle who shaped it briefly in the phrase:

«Ο άνθρωπος του ειδέναι ορέγεται φύσει»

"man is tempted to knowledge by his own nature", although we all know, eating from the tree of knowledge sometimes has lethal consequences.

Indeed if we believe in Paradise we are for sure then on the descending, in a free fall path, decaying through the miseries and the sins of knowledge. If however we have lost the Paradise from our sight, we might think of ourselves as the mariners of space, searching for meaning in our brief lives within the vast and mysterious extend of the universe.

Is the universe an enormous cosmic firework so dear in our imagination, giving rise to a sequential chain of multicolor fireworks, behaving in fact like a cosmic fractal, or is it a cage, the bars of which we cannot see. Or is it our efforts that

M. Plionis (ed.), Multiwavelength Cosmology, 3-4.

make them invisible, by deepening our understanding, broadening our horizons and gradually pushing further away the limits of our ignorance ?
Cosmology has something to say about it, that's for sure. And it is your particular endeavour of observational cosmology that makes the whole effort real, indeed establishing the very foundations of reality around us, which we are obliged to accept and confront.
Is however reality continuously under construction? Are the messages we are receiving from the universe clear enough to decipher its quintessence. Will it ever be possible for entities like us, insignificant manifestations of the Cosmos to perceive the origin and workings of our Creator ? Or even the best ideas we could ever generate are bound to be contradictory ? But then again, the Universe itself might be a Contradiction *per se*. I can sense it through the spiritual voice of that great American, that inflationary figure of Walt Whitman, big banging to us:

Do I contradict myself '?
Very well then, I contradict myself.
(I am large. I contain multitudes.).

At least some of these multitudes we are now able, thanks to your quest, to approach, understand and comprehend.
We are grateful to you for participating in this conference and we wish you further successes in your multi-wavelength drive to enlighten us all. Welcome and thank you again.

OPTICAL WAVELENGTHS

DEEP REDSHIFT SURVEYS: THE VIMOS VLT DEEP SURVEY (*INVITED*)

O. Le Fevre[1], G. Vettolani[2], D. Maccagni[3], J.P. Picat[4], C. Adami[1], M. Arnaboldi[5], S. Bardelli[10], D. Bottini[3], G. Busarello[5], A. Cappi[2], T. Contini[4], S. Charlot[7] B. Garilli[3], I. Gavignaud[4], L. Guzzo[8], O. Ilbert[1], A. Iovino[8], V. Le Brun[1], C. Marinoni[1], H.J. McCracken[2], G. Mathez[4], A. Mazure[1], Y. Mellier[6], B. Meneux[1] P. Merluzi[5], S. Paltani[1], R. Pelló[4], M. Radovich[5], R. Scaramella[9], M. Scodeggio[3], L. Tresse[1], G. Zamorani[10], A. Zanichelli[10], E. Zucca[10]

[1]*Laboratoire d'Astrophysique de Marseille - France,* [2]*Istituto di Radio-Astronomia - CNR, Bologna - Italy,* [3]*IASF - INAF, Milano - Italy,* [4]*Laboratoire d'Astrophysique de Toulouse,* [5]*Osservatorio Astronomico di Capodimonte, Naples, Italy* [6]*Institut d'Astrophysique de Paris, Paris, France,* [7]*Max Planck fur Astrophysik, Garching, Germany* [8]*Osservatorio Astronomico di Brera, Milan, Italy,* [9]*Osservatorio Astronomico di Roma, Italy,* [10]*Osservatorio Astronomico di Bologna, Bologna, Italy*

olivier.lefevre@oamp.fr, vettolani@ira.cnr.it,dario@iasf.mi.cnr.it, picat@obs-mip.fr

Abstract In this paper, the goals and methods of deep redshift surveys are reviewed and on-going projects are discussed. The requirements on instrumentation and observations methods are very stringent and call for a dedicated approach. Among the several on-going deep surveys, we are describing the VIMOS VLT Deep Survey (VVDS) which our team is conducting. It aims to obtain spectra in the redshift range $0 < z < 5$, for a complete magnitude limited sample of more than 150000 galaxies. This will allow to quantify the evolution of the galaxy and AGN population over more than 90% of the current age of the universe, as a function of galaxy type, luminosity, or local environment.

Keywords: Deep Redshift Surveys, galaxy and large scale structure evolution

1. Introduction

Understanding how galaxies and large scale structures formed and evolved is one of the major goals of modern cosmology. In order to identify the relative contributions of the various physical processes at play and the associated timescales, a comprehensive picture of the evolutionary properties of the con stituents of the universe is needed over a large volume and a large time base. Samples of high redshift galaxies known today reach less than two thousand at redshifts $1-3$, and statistical analysis suffer from small number statistics,

M. Plionis (ed.), Multiwavelength Cosmology, 7-13.

small explored volumes, selection biases, which prevent detailed analysis. In the local universe, large surveys like the 2dFGRS and the SDSS, will contain from 2.5×10^5 up to 10^6 galaxies to reach a high level of accuracy in measuring the fundamental parameters of the galaxies and AGNs populations. Similarly, we need to gather large numbers of galaxies at high redshifts to accurately measure the luminosity function, correlation function, star formation rate of various populations in environments ranging from low density to the dense cluster cores.

This can be only achieved through massive observational programs assembling galaxies and QSO samples representative of the universe at the various look-back times explored. As the galaxies observed are at great distances and thus necessarily faint, instruments have to be conceived to combine wide field, high throughput, and high multiplex gain. Multi-object spectrographs are routinely in operation since $\sim$15 years, and the new generation now in place on the $6-10$m telescopes, like DEIMOS on the Keck-10m or VIMOS on the VLT-8m, allows to explore large volumes of the distant universe through the observations of many tens of thousand of objects.

Because of this leap forward in instrument performances, several large deep surveys are now in progress. We are presenting here the VIMOS VLT Deep Survey (VVDS). It is combining a common observing strategy to assemble more than 150000 redshifts in 3 large galaxy datasets observed over more than 16deg^2 in 5 equatorial fields, selected in magnitude from $I_{AB} \leq 22.5$ to $I_{AB} \leq 25$.

2. The need for large deep redshift surveys

The current theoretical picture of galaxy formation and evolution and of large scale structure growth in the universe is well advanced. By contrast, while existing observations already provide an exciting glimpse of the properties and distribution of the high redshift objects (e.g. CFRS to redshift $\sim$ 1, Lilly et al., 1995, ApJ 455, 100, Le Fevre et al., 1996, ApJ 461, 534; Lyman-break galaxies, Steidel et al., 1996, ApJ, 462, L17, Steidel et al.,1998, ApJ, 492, 428; evolution of the star formation rate, Lilly et al., 1996, ApJ, 460, L1, Madau et al., 1998, ApJ, 498, 106, Cimatti, et al., A&A, 392, 395, Daddi et al., ApJ, astro-ph/0308456, Cuby, Le Fevre, et al., A&A, astro-ph/0303646) the samples remain too small to investigate the relationship between the evolution observed and the luminosity, type or local environment of the populations, relate the observations made with different selection criteria, and compare to theoretical predictions.

The massive efforts conducted by the 2dF and Sloan surveys (http://mso.anu.edu.au/2dFGRS/ and http://www.sdss.org/) are motivated to map the distribution and establish the properties of more than one million galaxies up to redshifts ~ 0.3. Deep surveys on volumes comparable to the 2dFGRS or SDSS at look back times spanning a large fraction of the age of the universe are required to give access to the critical time dimension and trace back the history of structure formation on scales from galaxies up to $\sim 100h^{-1}$ Mpc. This reasoning forms the basis of the VIMOS-VLT Deep Survey (VVDS): probing the universe at increasingly higher redshifts to establish the evolutionary sequence of galaxies, AGNs, clusters and large scale structure, and provide a statistically robust dataset to challenge future models.

3. Instrumentation and methods for deep redshift surveys

3.1 Multi-object spectrographs

Multi-object spectrographs (MOS) are a key element in the instrument complement of a modern observatory. Only 15 years ago, the first facility MOS on 4m class telescopes were delivering the first scientific results on deep redshift surveys. Since then, many facilities have appeared and allow to conduct e.g. the large redshift surveys at redshifts up to about 4 (Le Fevre et al., 1994, A&A, 282, 325, Allington-Smith et al., 1994, Oke et al, 1995, Szeifert et al., 1998, Steidel et al., 1996, ApJ, 462, 17).

On the new generation of 8-10m telescopes, several advanced multi-object spectrographs are now in operation DIEMOS on Keck-II, GMOS on Gemini, and VIMOS on the VLT, while several others are under construction (e.g. IMAX on Magellan). The VIMOS project has been conducted by our team under contract with ESO, and provides the European astronomical community with a highly efficient multi-object spectrograph covering the wavelength range 0.37 to 1 micron. VIMOS is installed on the VLT unit no.3. It is a 4 channel imaging spectrograph, each channel with a field of 7×8arcmin^2. Spectral resolutions from $R \sim 200$ to $R \sim 5000$ are available. The key element for a multi-object spectrograph is its multiplex gain: up to 1000 spectra can be observed simultaneously with VIMOS at low spectral resolution, and about 200 spectra at high spectral resolution. See Le Fevre et al. (1998, proc. SPIE, SPIE, 3355, 8) for more details on the general concept and optico-mechanical design.

3.2 Preparing and conducting a deep redshift survey

A deep redshift survey requires to handle the following steps: (1) build a photometric catalog with sufficient depth and astrometric accuracy to select the spectroscopic targets without instrumental bias, (2) define a field coverage strategy, (3) pre-select targets from the imaging catalog to prepare the slit-mask layout (4) conduct the survey observations with database follow-up of the observed objects, (5) process the raw spectra to correct for instrumental signature, flux and wavelength calibrate, (6) measure redshifts and spectral features (flux, EW,...), (7) store the data in a query-oriented database. Each of these steps require detailed planning, coding of specific software tools, thorough testing, and has to be prepared in a engineering-like environment to ensure proper quality control. This is the approach which has been followed in preparation for the VIMOS VLT Deep Survey.

4. VVDS: the VIMOS VLT Deep Survey

4.1 Survey goals

With the aim of studying galaxy and large scale structure evolution, the VVDS is assembling several redshift samples of faint galaxies: (1) a "wide" survey: $>$100000 galaxies and AGNs with measured redshifts in 16 deg^2 up to a limiting magnitude $I_{AB} = 22.5$, up to redshifts ~ 1.3; (2) a "deep" survey: >40000 galaxies and AGNs with measured redshifts in 2 deg^2 brighter than $I_{AB} = 24$, to map the early growth of large scale structures and establish the evolution of the star formation rate over 90% of the age of the universe; (3) an "ultra-deep" survey: > 1000 galaxies and AGNs brighter than $I_{AB} = 25$, this will probe deep into the luminosity function (3.5 magnitudes below M* at z=1). Natural follow-ups will include a high spectral resolution survey: 10000 galaxies selected from the "wide" and "deep" surveys for spectroscopy at $R = 2500 - 5000$ to study the evolution of the fundamental plane and spectrophotometric properties and a "high-z cluster" survey with clusters selected in the "wide" and "deep" surveys. The observing strategy that we have devised allows to carry these goals in an optimized and efficient approach, starting with the guaranteed time allocated to this program, which will then be expanded into a large program.

4.2 The VIRMOS-CFH12K deep imaging survey

A comprehensive photometric survey has been conducted on 16 deg^2 in four fields, 4 fields 2×2 deg^2 each selected at high galactic latitude around the celestial equator. Each field allows to probe scales up to $\sim$ 100h^{-1}Mpc. The 0226-04 field has been selected to coincide with the high visibility area of XMM, and, subsequently the NOAO deep imaging survey area and the Subaru deep field area have been set next to it. This field is also subject to deep VLA observations with a sensitivity down to 70μJy. In each field we have obtained BVRI data, as well as U-band data for a fraction of the fields (Radovich et al, 2003, in preparation) and K' data for a smaller 900 arcmin^2 area (Iovino et al., 2003, in preparation).

The depth is set to probe the luminosity function of galaxies to M*+3 at z=1 and to select lyman-break and QSO candidates out to redshifts $z > 3 - 5$.

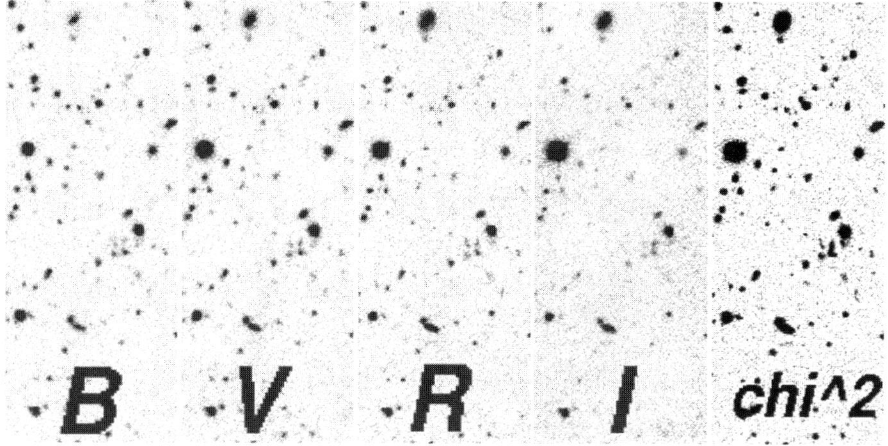

Figure 1. Deep images obtained with the CFH12K camera, it reaches a depth $I_{AB} = 25.3$ at 5σ in a 3 arcsec aperture. The χ^2 image is a combination of all the images and is used for source detection.

Particular care has been given to set a limiting magnitude deep enough to allow to select magnitude limited samples for the spectroscopic survey, free of bias and systematics. The limiting magnitude of $I_{AB} \leq 25$ at 5σ in a 3 arcsec aperture indeed represents the magnitude at which 95% of all galaxies with this magnitude are detected and measured. Therefore, at the limit $I_{AB} = 24$ of the deep spectroscopic survey, all galaxies including those with low surface brightness are expected to be detected, and no bias is expected. Details of this imaging survey are given in Le Fevre et al. (2003, astro-ph/0306252) and McCracken et al. (2003, astro-ph/0306254).

4.3 Spectroscopic survey status

The first observing campaign has been conducted in October and November 2002. A total of 9444 spectra have been observed in the "wide" survey mode, and 11297 in the "deep" survey mode. All data are processed at this time and the cross-checking of the redshift measurements is in progress (figure 2). The first step in the data processing is to produced 1D wavelength and flux calibrated spectra corrected from instrumental signature, and with the sky signal removed. This is done with the VIPGI pipeline environment (Scodeggio et al., 2003, in preparation), which allows to organize and process large amounts of data through the control of a few selection buttons in the user interface.

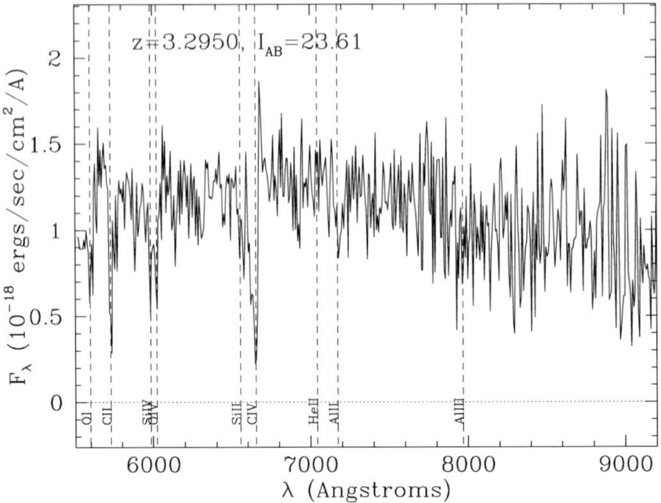

Figure 2. Example of a galaxy spectrum from the "Deep" survey, with magnitude $I_{AB} = 23.61$, and $z = 3.2950$

4.4 Measuring redshifts

Measuring redshifts is not completely trivial over a large redshift range. The basic technique calls for a cross correlation of observed spectra with a set of template spectra as representative as possible of the general population, combined with Principal Component Analysis (PCA) reconstruction of the spectral energy distribution. The KBRED tool (Scaramella et al., 2003) has been developed within the VVDS collaboration in the spirit to automate the redshift measurements. In this process, we have found that it is especially critical to assemble a good set of template spectra to compare against, to help the redshift measurement to converge. While templates ranging from early type to late type spectra can be found in the literature over most of the rest frame domain from Lyα to Hα, good high S/N templates with spectral resolution $\sim 10A$ are missing in the range $1600 - 3000A$. We are therefore taking an iterative approach to build from the VVDS data templates ranging down in the rest frame UV as far as possible. These templates are implemented in the KBRED templates list to improve significantly the measurement rate, especially in the redshift range $1.5 < z < 3$.

4.5 First results

In Figure 3, we show the redshift distribution N(z) for the $I_{AB} \leq 22.5$ and the $I_{AB} \leq 24$ samples assembled so far in our database. The mean redshift

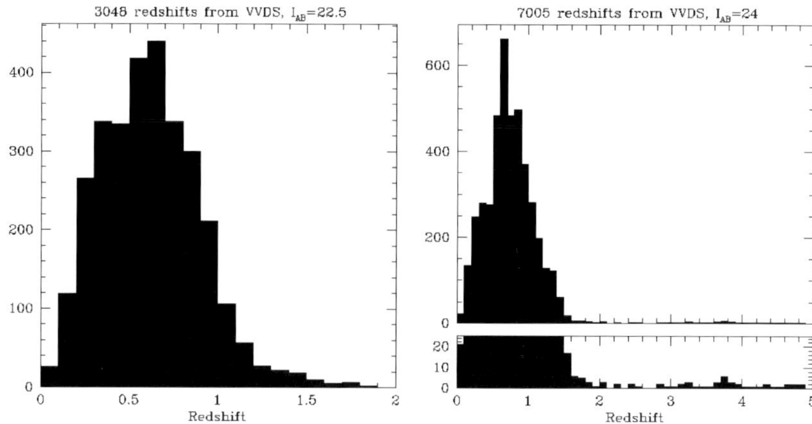

Figure 3. The redshift distribution N(z) for the VVDS $I_{AB} \leq 22.5$ and $I_{AB} \leq 24$ samples.

of the $I_{AB} \leq 22.5$ sample is $< z >= 0.63$, while it is $< z >= 0.70$ for the $I_{AB} \leq 24$ sample. In the latter, a high redshift tail is present, which is expected to be under-populated at the time of this writing, in particular in the range $1.5 < z < 3$: the $\sim 15\%$ spectroscopic incompleteness is expected to contain a significant fraction of galaxies in this redshift range which have escaped our redshift measurement so far, waiting for the use of the better UV templates described in the previous section.

We are now computing various statistical functions describing the galaxy population: the luminosity function, the correlation function, the star formation rate, and the clusters and groups content. Of particular interest will be to compute these indicators for various types of galaxies (using the PCA classification of spectra), and for various galaxy environments.

5. Conclusion

Deep redshift surveys are now feasible over more than 90% of the current age of the universe. This is due to a combination of powerful instruments on $6-10$m class telescopes and new methods developed to process large amounts of spectra and conduct redshift measurements in a quasi automated fashion. New results are now coming out from deep surveys like the VIMOS VLT Deep Survey or the DEEP2 survey, and will shed new light on our understanding of galaxies and large scale structure evolution.

CONSTRAINTS ON COSMOLOGY AND GALAXY FORMATION FROM THE NHDF *

Rodger I. Thompson
Steward Observatory
University of Arizona
Tucson, Arizona 85721, USA
rthompson@as.arizona.edu

Abstract Deep optical and near infrared images of the NHDF with HST provide information and place constraints on models of galaxy formation and evolution. Successful theories should match these observations. The constraints go from model and cosmology independent observations such as the apparent magnitude function to heavily cosmology and model dependent constraints such as the star formation history. All of the constraints are limited by large scale structure due to the small size of the NHDF.

Keywords: Cosmology, Galaxy Evolution, Star Formation

1. Introduction

The Northern Hubble Deep Field (NHDF) is currently the only area of sky that has both deep optical imaging with WFPC2 (Williams et al. 1996) and deep near infrared imaging with NICMOS (Dickinson 2000). As such it provides a unique opportunity for viewing the processes of galaxy formation and evolution at high redshifts. In particular the observations provide constraints on theories attempting to describe these processes. The constraints span a range from completely model and cosmology independent constraints such as the apparent magnitude function to constraints on star formation histories which are very cosmology and model dependent. The analysis methods are documented in Thompson, Weymann and Storrie-Lombardi 2001 and Thompson 2003 (TWS).

*Partial funding by NASA grant N-10843

M. Plionis (ed.), Multiwavelength Cosmology, 15-18.

2. Apparent Magnitude Function

The very high sensitivity of both WFPC2 and NICMOS provides accurate fluxes for even very faint galaxies. This extends the observed apparent magnitudes to limits beyond current ground based studies. The measured apparent magnitude function has the advantage of being completely free of any assumptions on cosmology, galaxy evolution model or even photometric redshift. Any galaxy formation and evolution theory must be able to reproduce the apparent magnitude function. The 1.6 and 0.814 micron apparent magnitude functions are shown in Figure 1. All magnitudes are in the AB system.

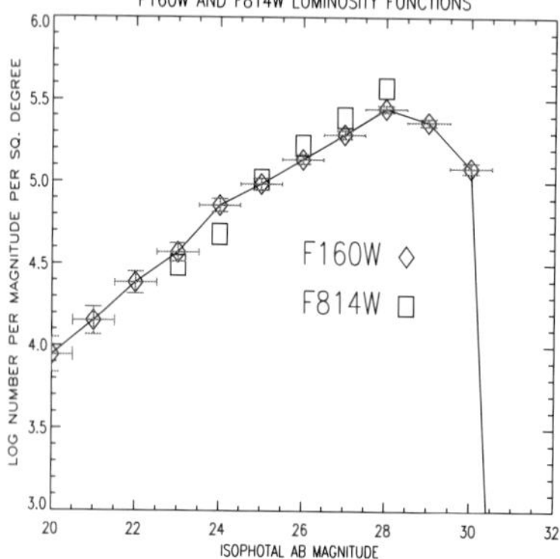

Figure 1. The 1.6 and 0.814 micron apparent magnitude functions in isophotal AB magnitudes.

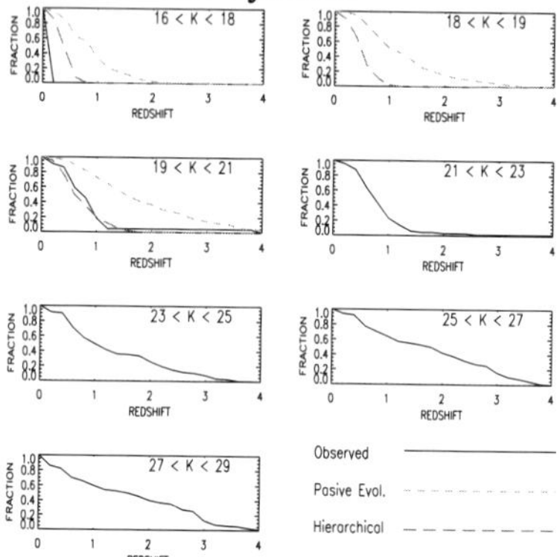

Figure 2. Comparison between the Kaufmann and Charlot predictions and observations.

The fall off of the 1.6 micron function at 28 mag. is due to incompleteness and is not a true feature of the function. The similarity between the 1.6 micron and 0.814 functions in the range of validity is striking. The main limitation on extending this constraint to the universe as a whole is large scale structure.

3. Kaufmann and Charlot Test

Kaufmann and Charlot (1998) proposed several tests to discriminate between hierarchical and passive luminosity evolution galaxy formation models. One of these tests predicts the redshift distribution of galaxies in constant apparent magnitude bins. In each magnitude bin the passive evolution model predicts a higher fraction of galaxies at high redshift than the hierarchical model. The predictions are for K band apparent magnitudes. The observed 1.6 micron apparent magnitudes were extended to K band magnitudes by the method given in Thompson (2003). The size of the difference between the two prediction and the width of the magnitude bins are much greater than any errors in the ex-

tension from H to K band magnitudes. Figure 2 shows the comparison between the predictions and the observations.
The first two magnitude bins have 1 and 0 observed galaxies and are shown only to indicate the trend of the predictions. The third magnitude bin has several galaxies and the observed distribution is indistinguishable from the hierarchical prediction. The observations extend to much fainter magnitudes than the predictions but it is clear that they follow the trend of the hierarchical predictions. Any revival of a model of pure passive luminosity evolution will have to account for this constraint. See, however, the contribution by Cimatti in this volume. This constraint depends on accurate photometric redshift determination. Given the brightness of the galaxies in the overlap bin between observation and prediction, this error should be small relative to the size of the difference between the predictions.

4. Luminosity Function

The deep NHDF optical and near infrared images provide an opportunity to probe the luminosity function at high redshift and produce a strong and important constraint on galaxy formation and evolution models. The combination of optical and near infrared measurements produce meaningful photometric redshifts beyond 6. The analysis techniques used in this work provide total luminosities by determining the extinction corrected SEDs for each galaxy. A comparison of the observed luminosity functions at redshifts 1 through 6 is shown in Figure 3.

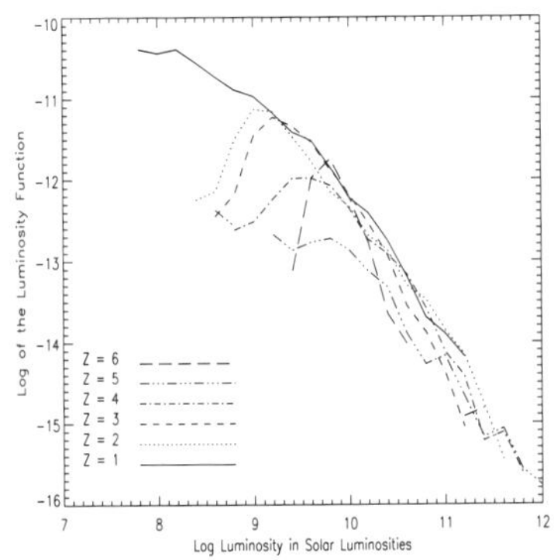

Figure 3. The observed luminosity function at redshifts 1 through 6.

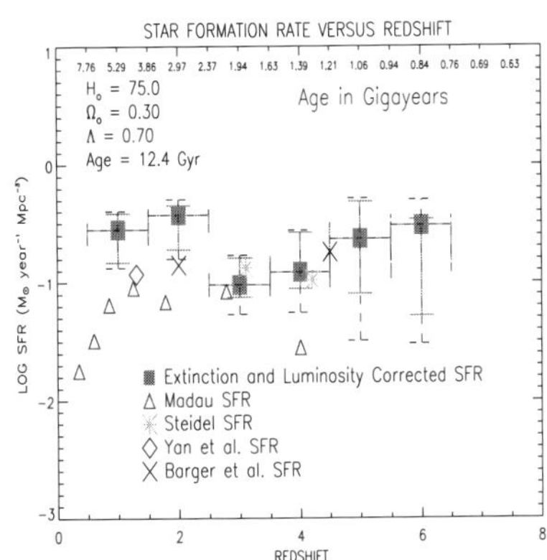

Figure 4. The star formation rate at redshifts 1 through 6.

The roll off of the luminosity function for high redshifts at low luminosities is just surface brightness dimming making the luminosity function incomplete at low luminosity. In the region of validity, however, the luminosity functions are remarkably similar in shape. This is a particularly strong constraint on galaxy formation theories. The luminosity functions in Figure 3 have not been normalized. The low absolute value of the luminosity function at redshift 5 may indicate a void at that redshift. The luminosity function depends on accurate redshifts, extinctions and SED determination from the photometric redshift and extinction analysis. It is also dependent on the cosmology which is taken as H_o = 75, Ω_{matter} = .3 and Ω_Λ = 0.7 in this analysis. The shape of the luminosity function however is independent of the cosmology.

5. Star Formation History

The final constraint is the star formation history. Accurate star formation histories are extremely strong constraints on galaxy formation and evolution, however, translating observations into star formation histories is subject to all of the uncertainties discussed above as well as the fundamental astrophysics of relating the measured ultra-violet flux to a star formation rate. An extensive description of the methods is given in TWS and Thompson (2003). The net result is an approximately constant star formation rate of a few tenths of a solar mass per year per Mpc^3 for a redshift range of 1 to 6 in redshift bins of 1 centered on integer redshifts. Figure 4 shows this result.
In figure 4 the solid error bars are for all errors except for large scale structure. The larger dashed error bars are representative of the error in translating the observations to the universe as a whole. It is clear that more extensive observations are needed to reduce the error due to large scale structure.

6. Conclusions

Theories of galaxy formation and evolution must confront the observational evidence gathered from both wide area and deep observations. The deep observations in this work provide constraints at high redshifts while wide area observations provide constraints that are less subject to large scale structure. The constraints from this work are a 1.6 micron apparent magnitude function, a redshift distribution of galaxies in apparent magnitude bins that appears to favor hierarchical galaxy assembly, a luminosity function that is independent of redshift over its range of validity, and a constant star formation rate at redshifts form 1 to 6.

References

Dickinson, M. 2000, Phil. Trans. Royal Soc. Lond. Ac., 358, 2001
Kaufmann, G. and Charlot, C. 1998, M.N.R.A.S., 297, L23
Thompson, R.I., Weymann, R.J. & Storrie-Lombardi, L.J. 2001, 546, 694
Thompson, R.I. 2003, Ap.J., submitted
Williams, R.E. et al. 1996, A.J., 112, 1335

THE END OF THE DARK AGES: PROBING THE REIONIZATION OF THE UNIVERSE WITH HST AND JWST

N. Panagia [ESA/STScI], M. Stiavelli & S.M. Fall [STScI]
panagia@stsci.edu; mstiavel@stsci.edu; fall@stsci.edu

Abstract Limiting the number of model-dependent assumptions to a minimum, we discuss the detectability of the sources responsible for reionization with existing and planned telescopes. We conclude that if reionization sources are UV-efficient, minimum luminosity sources, then it may be difficult to detect them before the advent of the James Webb Space Telescope (JWST). The best approach before the launch of JWST may be either to exploit gravitational lensing by clusters of galaxies, or to search for strong Ly-α sources by means of narrow-band excess techniques or slitless grism spectroscopy.

Keywords: cosmology: reionization; high-z galaxies

1. Introduction

Motivated by recent evidence that the epoch of reionization of hydrogen may have ended at a redshift as low as $z \approx 6$ (*e.g.,* Becker et al. (2001); Fan et al. (2002)), we have considered the detectability of the sources responsible for this reionization. The main idea is that reionization places limits on the mean surface brightness of the Population of reionization sources. We have defined a family of models characterized by two parameters: the Lyman continuum escape fraction f_c from the sources, and the clumpiness parameter C of the intergalactic medium. The minimum surface brightness model corresponds to a value of unity for both parameters. A maximum surface brightness is obtained by requiring that the reionization sources do not overproduce heavy elements. Our general approach is applicable to most types of reionization sources, but in specific numerical examples we focus on Population III stars, because they have very high effective temperatures and, therefore, are very effective producers of ionizing UV photons (*e.g.,* Panagia et al. (2003)). Our mean surface brightness estimates are compared to the parameter space that can be probed by existing and future telescopes, in order to help planning the

M. Plionis (ed.), Multiwavelength Cosmology, 19-22.

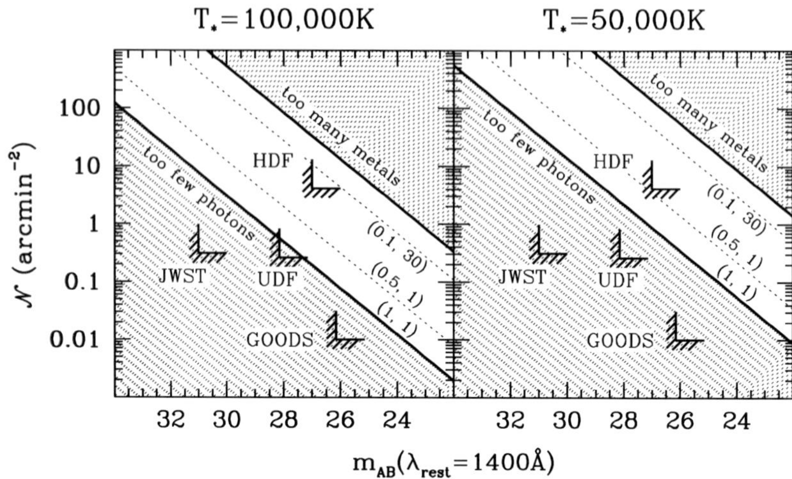

Figure 1. The loci of surface density vs apparent AB magnitude for identical reionization sources that are either Population III (left hand panel) or Population II stars (right hand panel).

most effective surveys. A full account of our work can be found in Stiavelli, Fall & Panagia (2003).

2. Results and Discussion

In Figure 1 we show the loci of the mean surface brightness of *identical* reionization sources as a function of their observable AB magnitude. The left panel refers to Population III sources with effective temperature of 10^5 K, the right panel to Population II reionization sources with effective temperature 5×10^4 K. In both panels, the lower solid line represents the minimum surface brightness model, (1,1), while the upper solid line represents the global metallicity constraint $Z \leq 0.01 Z_\odot$ at $z = 6$. The thin dotted lines represent the (0.5, 1) and (0.1, 30) models. The non-shaded area is the only one accessible to reionization sources that do not overproduce metals. The L-shaped markers delimit the quadrants (i.e., the areas above and to the right of the markers) probed by the GOODS/ACS survey (Dickinson and Giavalisco (2003)), the HDF/HDFS NICMOS fields (Thompson et al. (1999); Williams et al. (2000)), the Ultra Deep Field (UDF) and by a hypothetical ultra-deep survey with JWST.
In Figure 2 we show the expected cumulative surface density distributions for reionization sources with a variety of luminosity functions. In each panel, the upper solid line represents the global metallicity constraint. The lower solid line represents the minimum surface brightness model, (1,1). The thin dotted lines give the luminosity function for the (0.5,1) and the (0.1, 30) models. The

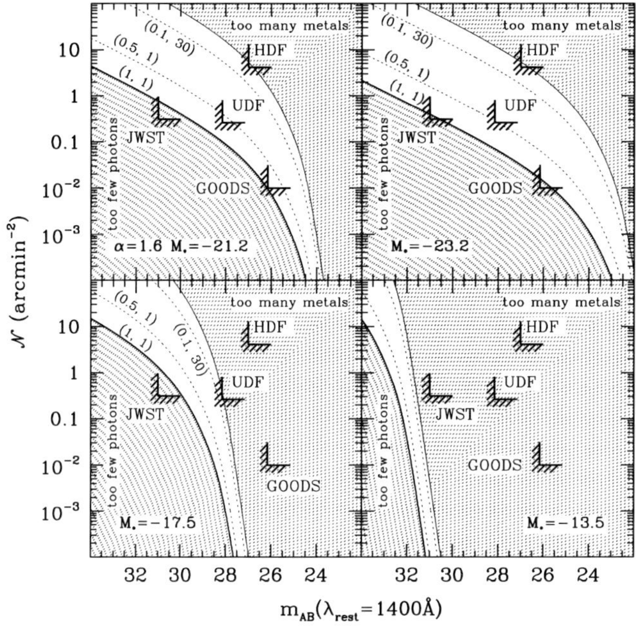

Figure 2. The cumulative distribution of the surface density vs apparent AB magnitude of reionization sources with luminosity functions with different knees.

L-shaped markers delimit the quadrants probed by the GOODS/ACS survey, the HDF and HDFS NICMOS fields, the UDF, and an ultra-deep survey with JWST, respectively.

From these results it appears that if reionization is caused by UV-efficient, minimum surface brightness sources, the non-ionizing continuum emission from reionization sources will be difficult to detect before the advent of JWST. On the other hand, if the sources of reionization were not extremely hot Population III stars but cooler Population II stars or AGNs, they would be brighter by 1-2 magnitudes and thus they would be easier to detect.

Finally, Figure 3 presents the required surface density as a function of the Ly-α line flux of reionization sources. The left hand panel shows the loci of identical sources for two different (f_c, C) models (thin dotted lines). The top solid line identifies the global metallicity constraint. The right hand panel illustrates two different luminosity functions. The solid line refer to a luminosity function identical to that of $z = 3$ Lyman break galaxies, (*i.e.*, $M_{*,1400} = -21.2$ and $\alpha = 1.6$, while the dashed line refers to a luminosity function with $M_{*,1400} = -17.5$ and a slope identical to the local Universe slope, $\alpha = 1.1$. In both panels, the L-shaped marker identifies the quadrant probed by a narrow-band excess survey at $z \geq 6$ (LALA survey, Rhoads and Malhotra (2001)). The

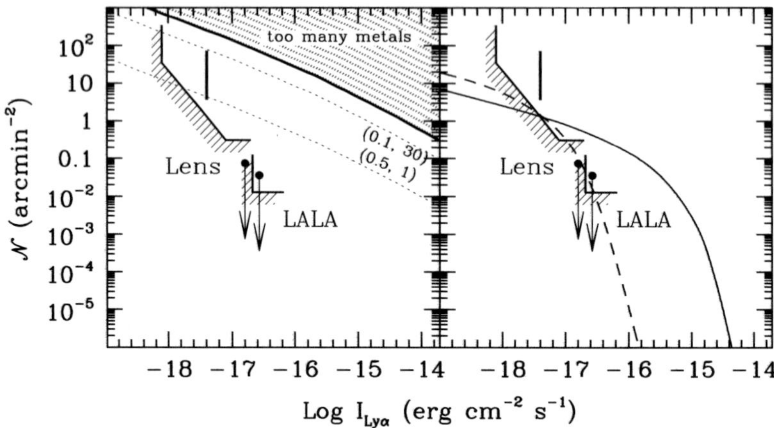

Figure 3. Surface density vs Ly-α line flux of reionization sources.

oblique marker labeled *Lens* represents a hypothetical 100-orbit survey with the ACS grism on a cluster of galaxies to exploit gravitational amplification. The solid bar represents the density estimated from the detection at $z = 6.56$ by Hu et al. (2002) while the two points with down-pointing arrows represent their upper limits.

It appears that searches based on narrow-band excess techniques or slitless grisms would be promising and might lead to the detection of the reionization sources within this decade.

References

Becker, R.H., et al. 2001, AJ, 122, 2850

Fan, X., et al. 2002, AJ, 123, 1247

Dickinson, M., & Giavalisco, M. 2003, in the proceedings of the ESO/USM Workshop "The Mass of Galaxies at Low and High Redshift", eds. R. Bender & A. Renzini, p. 324

Hu, E.M., et al., 2002, ApJ, 568, L75

Panagia, N., Stiavelli, M., Ferguson, H. & Stockman, H.S., 2003, in preparation [see also astro-ph/0209278 and /0209346]

Rhoads, J.E. & Malhotra, S., 2001, ApJ, 563, L5

Stiavelli, M., Fall, S.M., & Panagia, N., 2003, ApJ, in press

Thompson, R., Storrie-Lombardi, L., Weymann, R.J., Rieke, M.J., Schneider, G., Stobie, E., & Lytle, D., 1999, AJ, 117, 17

Williams, R.E., et al., 2000, AJ, 120, 2735

THE LENSES STRUCTURE & DYNAMICS SURVEY

The internal structure and evolution of E/S0 galaxies and the determination of H_0 from time-delay systems

L.V.E. Koopmans[1,2] & T.Treu[2]
[1]*Space Telescope Science Institute, 3700 San Martin Drive, Baltimore, MD 21218, USA*
[2]*California Institute of Technology, mailcode 105–24, Pasadena, CA 91125, USA*
koopmans@stsci.edu, tt@astro.caltech.edu

Abstract The Lenses Structure & Dynamics (LSD) Survey aims at studying the internal structure of luminous and dark matter – as well as their evolution – of field early-type (E/SO) galaxies to $z \sim 1$. In particular, E/S0 lens galaxies are studied by combining gravitational lensing, photometric and kinematic data obtained with ground-based (VLA/Keck/VLT) and space-based telescopes (HST). Here, we report on preliminary results from the LSD Survey, in particular on (i) the constraints set on the luminous and dark-matter distributions in the inner several $R_{\rm eff}$ of E/S0 galaxies, (ii) the evolution of their stellar component and (iii) the constraints set on the value of H_0 from time-delay systems by combining lensing and kinematic data to break degeneracies in gravitational-lens models.

Keywords: gravitational lensing — distance scale — galaxies: kinematics and dynamics — galaxies: fundamental parameters — galaxies: elliptical and lenticular, cD

Introduction

Even though massive E/S0 galaxies enclose most mass (luminous and dark) in the Universe on galactic scales, relatively little is observationally known about their their dark-matter halos or the evolution of their internal structure with time, and only recently studies have started to shed some light on the evolution of their stellar population with redshift.

The reasons for this are both observational and intrinsic to the mass modeling. First, since many of the studies of the mass distribution of E/S0 galaxies rely on stellar kinematics through spectroscopic observations, only with the advent of 8-10 m class telescopes (e.g. Keck and VLT) has it becomes possible to study E/S0 galaxies in any detail beyond the local Universe. Second, degeneracies in mass models that rely only on kinematic and photometric data often allow for multiple solutions and place limited or no constraints on the presence

M. Plionis (ed.), Multiwavelength Cosmology, 23-26.

and distribution of dark-matter. These problems exacerbate with increasing redshift due to poorer observational constraints. Model degeneracies are often due to the unknown mass of the galaxy, allowing one to freely play with stellar anisotropy, the mass-density slope and its normalization. For example, approximately constant velocity dispersion profiles can be explained with an isothermal, kinematically isotropic, luminous plus dark-matter distribution, as well as a constant stellar M/L model with a radially increasing tangential anisotropy. However, the latter model requires significantly less mass than the isothermal model within, say, several effective radii ($R_{\rm eff}$).
Hence, if the total mass of an E/S0 galaxy enclosed within some radius (around $\sim R_{\rm eff}$) is known, one can break the degeneracy between the stellar M/L, the stellar anisotropy and the radial mass distribution. Strong gravitational lensing by E/S0 galaxies provides exactly the required information!

The Lenses Structure & Dynamics (LSD) Survey

The 'clean' LSD Survey sample consists of 11 relatively isolated (e.g. no massive clusters nearby) E/S0 lens galaxies spread between redshifts $z = 0.04$ and 1.01. The galaxies have a mass range of ~1.5 orders of magnitude. Multi-color photometric data is available in the HST archive (mostly from the CASTLeS collaboration) for each system (typically V, I and H bands). In 2001–2002, we obtained stellar kinematic data using the Echelle Spectrograph and Imager (ESI) on the Keck–II telescope with typically $0.7''$ seeing and under photometric conditions. Some systems have extended kinematic profiles (along major and sometimes minor axes), others only luminosity weighted dispersions, depending on their brightness and extent. Besides the clean sample, we also observed several other systems (also with the VLT), including several disk-galaxy lenses and lens systems with measured time-delays.
We continue with some of the high-lights of the LSD survey and related studies (e.g. the determination of H_0 from lensing).

The evolution of the stellar mass-to-light ratio

Two LSD systems (MG2016+112 and 0047–285) have thus far been studied in detail (Koopmans & Treu 2002, 2003; Treu & Koopmans 2002a). Since we have available the (central) stellar velocity dispersion, effective radius and effective surface brightness (from the HST images; transformed to rest-frame B-band), each lens system can directly be compared with the local Fundamental Plane (FP). The offset from the FP is an indicator of the evolution of the effective surface brightness, due to fading of the stellar population with time (i.e. "passive evolution"). Both systems are consistent with passive evolution of field E/S0 galaxies, marginally faster than that of cluster E/S0 galaxies. We find that the stellar M/L determined from the FP evolution and local measurements, being close to the "maximum-bulge solution", agree with those measured *only* from lensing and dynamics, suggesting that no structural evolution has occurred in the FP below $z \sim 1$.

The luminous and dark-matter mass profile

The combination of stellar kinematics and gravitational lensing can also be used to place stringent constraints on the luminous plus dark-matter mass profiles of E/S0 galaxies to $z \sim 1$. The reason is that lensing determines the mass inside the Einstein radius to a few percent accuracy. Varying the inner (total) mass slope – but satisfying the stringent mass constraint – leads to a considerable change in the observed line-of-sight stellar velocity dispersion profile, as well as the luminosity weighted stellar dispersion, of the E/S0 galaxy, only weakly dependent on details such as anisotropy, etc (see KT). Thence, a comparison with the observed stellar kinematics allows the determination of an 'effective' slope (γ' for $\rho \propto r^{-\gamma'}$; i.e. the average luminous dark-matter power-law slope) inside the Einstein radius (typically 1–5 $R_{\rm eff}$).

Thus far, we have found that $\gamma' = 2.0 \pm 0.1 \pm 0.1$ for MG2016+112 (Treu & Koopmans 2002a) and $\gamma' = 1.9^{+0.05}_{-0.23} \pm 0.1$ for 0047−285 (Koopmans & Treu 2003; 68% C.L. and syst. error). For B1608+656–G1 and PG1115+080, both not part of the 'clean' LSD sample, we find $\gamma' = 2.03 \pm 0.14 \pm 0.03$ and $\gamma' = 2.35 \pm 0.1 \pm 0.05$, respectively (Koopmans et al. 2003; Treu & Koopmans 2002b). E/S0 galaxies in the clean sample are both consistent with isothermal mass profiles (i.e. $\gamma' = 2$). From all four systems studied thus far, the average is $\langle\gamma'\rangle = 2.1$ with an rms of 0.2; *E/S0 galaxies appear on-average isothermal to within ~ 10% inside several effective radii, eventhough some intrinsic scatter between systems is found, as expected.*

We note that this is a preliminary results and the analysis of the full sample is required to confirm/strengthen this conclusion. Even so, E/S0 galaxies require a considerable diffuse dark-matter component inside the stellar spheroid in order to explain the observed stellar kinematics; constant stellar M/L models are excluded with high confidence in all cases. The luminous plus dark-matter appears to conspire to form an isothermal profile in its inner regions, similar to that observed for spiral galaxies. Finally, upper limits have been set on the inner dark-matter profile of E/S0 as well as the (an)isotropy of the stellar component, but we defer a discussion until the entire sample has been analyzed.

The value of H_0 from lens time-delays

The most severe degeneracy in lens models is that of the (unknown) slope of the radial mass profile of the lens mass distribution. Different power-law slopes (other than e.g. isothermal) can often equally well fit the same lensing-only constraints (e.g. Wucknitz 2002). Different slopes, however, lead to different

inferred values of H_0 from time delays, roughly following $\Delta H_0/H_0^{\gamma'=2} = (\gamma' - 2)$; i.e. steeper (shallower) than isothermal mass profile lead to higher (lower) inferred values of H_0.
Since the combination of stellar kinematics and gravitational lensing can tightly constrain γ' (see above; assuming the mass profile indeed follows approximately a power-law), the usefulness of this to time-delay lenses and the determination of H_0 is apparent. We have thus far looked at two systems in detail and find $H_0 = 59^{+12}_{-7} \pm 3$ km s^{-1} Mpc^{-1} (PG1115+080) and $H_0 = 75^{+7}_{-6} \pm 4$ km s^{-1} Mpc^{-1} (B1608+656) for $(\Omega_m, \Omega_\Lambda) = (0.3, 0.7)$. The errors are the 68% C.L. and systematic errors and include a realistic uncertainty due to the slope of the radial mass profile. These values should therefore be relatively unbiased. The deviation of PG1115+080 from isothermal (see above) increases H_0 by $\sim$35% from 44 to 59 km s^{-1} Mpc^{-1}, exemplifying the need to measure the stellar kinematics of lens galaxies for each time-delay system. Both values are consistent with local determinations of H_0 (e.g. Freedman et al. 2001), but inconsistent with the sample studied by e.g. Kochanek & Schechter (2003; and references therein). We are currently observing more systems with Keck and the VLT to improve the statistics of the sample and further examine the nature of this apparent inconsistency between lens systems.

Conclusions

The combination of stellar kinematic with gravitational lensing provides a powerful new tool to study the internal structure and evolution of E/S0 galaxies. Some of the preliminary results from the LSD Survey indicate that E/S0 lens galaxies to $z \sim 1$ evolve passively and have nearly-isothermal luminous plus dark matter mass profiles inside several $R_{\rm eff}$. Application of this to lens systems with time-delays gives more accurate values of H_0, so far in agreement with local determinations. The study of more lens systems is required to confirm/strengthen these conclusion, but results so far have been encouraging.

References

Freedman, W. L. et al. 2001, ApJ 553, 47
Kochanek, C. K. & Schechter, P. L. [astro-ph/0306040].
Koopmans, L. V. E. & Treu, T. 2002, ApJ 568, L5
Koopmans, L. V. E. & Treu, T. 2003, ApJ 583, 606
Koopmans, L. V. E., et. al. 2003, ApJ submitted, [astro-ph/0306216]
Treu, T. & Koopmans, L. V. E. 2002a, ApJ 575, 87
Treu, T. & Koopmans, L. V. E. 2002b, MNRAS 337, L6
Wucknitz, O. 2002, MNRAS 332, 951

SEARCHES FOR HIGH REDSHIFT GALAXIES USING GRAVITATIONAL LENSING

J. Richard[1], R. Pelló[1], J.-P. Kneib[1,2], D. Schaerer[1,3], M.R. Santos[2], R. Ellis[2]
[1]*O.M.P., Laboratoire d'Astrophysique, UMR 5572, 14 Avenue E. Belin, 31400 Toulouse, France*
[2]*California Institute of Technology, 105-24 Caltech, Pasadena, CA 91125, USA*
[3]*Geneva Observatory, 51 Ch. des Maillettes, CH-1290 Sauverny, Switzerland*
jrichard@ast.obs-mip.fr

Abstract We present different methods used to identify high redshift ($z > 5$) objects in the high-magnification regions of lensing galaxy clusters, taking advantage of very well constrained lensing models. The research procedures are explained and discussed. The detection of emission lines in the optical/NIR spectra, such as Lyman-alpha, allows us to determine the redshift of these sources. Thanks to the lensing magnification, it is possible to identify and to study more distant or intrinsically fainter objects with respect to standard field surveys.

Keywords: galaxies: formation, evolution, high-redshift, luminosity function, clusters : lensing —cosmology: observations

1. Introduction

The main purpose of looking at high redshift ($z > 5$) objects is to get constraints about the nature and the formation epoch of the first sources in the Universe. The advent of 8-10 m class telescopes, such as VLT and Keck, has opened up this field of study. Moreover, the use of clusters of galaxies as gravitational telescopes can help a lot for this. Strong lensing effect in clusters has already enabled the detection of one of the most distant galaxies known up to now (Hu et al. 2002), thanks to the gravitational magnification and despite the decrease in effective area of the survey.

We present here two methods aimed at the detection of lyman-alpha sources behind galaxy clusters : first a spectroscopic search along the critical lines of clusters, and then a photometric selection technique for very low metallicity starbursts (the so-called Population III objects, Loeb & Barkana 2001), using ultra-deep near infrared imaging.

M. Plionis (ed.), Multiwavelength Cosmology, 27-30.

2. Critical lines survey

Using the LRIS spectrograph at Keck, we searched for $Ly\alpha$ emitters at redshift 2.5 to 6.8 in the most magnified parts of a sample of lensing clusters, selected for having well-constrained mass models. We scanned the regions located near the critical lines (Figure 1), defined as the lines of infinite magnification for a given redshift, using a 175"-long slit. Half of the area covered (4.2 arcmin^2) is at least magnified by a factor of 10 at $z = 5$.

We systematically looked for every single emission line in the spectra, and we confirmed $Ly\alpha$ candidates using HST images available for these clusters, optical photometry, and further spectroscopy at higher resolution, using ESI at Keck that can easily resolve the [OII] doublet, thus preventing this contamination. We identified 12 $Ly\alpha$ candidates, three of them lying in the redshift range $\sim 4.6 - 5.6$. One is a double image at $z \sim 5.6$ which was analysed with more details by Ellis et. al (2001). The two-dimensional LRIS spectra, showing $Ly\alpha$ emission lines, are presented in figure 1.

Thanks to the strong lensing magnification, these results can give us constraints on the luminosity function of emitters at $4.6 < z < 5.6$ with $Ly\alpha$ luminosity $10^{40} < L < 10^{42}\, erg/s$, which is a depth that was not reached by other surveys of $Ly\alpha$ emitters or Lyman Break Galaxies. This will be presented in Santos et al. (2003, *ApJ* submitted).

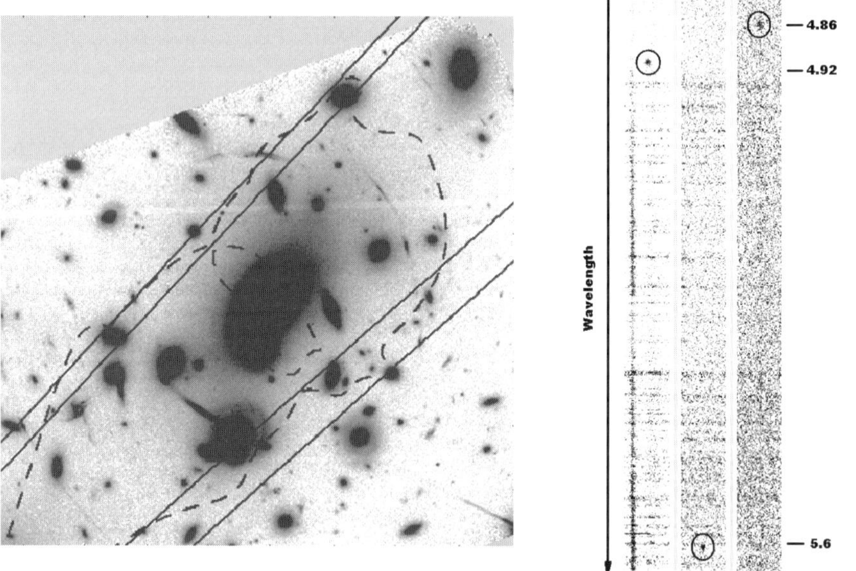

Figure 1. On the left: zoom on the center part of the galaxy cluster Abell 2218. The critical lines for z=5 are shown as dashed lines, and the two regions scanned by the survey as rectangles. On the right: composite spectra of the three $Ly\alpha$ emitters (circled) found at $z \sim 5 - 6$.

3. Looking for Population III objects

Recent models by D. Schaerer (2002, 2003) for the Spectral Energy Distributions (SED) of Population III objects show that they may be currently observable using 8 - 10 m telescopes, at the limits of conventional spectroscopy. The identification of such objects should be possible thanks to their very strong emission lines, mainly $Ly\alpha$ and HeII λ1640. In order to find these objects, the colors predicted by the same models can allow us to define a color-color region in the near-infrared diagram (J-H) vs (H-K'), where we can pick up candidates (Fig. 2). By doing simulations with existing models, we found that we should not be contaminated by stars or $z < 8$ galaxies, even in the case of important reddening.
As a first test of these selection criteria, we did very deep imaging (limiting magnitudes of J=25, H=24.5, K'=24, Vega system), with ISAAC on VLT, of two lensing clusters, taking advantage of the lensing magnification to help us detecting these faint objects.

4. Preliminary results

We selected several ($\sim$ 10) candidates per cluster, satisfying our selection criteria in the near-IR, and being undetected on available optical images. These objects have the expected magnitudes and SEDs of $8 < z < 10$ Population III objects (Figure 2), and are magnified by 2 to 4 magnitudes thanks to the strong lensing effect. We used a modified version of the photometric software **hyperz** (Bolzonella et al. 2000) to find the redshift distribution probability of our candidates with the spectra models quoted above.

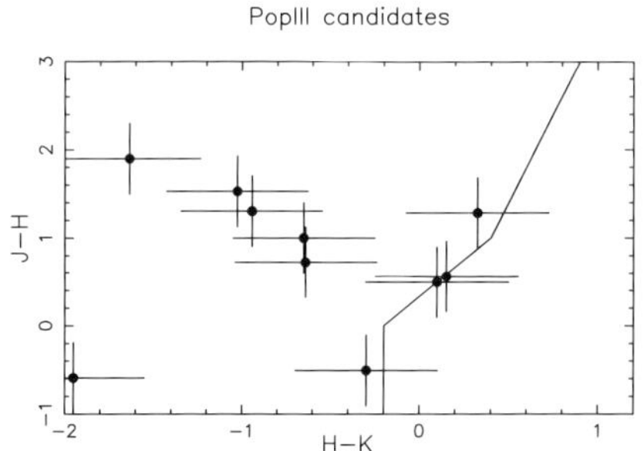

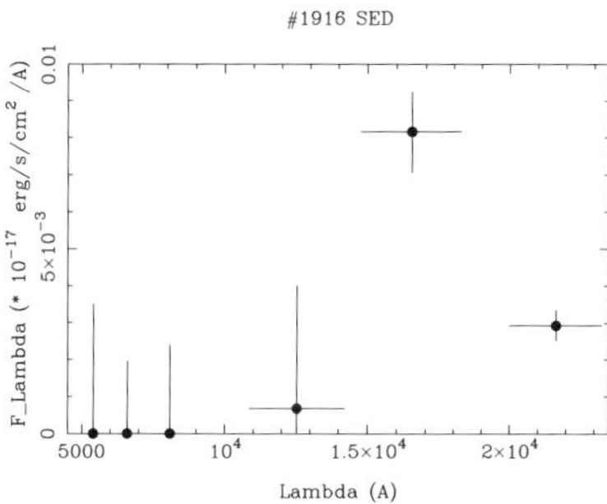

Figure 2. Left: location, on the NIR color-color diagram (J-H) vs (H-K'), of the candidates found with the typical photometric errors (Vega system). The selection region for Population III objects is delimited by a solid line. Right : example of SED, combining infrared and optical photometry, for one of the candidates. Photometric redshift gives $z \sim 9$.

As a preliminary result, we can try to compare the number of Population III objects per redshift that was expected to be detected in our field with the upper limit corresponding to our candidates, using a simple model of dark-matter halos distribution (Press & Schechter, 1974), and 4 different models of IMF for PopIII galaxies (Figure 3). Furthermore, we can estimate the efficiency of using strong lensing in this field by plotting the expected number counts in a blank field of same size and depth. We find that lensing is more efficient at high redshifts ($z > 8$), and that the number of candidates we found is consistent with some of the models we used.

5. Conclusions

The use of gravitational lensing is efficient to detect more distant or intrinsically fainter galaxies lying behind galaxy clusters : we can have constraints on luminosity functions at fainter scales, and the expected number of primordial objects in a cluster field is boosted at high redshifts.
Even if the candidates we found with our selection criteria are very faint, we should try to perform spectroscopy of the best ones with present day facilities. The detection of these sources is one of the major science cases for the next generation NIR instruments, like EMIR/GTC, KMOS/VLT or KIRMOS/Keck.

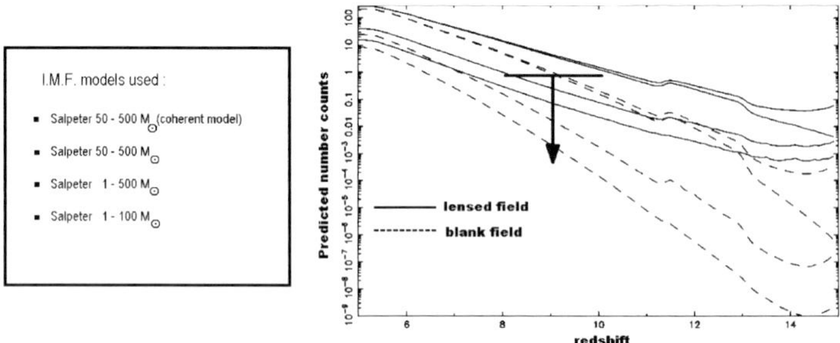

Figure 3. Number counts of PopIII objects per interval of 0.1 in z, expected to have K'<24 in the ISAAC fov. The different curves correspond to different IMF models. The values obtained with gravitational lensing (solid curves) are boosted by a factor of 10 at $z \sim 8 - 10$ regarding the one expected for a blank field (dashed curves). Overplotted is the upper limit of our survey.

References

Bolzonella, M., Miralles, J.M., Pelló, R., 2000, A & A, **363**, 476
Ellis, R., Santos, M. R., Kneib, J.-P., Kuijken, K., 2001, *ApJ*, **560**, L119
Hu, E., et al. 2002a, *ApJ*, **568**, L75
Loeb, A. & Barkana, R., 2001, ARA& A, **39**, 19
Press, W. H. & Schechter, P., 1974, *ApJ*, **187**, 425
Santos, M. R., Ellis, R., Kneib, J.-P., Richard, J., Kuijken, K., *ApJ* submitted
Schaerer, D. 2002, A & A, **382**, 28
Schaerer, D. 2003, A & A, **397**, 527

THE 2DF QSO REDSHIFT SURVEY

P. J. Outram[1], T. Shanks[1], B. J. Boyle[2,3], S. M. Croom[3], F. Hoyle[4], N. S. Loaring[5,6], L. Miller[6], A. D. Myers[1], and R. J. Smith[7]

[1]*Dept. of Physics, University of Durham, South Road, Durham DH1 3LE, UK*

[2]*Australia Telescope National Facility, PO Box 76, Epping, NSW 1710, Australia*

[3]*Anglo-Australian Observatory, PO Box 296, Epping, NSW 2121, Australia*

[4]*Dept. of Physics, Drexel University, 3141 Chestnut Street, Philadelphia, PA 19104, USA*

[5]*Mullard Space Science Laboratory, University College London, Holmbury St. Mary, Dorking, Surrey, RH5 6NT, UK*

[6]*Dept. of Physics, University of Oxford, Nuclear & Astrophysics Laboratory, Keble Road, Oxford, OX1 3RH, UK*

[7]*Astrophysics Research Institute, Liverpool John Moores University, 12 Quays House, Egerton Wharf, Birkenhead, CH41 1LD, UK*

Abstract With around 23000 QSOs, the recently completed 2dF QSO Redshift Survey (2QZ) is currently the largest individual QSO survey. Here we present a measurement of the QSO power spectrum, and investigate new constraints on Ω and Λ from redshift-space distortions in the power spectrum, and weak lensing of 2QZ QSOs by foreground galaxy groups.

1. Introduction

We have used the AAT 2dF facility to make a redshift survey of some 23000 $b_J < 20.85$ QSOs with redshifts $0.3 < z < 2.5$, probing a volume of $\sim 4 \times 10^9$ Mpc^3. Here we utilise the huge volume probed by the 2dF QSO Redshift Survey (2QZ) to determine the form of QSO clustering out to scales of $\sim 500h^{-1}$ Mpc and derive new cosmological constraints. Further details about the survey, including the catalogue and QSO spectra, can be obtained at `http://www.2dfquasar.org`.

2. The QSO Power Spectrum

QSOs are highly effective probes of the large-scale structure of the Universe over a wide range of scales and can trace clustering evolution from the present day out to high redshift. In the linear regime ($10 < r < 1000\mathrm{h}^{-1}$Mpc), they are clearly superior to galaxies as probes of large scale structure by virtue of both the large volumes they sample and their flat $n(z)$ distribution. Thus a

M. Plionis (ed.), Multiwavelength Cosmology, 31-34.

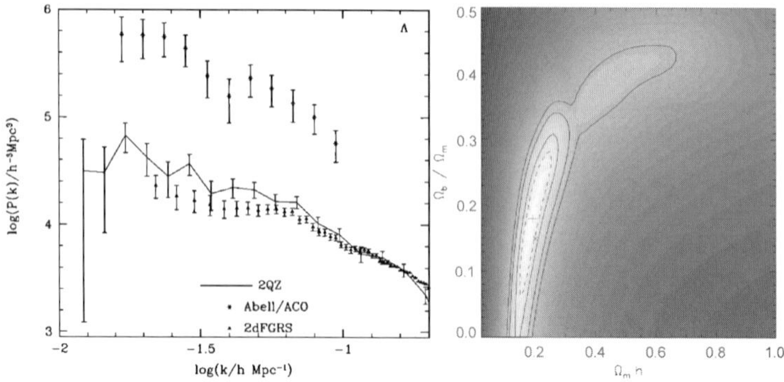

Figure 1. On the left we show the QSO power spectra derived assuming a standard Λ cosmology, compared with the 2dFGRS galaxy (Percival et al 2001), and ACO cluster (Miller & Batuski 2001) power spectra. On the right we show filled contours of decreasing likelihood in the $\Omega_m h$ – Ω_b/Ω_m plane, marginalizing over h and the power spectrum amplitude, for fits to the 2QZ power spectra. Countours are plotted for a one-parameter confidence of 68%, and two-parameter confidence of 68%, 95% and 99%. + marks the best fit models to the 2QZ data ($\Omega_b/\Omega_m = 0.18$, $\Omega_m h = 0.19$). For comparison, $\times$ marks the best fit model ($\Omega_b/\Omega_m = 0.15$, $\Omega_m h = 0.2$) determined from the 2dFGRS data.

measurement of the QSO power spectrum on these scales provides a powerful constraint on cosmological models.

In Fig. 1a we show the power spectrum derived from the final 2dF QSO Redshift Survey catalogue containing 23000 QSOs (Outram et al. 2003). Utilising the huge volume probed by the QSOs, we have accurately measured the power over the range $-2 < \log(k/h\,\mathrm{Mpc}^{-1}) < -0.7$, thus covering over a decade in scale, and reaching scales of $\sim 500h^{-1}$Mpc. We consider either an Einstein-de Sitter (Ω_m=1.0, Ω_Λ=0.0) cosmology (EdS hereafter) or an Ω_m=0.3, Ω_Λ=0.7 cosmology (Λ hereafter) to convert from the observed redshift to comoving distance. The QSO power spectrum can be well described by a model with shape parameter $\Gamma = 0.13 \pm 0.04$ (assuming a Λ $r(z)$; if an EdS $r(z)$ is instead assumed, a slightly higher value of $\Gamma = 0.17^{+0.07}_{-0.04}$ is obtained). These low measurements of Γ indicate significant large-scale power and assuming either cosmology we can strongly rule out a turnover in the power spectrum at scales below $\sim$300h^{-1}Mpc (log$k \sim -1.7$). The amplitude of clustering of the QSOs at $z \sim 1.4$ is similar to that of present day galaxies (Percival et al 2001), and an order of magnitude lower than present day clusters (Miller & Batuski 2001).

To derive constraints on the matter and baryonic contents of the Universe, we fit CDM model power spectra (assuming scale-invariant initial fluctuations). Assuming a Λ $r(z)$, we find that models with baryon oscillations are slightly prefered, with the baryon fraction $\Omega_b/\Omega_m = 0.18 \pm 0.10$. The overall shape of the power spectrum provides a strong constraint on $\Omega_m h$ (where h is the Hubble parameter), with $\Omega_m h = 0.19 \pm 0.05$, consistent with a standard ΛCDM model.

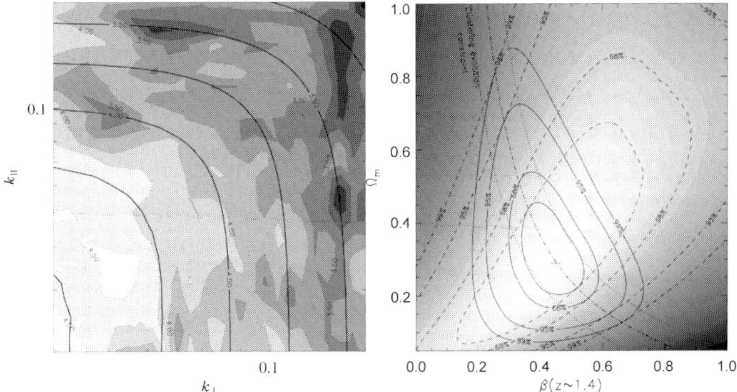

Figure 2. On the left we show $P^S(k_{\parallel}, \mathbf{k}_{\perp})$ determined from the 2QZ Catalogue. Filled contours of constant $log(P(k)/h^{-3}\text{Mpc}^3)$ are shown as a function of $k_{\parallel}/h\,\text{Mpc}^{-1}$ and $\mathbf{k}_{\perp}/h\,\text{Mpc}^{-1}$, appearing elongated along the line of sight. Overlaid is the best fit model with $\beta = 0.45$ and $\Omega_{\text{m}} = 0.29$. On the right we show filled contours of decreasing likelihood in the $\Omega_{\text{m}} - \beta$ plane for fits to $P^S(k_{\parallel}, \mathbf{k}_{\perp})$. Contours are plotted for χ^2 values corresponding to a one-parameter confidence of 68 per cent, and two-parameter confidence of 68, 95 and 99 per cent (dashed contours), calibrated using Monte Carlo simulations. Overlaid are the best-fit (dot-dash) and 1-σ (dot) values of β determined through consideration of mass clustering evolution. Significance contours given by joint consideration of the two constraints are also plotted for a one-parameter confidence of 68 per cent, and two-parameter confidence of 68, 95 and 99 per cent (solid contours).

Assuming instead an EdS $r(z)$, we find a degeneracy between the two parameters. The best fit model has $\Omega_b/\Omega_m = 0.16 \pm 0.09$ and $\Omega_m h = 0.23 \pm 0.04$, however a second solution with $\Omega_b/\Omega_m = 0.42$ and $\Omega_m h = 0.69$ cannot be rejected at 95 per cent confidence from this analysis alone. However, with $\Omega_m = 1$, the first solution would require a very low value of $h \sim 0.2$, whilst second solution requires a very high value of $\Omega_b \sim 0.4$. Neither of these solutions is therefore compatible with other observations.

2.1 Redshift-space distortions in the Power Spectrum

Alcock & Paczynski (1979) suggested that Λ might be measured from distortions in the shape of large-scale structure. Geometrical distortions occur if the wrong cosmology is assumed, due to the different dependence on cosmology of the redshift-distance relation along and across the line of sight. By making the simple assumption that clustering in real-space is on average spherically symmetric, we can attempt to detect these geometrical distortions by measuring the shape of clustering of 2QZ QSOs parallel and perpendicular to the observer's line of sight, through the power spectrum, $P^S(k_{\parallel}, \mathbf{k}_{\perp})$ (Outram et al. 2003b), shown in Fig. 2.

Unfortunately redshift-space distortions are also caused by peculiar velocities. On the large, linear scales probed by the power spectrum coherent peculiar velocities due to infall into overdense regions are the main cause of anisotropy. According to gravitational instability theory these distortions depend only on the parameter $\beta \approx \Omega_{\rm m}^{0.6}/b$. The effects of infall and geometry on the clustering distribution are very similar, and this approach alone could therefore only provide a degenerate constraint on Λ and β. This degeneracy can be broken, however, by jointly considering a second constraint based on evolution of the amplitude of mass clustering. Assuming a flat ($\Omega_m + \Omega_\Lambda = 1$) cosmology and a Λ $r(z)$, we find best fit values of $\Omega_m = 0.29^{+0.17}_{-0.09}$ and $\beta_q(z \sim 1.4) = 0.45^{+0.09}_{-0.11}$. Assuming instead an EdS $r(z)$ we find that the best fit model obtained, with $\Omega_m = 0.36^{+0.16}_{-0.11}$ and $\beta_q(z \sim 1.4) = 0.40^{+0.09}_{-0.09}$, is consistent with the Λ $r(z)$ results, and inconsistent with a $\Omega_\Lambda = 0$ flat cosmology at over 95 per cent confidence.

3. Weak Lensing

There is a 3σ anti-correlation between 2QZ QSOs and galaxy groups (Myers et al. 2003). If attributed to gravitational lensing by foreground structure, simple models suggest that more mass is present in these groups than can be accounted for in a Universe with $\Omega_m = 0.3$. An alternative explanation is dust in the galaxy groups obscuring background QSOs. The colours of 2QZ QSOs near to galaxy groups do not change relative to the field, however, meaning such dust would have to be significantly 'greyer' than even a Calzetti law.

4. Conclusions

The 2dF QSO Redshift Survey has obtained redshifts for 23,000 $B_J < 20.85$ $z < 2.5$ QSOs. Using the 2QZ, we are measuring large scale structure on unrivaled scales. Results from the QSO power spectrum, including redshift-space distortions in the power spectrum, appear consistent with ΛCDM, however lensing results suggest a higher value for Ω_m.

References

Alcock, C., Paczynski, B. 1979, Nature, 281, 358

Miller, C. J. & Batuski, D. J. 2001, ApJ, 551, 635

Myers, A. D., Outram, P. J., Shanks, T., Boyle, B. J., Croom, S. M., Loaring, N. S., Miller, L., & Smith, R. J. 2003, MNRAS, 342, 467

Outram, P. J., Hoyle, F., Shanks, T., Croom, S. M., Boyle, B. J., Miller, L., Smith, R. J., & Myers, A. D. 2003, MNRAS, 342, 483

Outram, P. J., Shanks, T., Boyle, B. J., Croom, S. M., Hoyle, F., Loaring, N., Miller, L., & Smith, R. J. 2003, MNRAS, submitted

Percival, W. J. et al. 2001, MNRAS, 327, 1297

REST-FRAME UV SPECTRA OF STAR-FORMING GALAXIES: FROM $Z \sim 3$ TO THE REDSHIFT DESERT

Alice E. Shapley,[1] Charles C. Steidel,[1] Kurt L. Adelberger,[2] Max Pettini,[3] Matthew P. Hunt,[1] Dawn K. Erb, [1] & Naveen Reddy [1]

[1]*Palomar Observatory, California Institute of Technology 105-24, Pasadena, CA 91125, USA*

[2]*Harvard-Smithsonian Center for Astrophysics, 60 Garden St., Cambridge, MA 02138, USA*

[3]*Institute of Astronomy, Madingley Road, Cambridge, CB3 0HA, UK*

Abstract We present the results of detailed studies of the astrophysical conditions in $z \sim 3$ Lyman Break Galaxies (LBGs), placing particular emphasis on what is learned from LBG rest frame UV spectra. By drawing from our database of ~ 1000 spectra, and constructing higher S/N composite spectra from galaxies grouped according to various parameters, we can show how the rest-frame UV spectroscopic properties systematically depend on other galaxy properties. Such information is crucial to understanding the detailed nature of LBGs, and their impact on the surrounding IGM. We then turn to a new survey of UV-selected star-forming galaxies at $z \sim 1.5 - 2.5$, in the so-called "redshift desert." Adjusting the $z \sim 3$ LBG color criteria to find similar types of galaxies at $z \sim 2$, and using a UV-optimized spectroscopic set-up, we have assembled nearly 700 galaxies spectroscopically confirmed to lie in this virtually unexplored, yet critical, period in the history of the universe. Ongoing and future work will probe the detailed astrophysical conditions in $z \sim 2$ galaxies.

Keywords: galaxies: evolution — galaxies: formation — galaxies: starburst

1. Introduction

Until now, the rest-frame UV spectra of Lyman Break Galaxies (LBGs) have been used primarily to measure redshifts (Steidel et al. 1996). Given the faint nature of LBGs (most have $\mathcal{R}_{AB} = 24 - 25.5$), the desire to observe a large sample results in individual spectra with low signal-to-noise ratios and spectral resolution. In most cases, the low signal-to-noise of LBG spectra precludes any analysis more detailed than the determination of redshifts.

One notable exception is the galaxy MS1512-cB58, an L^* LBG at $z = 2.73$ with an apparent magnitude of $V = 20.6$ due to lensing-magnification by a

M. Plionis (ed.), Multiwavelength Cosmology, 35-38.

factor of ~ 30 from a foreground cluster at $z = 0.37$ (Yee et al. 1996). Many detailed properties of cB58 have been determined from observations ranging in wavelength from the submillimeter to the X-ray. From high-resolution ($R \simeq 5000$) studies of its rest-frame UV spectrum alone, we have learned about this galaxy's stellar population, initial mass function (IMF), dust-content, chemical abundance pattern, and large-scale outflow velocity field (Pettini et al. 2000; Pettini et al. 2002). For the vast majority of unlensed LBGs, it is unfortunately not possible to obtain individual spectra which contain the same high-quality information about physical conditions. Since cB58 is only one object, we need to worry about how "typical" its continuum and spectroscopic properties are, relative to the range seen in the entire sample of LBGs.

While there is no hope of collecting data of comparable quality to the cB58 spectra for individual unlensed LBGs, we have assembled a sample of almost 1000 spectroscopically confirmed $z \sim 3$ galaxies over the past six years. By dividing our spectroscopic database into subsamples according to specific criteria, and creating high S/N composite spectra of each subsample, we hope to understand how the LBG spectroscopic properties depend in a systematic way on other galaxy properties. In section 2, we therefore describe our efforts to understand the physical conditions in the interstellar medium of LBGs using high S/N composite spectra. We would like to characterize the nature of both star-forming regions and the large-scale outflows of interstellar gas caused by the kinetic energy input from frequent supernova explosions. In section 3, we describe a new survey of UV-selected star-forming galaxies at $z \sim 1.5 - 2.5$. We have spectroscopically confirmed almost 700 galaxies in this critical redshift range, and will also attempt to characterize the detailed properties of the sample, using deep rest-frame UV spectra, near-IR imaging, and near-IR spectroscopy.

2. Detailed Astrophysical Properties of $z \sim 3$ Lyman Break Galaxies

Figure 1 shows a composite spectrum which is the average of our entire spectroscopic sample of 811 LBGs. Rest-frame UV spectra of LBGs are dominated by the emission from O and B stars with masses higher than $10 M_\odot$ and $T \geq 25000 K$. Marked in Figure 1 are spectroscopic features that trace the photospheres and winds of massive stars, neutral and ionized gas associated with large-scale outflows, and ionized gas in H II regions where star formation is taking place.

By grouping the database of LBG spectra according to galaxy parameters such as Lyα equivalent width, UV spectral slope, and interstellar kinematics, we isolate some of the major trends in LBG spectra which are least compromised by selection effects. We find that LBGs with stronger Lyα emission have bluer

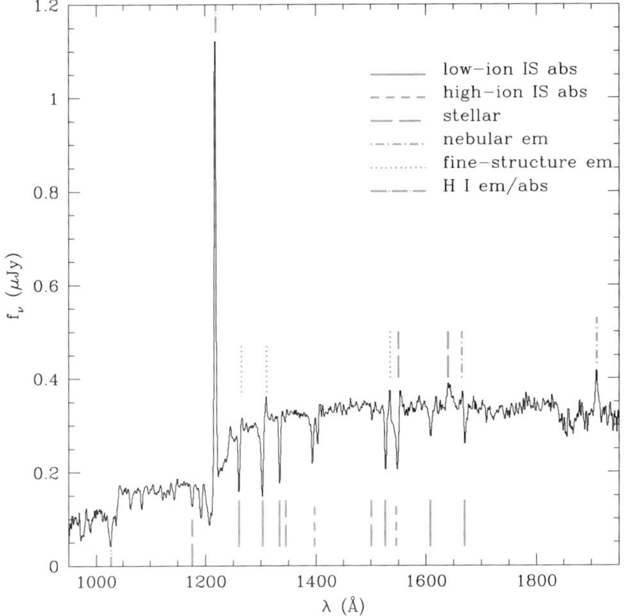

Figure 1. A composite rest-frame UV spectrum constructed from 811 individual LBG spectra.

UV continua, weaker low-ionization interstellar absorption lines, smaller kinematic offsets between Lyα and the interstellar absorption lines, and lower star-formation rates. There is a decoupling between the dependence of low- and high-ionization outflow features on other spectral properties. Most of the above trends can be explained in terms of the properties of the large-scale outflows seen in LBGs. According to this scenario, the appearance of LBG spectra is determined by a combination of the covering fraction of outflowing neutral gas which contains dust, and the range of velocities over which this gas is absorbing (Shapley et al. 2003). Higher sensitivity and spectral resolution observations are still required for a full understanding of the covering fraction and velocity dispersion of the outflowing neutral gas in LBGs, and its relationship to the escape fraction of Lyman continuum radiation in galaxies at $z \sim 3$.

3. Star-forming galaxies in the "Redshift Desert"

Figure 2a shows the region of $U G \mathcal{R}$ color-space inhabited by high-redshift star-forming galaxies. With the same intrinsic colors as $z \sim 3$ LBGs, star-forming $z \sim 2$ galaxies should inhabit the light-grey box immediately below the darker-grey $z \sim 3$ selection box along the $U - G$ axis. Applying these color criteria to select objects detected in deep ground-based optical ($U G \mathcal{R}$) images,

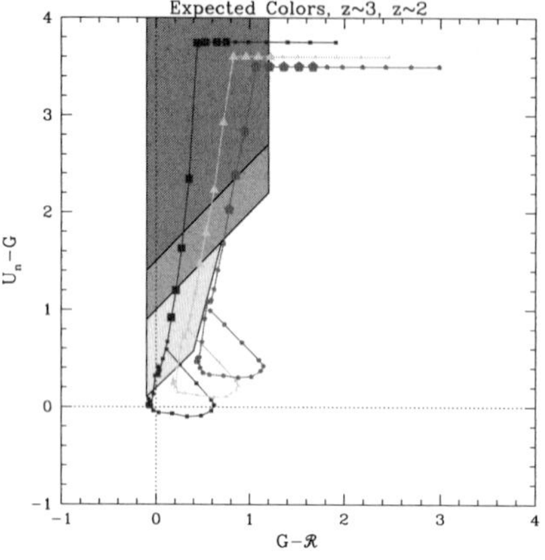

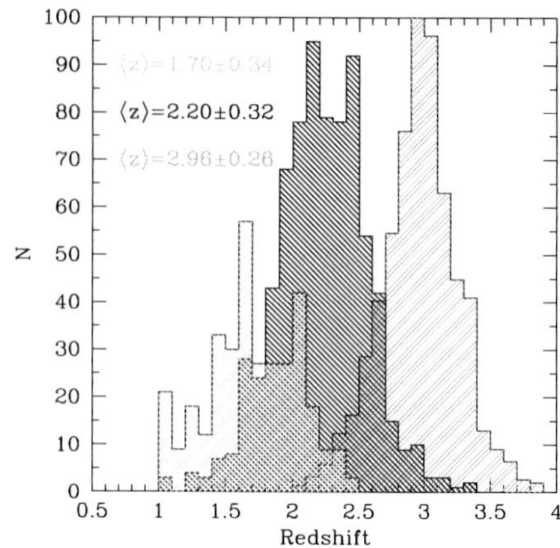

Figure 2. a) A plot of $UG\mathcal{R}$ color space showing the regions inhabited, respectively, by $z \sim 3$ galaxies (dark and medium grey regions) and $z \sim 2$ galaxies (light grey region). The tracks show how star-forming galaxies at different redshifts move through color space. The three tracks indicate different amounts of interstellar extinction. b) The redshift histograms of our "BM" ($\langle z \rangle = 1.70$) and "BX" ($\langle z \rangle = 2.20$) samples alongside the (normalized) $z \sim 3$ LBG sample.

we then follow up our $z \sim 2$ photometric candidates using the UV-sensitive Keck I/LRIS-B instrumental configuration. As shown in Figure 2b, we have spectroscopically confirmed almost 700 galaxies in the range $z = 1.4 - 2.5$, the so-called "redshift desert." Early rest-frame UV spectroscopic results for our $z \sim 2$ sample indicate that with integration times of 10 hours using Keck I/LRIS-B (combined with ancillary near-IR imaging and spectroscopic data), we will learn a vast amount of information about these galaxies including the nature of their stellar IMFs, stellar and interstellar metallicities, dynamical masses, stellar populations, and the nature of the large scale outflows which are also characteristic of $z \sim 2$ star-forming galaxies. Differential comparisons with the $z \sim 3$ sample will be important for studying the evolution of high-redshift star-forming galaxies.

References

Pettini et al. 2000, ApJ, 528, 96
Pettini et al. 2002, ApJ, 569, 742
Shapley et al. 2003, ApJ, 588, 65
Steidel et al. 1996, ApJ, 462, L17
Yee et al. 1996, AJ, 111, 1783

EVOLUTION OF THE GALAXY LUMINOSITY FUNCTION IN THE FORS DEEP FIELD (FDF)

A. Gabasch[1], R. Bender[1], U. Hopp[1], R.P. Saglia[1], S. Seitz[1], J. Snigula[1], I. Appenzeller[2], J. Heidt[2] D. Mehlert[2], S. Noll[2], A. Boehm[3], K.J. Fricke[3], K. Jaeger[3], B. Ziegler[3]

[1] *Universitaetssternwarte Muenchen, Scheinerstr. 1, D−81679 Muenchen, Germany*

[2] *Landessternwarte Heidelberg, Koenigstuhl, D−69117 Heidelberg, Germany*

[3] *Universitaets-Sternwarte Goettingen, Geismarlandstr. 11, D−37083 Goettingen, Germany*

gabasch@usm.uni-muenchen.de

Abstract With the large homogeneous dataset and the accurate photometric redshifts ($\Delta z/(z+1) \approx 0.03$) of the FORS Deep Field (FDF) we are able to study the evolution of the luminosity function in various bands from low ($\langle z \rangle = 0.3$) to high redshift ($\langle z \rangle = 4.5$). We find a strong brightening of M^* and decrease of ϕ^* with redshift in the rest-frame UV. We also find an increase of the star formation rate (SFR) between $0.5 < z < 1.5$, whereas the SFR remains approximately constant between $1.5 < z < 4.0$.

Keywords: galaxies: photometric redshift, luminosity function, star formation rate

1. The FORS Deep Field in Brief

The FORS Deep Field (Heidt et al., 2003) is a 9-band pencil beam survey observed with FORS at the VLT and SOFI at the NTT (NIR - data). It is about 8 times larger in area and nearly as deep as the HDFN. The 50% completeness limits for point sources are 26.5, 27.6, 26.9, 26.9, 26.8, ~ 25.5, ~ 25.8, 24.5, 23.5 in U, B, g, R, I, 834 nm, z, J and Ks (AB-system), respectively. The seeing varies from $0.5''$ in the I and z band to $1.0''$ in the U band. Furthermore the spectroscopic follow-up comprises more than $\sim$ 400 sources (Noll et al. 2003, in preparation). The following results are all based on an I band selected catalogue which contains about 5560 galaxies within the overlap area of all 9 filters. Throughout the following AB magnitudes and a flat cosmology ($\Omega_M = 0.3, \Omega_\Lambda = 0.7$) with $H_0 = 70\,\mathrm{km\,s^{-1}\,Mpc^{-1}}$ are used.

M. Plionis (ed.), Multiwavelength Cosmology, 39-42.

2. Photometric redshift

The 9 bands of the FDF and the large spectroscopic sample allow us to derive precise distances using the photometric redshift technique (Bender et al., 2001). This method seeks optimal agreement between redshifted spectral templates and the observed photometry. The semi-empirical templates were constructed by fitting stellar population models of Maraston (1998) of different dust extinction and age to the observed broad-band energy distribution of galaxies with spectroscopic redshift.

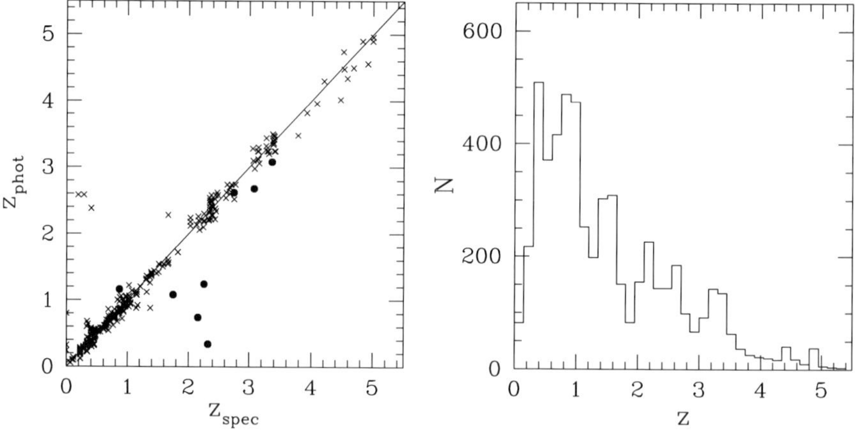

Figure 1. Comparison of spectroscopic and photometric redshifts of 362 objects. Crosses represent galaxies whereas filled circles represent QSOs.

Figure 2. Redshift distribution of all 5558 objects.

A comparison of the photometric and spectroscopic redshifts of 362 galaxies in our sample is shown in Figure 1. The agreement is very good with a typical error of only $\Delta z/(z+1) \approx 0.03$. Furthermore we have only 7 outliers out of 362. 4 of these outliers are QSOs for which a determination of a photometric redshift is difficult because of their power-law SED. A detailed description of the photometric redshifts in the FDF can be found in Bender et al. (2003, in preparation).

In Figure 2 the redshift distribution of the galaxies in the FDF used to derive the luminosity function and SFR is shown. Even at high redshift ($z > 4.0$) we have more than 150 objects.

3. Luminosity function

We use the best fitting SED as determined by the photometric redshift code and the SED itself to derive the rest-frame luminosity of the galaxies. This procedure reduces systematic influences of measurements in a single observer frame waveband. The error budget of the luminosity functions include both the photometric redshift error as well as the statistical error. We perform Monte Carlo simulations including the redshift probability distribution P(z) of the best fitting SED of every single object to account for the photometric redshift error. A detailed description of the evolution of the luminosity functions in different wavebands with redshift as derived from the FDF can be found in Gabasch et al. (2003, in preparation).

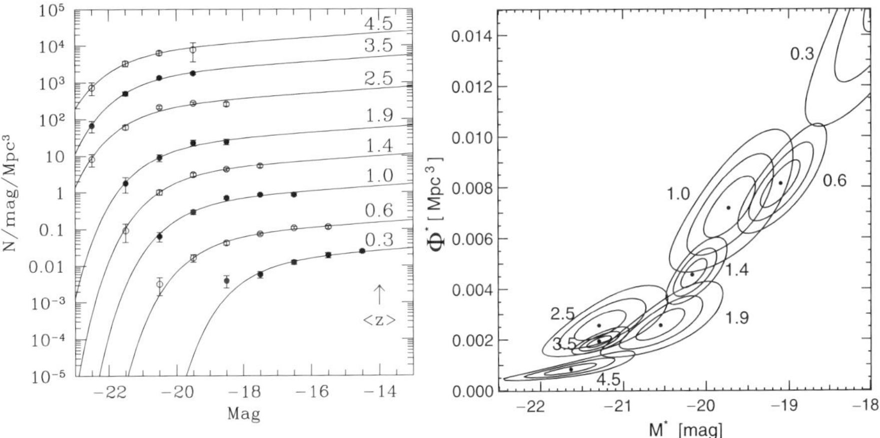

Figure 3. Luminosity functions at 280 nm from low ($\langle z \rangle = 0.3$) to high redshift ($\langle z \rangle = 4.5$). ϕ^* has been arbitrarily shifted by 1 dex per redshift bin for clarity.

Figure 4. 1σ, 2σ and 3σ confidence levels in Schechter parameter space for a fit to the luminosity function at 280 nm from $\langle z \rangle = 0.3$ to $\langle z \rangle = 4.5$ with a fixed slope $\alpha = -1.15$.

As an example we show the evolution of the luminosity function at 280 nm from $\langle z \rangle = 0.3$ to $\langle z \rangle = 4.5$ for a fixed slope $\alpha = -1.15$ in Figure 3 and Figure 4. We find a strong brightening of M^* and decrease of ϕ^* from low to high redshift. One can see from Figure 4 that we are able to derive M^* and ϕ^* with small errors, because we have about 700 galaxies in all but the last redshift bin. Although the variation of M^* and ϕ^* between adjacent redshift bins is in part influenced by large scale structure, the overall trend in the evolution of M^* and ϕ^* seems robust.

4. Star formation rate

From the luminosity density at 280 nm we derive the star formation rate (SFR) following Madau et al. (1998). The error budget of the SFR includes again the photometric redshift error as well as the statistical error.

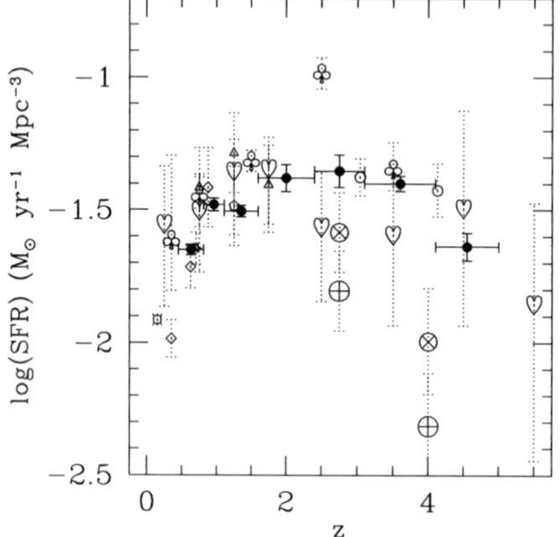

Figure 5. Evolution of the star formation rate derived from the luminosity density at 280 nm corrected for incompleteness, but not for dust.

In Figure 5 we show the evolution of the SFR assuming a Salpeter IMF. All points are corrected for incompleteness but, not for dust. The comparison with the literature (see Somerville et al. 2001 for the references) shows good agreement. Because of the large data sample of the FDF our errorbars are substantially smaller. A detailed description of the SFR in the FDF will be presented in Gabasch et al. (2003, in preparation).

Acknowledgments

This work was supported by the *Deutsche Forschungsgemeinschaft DFG*, SFB 375 *Astroteilchenphysik* and SFB 439 *Galaxien im jungen Universum.*

References

Bender, R. et al. (2001). In Christiani, S. editor, *ESO/ECF/STScI Workshop on Deep Fields*, pages 327, Berlin. Springer.

Heidt, J. et al., 2003 *A&A*, 398:49.

Madau, P., Pozzetti, L., and Dickinson, M. (1998). *ApJ*, 498:106.

Maraston, C. (1998). *MNRAS*, 300:872.

Somerville, R. S., Primack, J. R., and Faber, S. M. (2001). *MNRAS*, 320:504.

THE HALO OCCUPATION NUMBER AND SPATIAL DISTRIBUTION OF 2DF GALAXIES

Manuela Magliocchetti[1] & Cristiano Porciani[2]
[1]*SISSA, Via Beirut 4, 34014, Trieste, Italy*
[2]*Institute of Astronomy, HPF G3.2, ETH Hoenggerberg, 8093 Zuerich, Switzerland*

Abstract We use the clustering results obtained by Madgwick et al. (2003) for a sample of 96,791 2dF galaxies with redshift $0.01 < z < 0.15$ to study the distribution of late-type and early-type galaxies within dark matter haloes of different mass. The adopted method relies on the connection between the distribution of sources within haloes and their clustering properties by focusing on the issue of the halo occupation function i.e. the probability distribution of the number of galaxies brighter than some luminosity threshold hosted by a virialized halo of given mass. Within this framework, the distribution of galaxies within haloes is shown to determine galaxy-galaxy clustering on small scales, being responsible for the observed power-law behaviour at separations $r \lesssim 3$ Mpc. For a more extended analysis, we refer the reader to Magliocchetti & Porciani (2003).

Keywords: galaxies: clustering - cosmology: theory - cosmology: observations

1. Analysis and Results

Our approach follows the one adopted by Scoccimarro et al. (2001); in this framework, the galaxy-galaxy correlation function can be written as

$$\xi_g(\mathbf{x}-\mathbf{x}') = \xi_g^{1h}(\mathbf{x}-\mathbf{x}') + \xi_g^{2h}(\mathbf{x}-\mathbf{x}'), \quad (1)$$

where the first term ξ_g^{1h} accounts for pairs of galaxies residing within the same halo, while the ξ_g^{2h} represents the contribution coming from galaxies in different haloes. The above quantities can be written as a function of the mean number of galaxies per halo of mass m, $\langle N_{\rm gal}(m)\rangle$, of the spread about this mean value (also dependent on the mass of the halo hosting the galaxies), $\langle N_{\rm gal}^2(m)\rangle$, of the mean comoving number density of galaxies, $\bar{n}_g = \int n(m)\langle N_{\rm gal}(m)\rangle dm$ (where $n(m)$ is the halo mass function which gives the number density of dark matter haloes per unit mass and volume), of the two-point cross-correlation function between haloes of mass m_1 and m_2, $\xi(r, m_1, m_2)$ and finally of the (spatial) density distribution of galaxies within the haloes, $\rho_m(r)$.

M. Plionis (ed.), Multiwavelength Cosmology, 43-46.

$\langle N_{\rm gal}(m)\rangle$ and $\langle N^2_{\rm gal}(m)\rangle$ are the first and second moment of the halo occupation function $p(N_{\rm gal}|m)$ which gives the probability for a halo of specified mass m to contain $N_{\rm gal}$ galaxies. These can be parameterized as $\langle N_{\rm gal}\rangle = 0, (m/m_0)^{\alpha_1}, (m/m_0)^{\alpha_2}$ and $\langle N^2_{\rm gal}\rangle = \alpha(m)^2(\langle N_{\rm gal}\rangle)^2$ with $\alpha(m) = 0$, $\log(m/m_{\rm cut})/\log(m_0/m_{\rm cut}), 1$ in the different mass ranges $m < m_{\rm cut}$, $m_{\rm cut} \leq m < m_0$ and $m \geq m_0$. According to this approach, $m_{\rm cut}$, m_0, α_1 and α_2 are parameters to be determined by comparison with observations.

The last ingredient needed for the description of 2-point galaxy clustering is the spatial distribution of galaxies within their haloes. The first, easiest approach one can take is to assume that galaxies follow the dark matter profile (Navarro, Frenk & White 1997, hereafter NFW). However, since this assumption is not necessarily true, in the present analysis we also consider spatial distributions of the form $\rho_m(r) \propto (r)^{-\beta}$, with $\beta = 2, 2.5, 3$, where the first value corresponds to the singular isothermal sphere case. All the above profiles have been initially truncated at the virial radius $r_{\rm vir}$, as one expects galaxies to form within virialized regions, where the overdensity is greater than a certain threshold. However, this might not be the only possible choice since for instance – as a consequence of halo-halo merging – galaxies might also be found in the outer regions of the newly-formed halo, at a distance from the center greater than $r_{\rm vir}$.

In order to perform our analysis, we have considered the 4-dimensional grid: $-1 \leq \alpha_1 \leq 2$; $-1 \leq \alpha_2 \leq 2$; $10^9\ m_\odot \leq m_{\rm cut} \leq 10^{13}$; $m_{\rm cut} \leq m_0 \leq m_{\rm cut} \cdot 10^3$. Combinations of these four quantities have then been used to evaluate the mean number density of galaxies $\bar{n}_g$. Only values for $\bar{n}_g$ within 2σ from the observed ones (derived from the luminosity functions obtained by Madgwick et al. 2002) were accepted and the corresponding values for α_1, α_2, $m_{\rm cut}$ and m_0 have subsequently been plugged into equation (1) to produce – for a specified choice of the distribution profile – the predicted galaxy-galaxy correlation function to be compared with the Madwick et al. (2003) results on late-type and early-type galaxies by means of a least squares (χ^2) fit.

The main conclusions of this work are as follows:

1 Early-type galaxies (see Figure 1) are well described by a halo occupation number of the form broken power-law with $\alpha_1 \simeq -0.2$, $\alpha_2 \simeq 1.1$, $m_{\rm cut} \simeq 10^{12.6} m_\odot$ and $m_0 \simeq 10^{13.5} m_\odot$, where the two quantities which determine the intermediate-to-high mass behaviour of $\langle N_{\rm gal}\rangle$ are measured with a good accuracy.

2 No model can provide a reasonable fit to the correlation function of late-type galaxies since they all show an excess of power with respect to the data on scales $0.5 \lesssim r/[\rm Mpc] \lesssim 2$. In order to obtain an acceptable description of the observations, one has to assume that star-forming galaxies are distributed within haloes of masses comparable to those of groups

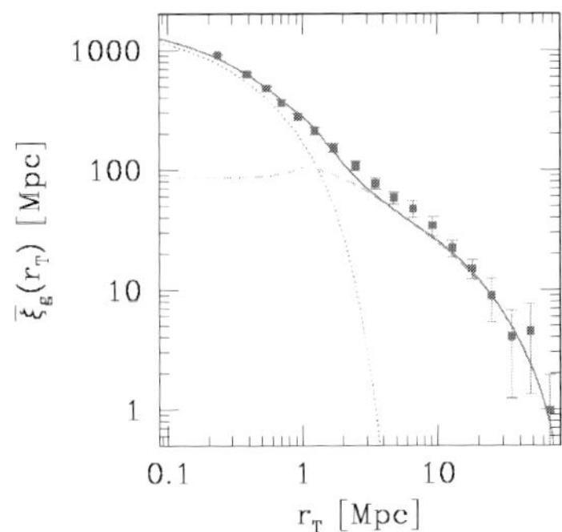

Figure 1. Projected correlation function of early-type galaxies. Data-points represent the results from Madgwick et al. (2003), while the solid curve is the best fit to the measurements obtained for a halo number density of the form broken power-law, with $\alpha_1 = -0.2$, $\alpha_2 = 1.1$, $m_{\rm cut} = 10^{12.6} m_\odot$, $m_0 = 10^{13.5} m_\odot$ and for galaxies distributed within their dark matter haloes according to a NFW profile. Dashed and dotted lines respectively indicate the contribution ξ_g^{2h} from galaxies residing in different haloes and the ξ_g^{1h} term originating from galaxies within the same halo.

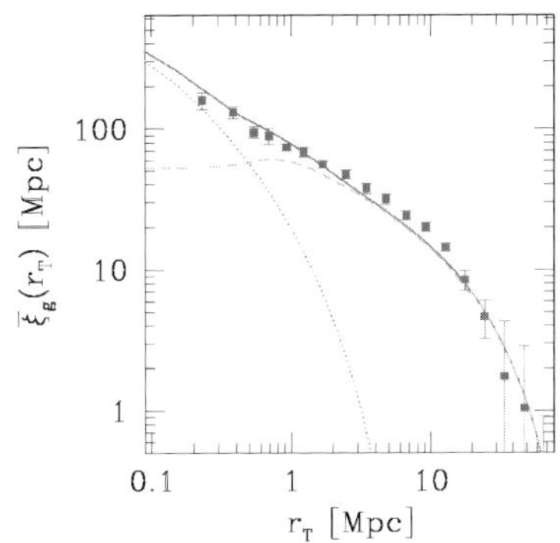

Figure 2. Projected correlation function of late-type galaxies. Data-points represent the results from Madgwick et al. (2003), while the solid curve is the best fit to the measurements obtained for a halo number density of the form broken power-law, with $\alpha_1 = -0.4$, $\alpha_2 = 0.7$, $m_{\rm cut} = 10^{11} m_\odot$, $m_0 = 10^{11.4} m_\odot$ and for galaxies distributed within their dark matter haloes according to a NFW profile with $r_{\rm cut} = 2 \cdot r_{\rm vir}$. Dashed and dotted lines respectively indicate the contribution ξ_g^{2h} from galaxies residing in different haloes and the ξ_g^{1h} term originating from galaxies within the same halo.

and clusters up to two virial radii. This result is consistent with the phenomenon of morphological segregation whereby late-type galaxies are mostly found in the outer regions of groups or clusters (extending well beyond their virial radii), while passive objects preferentially sink into their centres.

3 With the above result in mind, one finds that late-type galaxies (see Figure 2) can be described by a halo occupation number of the form single power-law with $\alpha_2 \simeq 0.7$, $m_{\rm cut} \simeq 10^{11} m_\odot$ and $m_0 \simeq 10^{11.4} m_\odot$, where the quantities which describe $\langle N_{\rm gal} \rangle$ in the high-mass regime are determined with a high degree of accuracy.

4 Within the framework of our models, galaxies of any kind seem to follow the underlying distribution of dark matter within haloes as they present the same degree of spatial concentration. In fact the data indicates both early-type and late-type galaxies to be distributed within their

host haloes according to NFW profiles. We note however that, even though early-type galaxies can also be described by means of a shallower distribution of the form $\rho(r) \propto r^{-\beta}$ with $\beta = 2$, this cannot be accepted as a fair modeling of the data in the case of late-type galaxies which instead allow for somehow steeper ($\beta \simeq 2.5$) profiles. In no case a $\beta = 3$ density run can provide an acceptable description of the observed correlation function. These conclusions depend somehow on assuming a specific functional form for the second moment of the halo occupation distribution. However, Magliocchetti & Porciani (2003) have shown that there is not much freedom in the choice of this function if one wants to accurately match the observational data.

An interesting point to note is that results on the spatial distribution of galaxies within haloes and on their halo occupation number are independent from each other. There is no degeneracy in the determination of $\langle N_{\rm gal}\rangle$ and $\rho(r)$ as they dominate the behaviour of the two-point correlation function ξ_g at different scales. Different distribution profiles in fact principally determine the slope of ξ_g on small enough ($r \lesssim 1$ Mpc) scales which probe the inner regions of the haloes, while the halo occupation number is mainly responsible for the overall normalization of ξ_g and for its slope on large-to-intermediate scales.
Our analysis shows that late-type galaxies can be hosted in haloes with masses smaller than it is the case for early-type objects. This is probably due to the fact that early-type galaxies are on average more massive (where the term here refers to stellar mass) than star-forming objects, especially if one considers the population of irregulars, and points to a relationship between stellar mass of galaxies and mass of the dark matter haloes which host them.

References

Madgwick D.S., et al. (the 2dFGRS Team), 2002, MNRAS, 333, 133
Madgwick D.S., et al. (the 2dFGRS Team), 2003, MNRAS, submitted (astro-ph/0303668)
Magliocchetti M., Porciani C., 2003, to appear on MNRAS, astro-ph/0304003
Navarro J.F., Frenk C.S., White S.D.M., 1997, ApJ, 490, 493
Scoccimarro R., Sheth R.K., Hui L.,Jain B., 2001, ApJ, 546, 20

THE FORMATION OF THE HUBBLE SEQUENCE

Christopher J. Conselice
California Institute of Technology
cc@astro.caltech.edu

Abstract Understanding when galaxies form via star formation histories and stellar mass assembly rates is becoming known with some certainty, yet the connection between high redshift and low redshift galaxy populations is not yet clear. By identifying and studying individual massive galaxies at high-redshifts, $z > 1.5$, we can uncover the physical effects that drove galaxy formation. Using the structures of high-z galaxies, as imaged with the Hubble Space Telescope, we argue that it is now possible to directly study the progenitors of ellipticals and disks. We also briefly describe early results that suggest many massive galaxies are forming at $z > 2$ through major mergers.

1. Introduction

Astronomers have come a long way in the last decade towards detecting and studying galaxies out to redshifts $z \sim 6$. However, the principle goal of observational galaxy formation studies, understanding how high redshift and low redshift populations are connected, and thus *how* galaxies form, remains largely unknown. The traditional method of studying galaxy formation is to try and answer *when* galaxies formed, e.g., through star formation and stellar mass assembly histories and by detecting various galaxy populations such as extremely red objects, Lyman-alpha emitters, and Lyman-break galaxies. These galaxies, and their stellar populations, tell us when galaxies formed, but they do not, and empirically cannot, tell us how they formed.

An approach to this problem is to use ancillary information about galaxies not available from spectroscopy or integrated photometry. One method is to utilize structural information from high-z galaxies, imaged with HST, to determine when normal galaxies (hereafter defined as giant ellipticals and spirals) formed, as well as how. Techniques for doing this are now developed (Conselice 2003) and early results demonstrate that we can identify and trace the physical processes responsible for the formation of galaxies at $z > 1.5$. In this article we briefly describe these approaches for understanding the formation of the Hubble sequence, or normal galaxies, and give some preliminary results.

M. Plionis (ed.), Multiwavelength Cosmology, 47-50.

2. Galaxies Near and Far

Bright and massive galaxies in the nearby universe are disks and ellipticals. At higher redshifts, the brightest galaxies in rest-frame UV selected samples nearly all have structural peculiarities (Conselice et al. 2003a). When did normal disks and ellipticals form? Figure 1a demonstrates through an I < 27 selected sample in the Hubble Deep Field North (HDF-N) that at high redshift the galaxy population is dominated by peculiars, while at low-z ellipticals and spirals are common. This transition is robust to effects of resolution, noise, and morphological k-corrections (Conselice et al. 2004, in prep.)

A fundamental question to ask is whether these high redshift peculiars, often called Lyman-break galaxies (LBGs), transform into disks and ellipticals. LBGs are likely the progenitors of massive nearby systems, based on their clustering properties (Giavalisco et al. 1998). However, LBGs are observed to have lower stellar masses, generally $<$ M* (e.g., Papovich et al. 2001). If LBGs passively evolve, they will not contain enough mass to become massive, $>$ M*, galaxies at $z \sim 0$. If we can determine the future evolution of LBGs, and other high redshift galaxy populations, we can begin to piece together the history of galaxy formation.

3. The Galaxy Merger History

There are a few major methods by which galaxies can form. The first is an early collapse when stars form in rapid bursts which then passively evolve. Other methods are due to hierarchical accretion of intergalactic material, or mergers with other galaxies. The formation methods of galaxies are hard to trace, although based on the star formation history of the universe it is unlikely that most galaxies formed rapidly and early (e.g., Madau et al. 1998).

One method that can be traced is the incidence of major mergers. Major mergers in the nearby universe create distinct disturbed asymmetric morphologies that can be distinguished from pure star forming systems through the use of indices that measure asymmetries and the clumpiness of light distributions (Conselice et al. 2000; Bershady et al. 2000; Conselice 2003). The basic idea is that galaxies which are asymmetric, without a corresponding high degree of light clumpiness, are likely to be involved in major mergers (Conselice 2003). How well does the asymmetry index (A) identify known mergers? Figure 1b shows the deviation in sigma units between asymmetry (A) and clumpiness (S) values for nearby normal galaxies and merging ULIRGs. While normal galaxies have mostly low σ deviations from the A-S correlation, ULIRGs span a much larger range, including the most highly asymmetric ones.

We can use asymmetry values of HDF-N galaxies to measure the evolution of implied major merger fractions out to $z \sim 3$. We avoid morphological K-corrections by using the rest-frame B-band asymmetries of galaxies in the

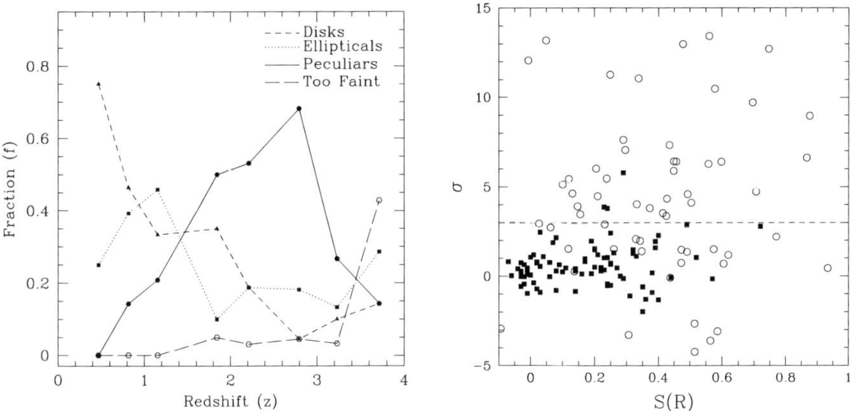

Figure 1. Left panel - the morphological distribution as a function of type in the HDF, normalized to I< 27 counts. Right panel - the sigma deviation from the asymmetry-clumpiness correlation, showing that major mergers deviate by more than 3σ from the relationship. Normal galaxies are solid boxes and major mergers (ULIRGs) are open circles.

HDF-N (Conselice et al. 2003a). We define major mergers as galaxies with rest-frame B-band asymmetry larger than $A_{\rm merger} = 0.35$. By taking the ratio of galaxies with asymmetries $A > A_{\rm merger}$ to the total number of galaxies in a given parameter range, implied merger fractions out to $z \sim 3$ can be computed as a function of absolute magnitude (M_B), stellar mass (M_*), and redshift (z). This allows us to determine how and when galaxies formed as a function of stellar mass and time.

Major merger fractions for galaxies brighter than $M = -21$ and -19 in the HDF-N are plotted in Figure 2a, along with semi-analytical model predictions of the same quantities. From this it appears that a large fraction of the brightest galaxies are undergoing mergers, while fainter systems generally do not. The fraction of bright galaxies undergoing major mergers drops quickly at lower redshifts, while the fainter systems have merger fractions that remain largely constant with redshift (Conselice et al. 2003a). Semi-analytic model predictions of the same quantities, based on Durham group simulations, do a relatively good job of predicting merger fractions at high redshift, but over predict the degree of major merging at lower redshifts for the most luminous and massive galaxies (Figure 2a).

4. Merger and Stellar Mass Accretion Rates

Using further information we can investigate the merger and stellar mass accretion rates of galaxies from $z = 0$ out to $z \sim 3$, or to when the universe was only ~ 2 Gyr old. By assuming a merger time scale of 1 Gyr, and that each merger consists of two galaxies of equal mass, we can measure the rate of merging, and the stellar mass accretion rate due to major mergers. We can

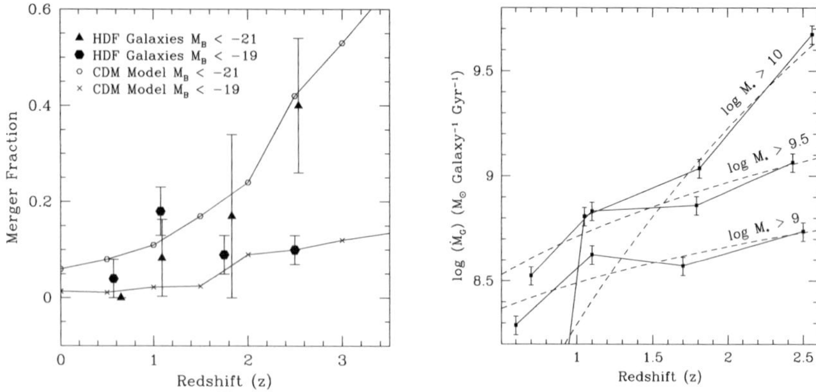

Figure 2. Left panel - major merger fractions to $z \sim 3$ at magnitude limits $M_B = -21$ and -19. Semi-analytical model predictions are also shown. Right panel - stellar mass accretion history from major mergers as a function of initial mass (see Conselice et al. 2003a).

also determine how this accretion history varies with initial stellar mass (Figure 2b; Conselice et al. 2003a,b). Based on these empirical measurements, the amount of stellar mass added to a M $> 10^{10}$ $M_{\odot}$ galaxy due to star formation and merging is enough to create a massive $> M^*$ galaxy at $z \sim 0$ (Conselice 2004, in prep).

Disk galaxies likely cannot form through these merger processes as they would have a hard time surviving mergers at high redshift. These systems are therefore likely forming at about the same time they appear morphologically, at $z \sim 1.5$ (Figure 1). These galaxies have now possibly been identified at $z > 1.5$ by their low light concentrations in GOODS ACS images (Conselice et al. 2003c). These luminous diffuse objects (LDOs) are fairly common at redshifts $1 < z < 2$ and have co-moving volumes similar to nearby massive disks. Follow up on these systems is now in progress.

References

Bershady, M.A., Jangren, A., & Conselice, C.J. 2000, AJ, 126, 1183

Conselice, C.J., et al. 2000, ApJ, 529, 886

Conselice, C.J. 2003, ApJS, 147, 1

Conselice, C.J., et al. 2003a, AJ, 126, 1183

Conselice, C.J., Chapman, S.C., & Windhorst, R.A. 2003b, ApJ, 596, 5L

Conselice, C.J., et al. 2003c, ApJ in press, astro-ph/0309039

Giavalisco, M., et al. 1998, ApJ, 503, 543

Madau, P., Pozzetti, L., & Dickinson, M. 1998, ApJ, 498, 106

Papovich, C., Dickinson, M., & Ferguson, H.C. 2001, ApJ, 559, 620

CLUSTERING OF HIGH REDSHIFT GALAXIES IN THE CANADA-FRANCE DEEP FIELDS SURVEY AND VIRMOS DEEP IMAGING SURVEY

S.Foucaud[1,3], H.J. McCracken[2,3], O. Le Fèvre[3], M.Brodwin[4], Y.Mellier[5,6], S.Arnouts[3,7], E.Bertin[5,6], D.Crampton[8], J.-C. Cuillandre[9], M.Dantel-Fort[6], S.D.J. Gwyn[3], S.J. Lilly[10] & M.Radovich[5,11] for the CFDF & VIRMOS collaboration

[1] *IASF-MI, via Bassini 15, 20133 Milano, Italy*
[2] *Univ. of Bologna, Dept. of Astronomy, via Ranzani 1, 40127 Bologna, Italy*
[3] *LAM, Traverse du Siphon, 13376 Marseille Cedex 12, France*
[4] *Univ. of Toronto, Dept. of Astronomy, 60 St. George Str., Toronto, Ontario, Canada M5S 3H8*
[5] *IAP, 98 bis Boulevard Arago, 75014 Paris, France*
[6] *Observatoire de Paris, LERMA, 61 Avenue de l'Observatoire, 75014 Paris, France*
[7] *ESO, Karl-Schwarzschild-Str. 2, 85748 Garching bei Munchen, Germany*
[8] *HIA, 5071 West Saanich Rd., Victoria, British Colombia, Canada V9E 2E7*
[9] *CFHT, 65-1238 Mamalahoa Highway, Kamuela, HI 96743, USA*
[10] *Institute of Astronomy - ETH Hoenggerberg, HPF D8, 8093 Zurich, Switzerland*
[11] *Osservatorio di Capodimonte, via Moiariello 16, 80131 Napoli, Italy*

Abstract The Canada-France Deep Fields (CFDF) and the VIRMOS Deep Imaging Survey (VDIS) are deep, wide angle, multi-colour imaging surveys. The CFDF covers ~ 0.65 deg^2 in $UBVI$ and reaches a limiting magnitude of $U_{AB}(3\sigma, 3'') \sim 27.5$. The VDIS deep field covers ~ 1.2 deg^2 in $BVRI$ and reaches a limiting magnitude of $B_{AB}(3\sigma, 3'') \sim 26.5$. With these data, we are able to identify some of the largest samples of photometrically selected Lyman-break galaxies (LBGs) at $z \sim 3$ to a limiting magnitude of $I_{AB}(3\sigma, 3'') \sim 24.5$ and at $z \sim 4$ to a limiting magnitude of $I_{AB}(3\sigma, 3'') \sim 24$. We select 1294 $z \sim 3$ LBGs for $20 < I_{AB} < 24.5$, and 194 $z \sim 4$ LBGs for $20 < I_{AB} < 24$. We determine the correlation length in comoving space, r_0 and, using Cold Dark Matter (CDM) models, we estimate a linear bias, b, for LBGs. For Λ-flat cosmological parameters, we measure $r_0 = (5.9 \pm 0.5)h^{-1}$ Mpc and $b = 3.5 \pm 0.3$ at $z \sim 3$ and $r_0 = (7.4 \pm 1.2)h^{-1}$ Mpc and $b = 5.4 \pm 0.8$ at $z \sim 4$. Although, the CFDF data indicates in one field (at a 3σ level) that r_0 depends on sample limiting magnitude: brighter LBGs are more clustered than fainter ones. The strong clustering and the large bias of high-redshift LBGs are consistent with models of biased galaxy formation and show that they are associated with massive CDM halos.

Keywords: Galaxies: observations: high-redshift – evolution. Cosmology: LSS of Universe

M. Plionis (ed.), Multiwavelength Cosmology, 51-54.

1. Lyman-break galaxy samples

We used the colour-colour method of high redshift galaxy selection developed by Steidel et al. (1999) using the Lyman break feature, to select $z \sim 3$ and $z \sim 4$ galaxies in two surveys. From catalogues created from our set of CFDF $UBVI$ data (covering ~ 0.65 deg^2 and described in McCracken et al., 2001 and Foucaud et al., 2003), using a selection in a $(U - B)_{AB}$ vs. $(B - I)_{AB}$ colour-colour diagram, we extracted samples of LBGs at redshift $z \sim 3$. From our set of VDIS $BVRI$ data (covering ~ 1.2 deg^2 and described in Le Fèvre et al., 2003 and McCracken et al., 2003), using a $(B - V)_{AB}$ vs. $(V - I)_{AB}$ colour-colour diagram, we extracted a sample at redshift $z \sim 4$. Simulated catalogues indicate that the redshift ranges of our LBGs are $2.9 < z < 3.5$ in the CFDF and $3.9 < z < 4.5$ in the VDIS. As the CFDF and the VDIS are reaching limiting magnitudes of $U_{AB}(3\sigma, 3'') \sim 27.5$ and $B_{AB}(3\sigma, 3'') \sim 26.5$ respectively, we were able to select samples consisting of 1294 LBGs over three different fields for $20 < I_{AB} < 24.5$, sub-samples in one field of 49 LBGs for $20 < I_{AB} < 23.5$ and 358 LBGs for $23.5 < I_{AB} < 24.5$ at $z \sim 3$, and a sample consisting of 194 LBGs over one field for $20 < I_{AB} < 24$ at $z \sim 4$ respectively. We measure a surface density of LBGs of $n = (0.8 \pm 0.2)$ arcmin^{-2} at $z \sim 3$ for $20 < I_{AB} < 24.5$ (mean over fields; errors from field-to-field variance) and $n = (0.05 \pm 0.005)$ arcmin^{-2} at $z \sim 4$ for $20 < I_{AB} < 24$ (poissonian error).

2. Clustering results

Using the estimator of Landy & Szalay (1993), we found that for our $z \sim 3$ and $z \sim 4$ samples the two-point projected galaxy correlation functions are well fitted by power-laws with a slope of $\delta = 0.8$. For two of our CFDF fields, we were able to fit simultaneously for both the slope δ and the amplitude A_ω, and as shown in figure 1a, the best fitted slopes are confirmed to be consistent with $\delta = 0.8$. Furthermore, in one of our CFDF field, measurements indicate that samples with fainter limits magnitudes have lower values of clustering amplitude. Comparing our brightest $20.0 < I_{AB} < 23.5$ and our faintest $23.5 < I_{AB} < 24.5$ samples, we detect this effect with a 3σ confidence level, as shown in the figure 1b. For these two sub-samples we also find approximately the same values of the slope. This segregation of clustering strength with apparent magnitudes is expected in the framework of galaxy biased formation, and is for the first time detected within a given high redshift galaxy sample.

From our measurements of correlation function amplitudes, we determine correlation lengths in comoving space, r_0, for a Λ-flat cosmology. At $z \sim 3$, we find $r_0 = (5.9 \pm 0.5)h^{-1}$ Mpc for $20 < I_{AB} < 24.5$ and at $z \sim 4$, $r_0 = (7.4 \pm 1.2)h^{-1}$ Mpc for $20 < I_{AB} < 24$. All errors are computed from

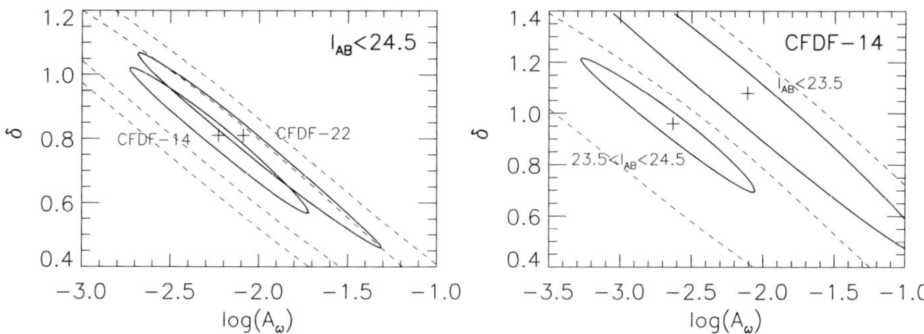

Figure 1. (a) Contours of χ^2 for the mean $\omega(\theta)$ computed for Lyman-break galaxies selected with $I_{AB} < 24.5$ in the CFDF-14hr and CFDF-22hr fields. (b) Contours of χ^2 for $\omega(\theta)$ for two subsamples of Lyman-break galaxies with $20.0 < I_{AB} < 23.5$ and $23.5 < I_{AB} < 24.5$ in the CFDF-14hr field. In those two plots, the plus symbol shows the best-fitting amplitudes and slope, and two contours correspond to the 1σ (thick contours) and 3σ confidence levels.

the poissonian errors.

In figure 2a, we present the comoving correlation length r_0 as a function of redshift for Λ-flat cosmological parameters. For comparison we show different values obtained from clustering found in the literature computed using the same cosmological parameters. The lines represents predicted evolution of distribution of the underlying dark matter. This implies that, as bias depends on redshifts and scales, the evolution of galaxy clustering is related of dark matter mass in a complex way.

In figure 2b, we present the comoving correlation length r_0 as a function of surface density (computed from the predicted number of LBGs per arcmin^2) for Λ-flat cosmological parameters. The line refers to the prediction from ΛCDM "transient" model (Moscardini et al., 1998). For comparison we show results obtained from clustering of LBGs by Adelberger et al. (1998) and from clustering of galaxies selected with photometric redshift by Arnouts et al. (1999, 2002) for the same sample of redshift. It reinforces that LBGs indicate locations of halos with a good efficiency. Furthermore, this figure shows that LBGs are strongly clustered, which implies that LBGs trace massive CDM halos, and the more massive are the halos, the more the LBGs are clustered and the less they are abundant. This decrease of the clustering strength with density is predicted by hierarchical models of structure formation and is consistent with the theory of biased galaxy formation. Using CDM models, we estimate a linear bias, b, for LBGs at $z \sim 3$ of $b = 3.5 \pm 0.3$ and at $z \sim 4$ of $b = 7.4 \pm 1.2$ in a Λ-flat cosmology. This study, in agreement with previous ones, provide evidence for a picture in which structures form hierarchically and massive objects form at highest peaks in the underlying density field (Bardeen et al., 1986).

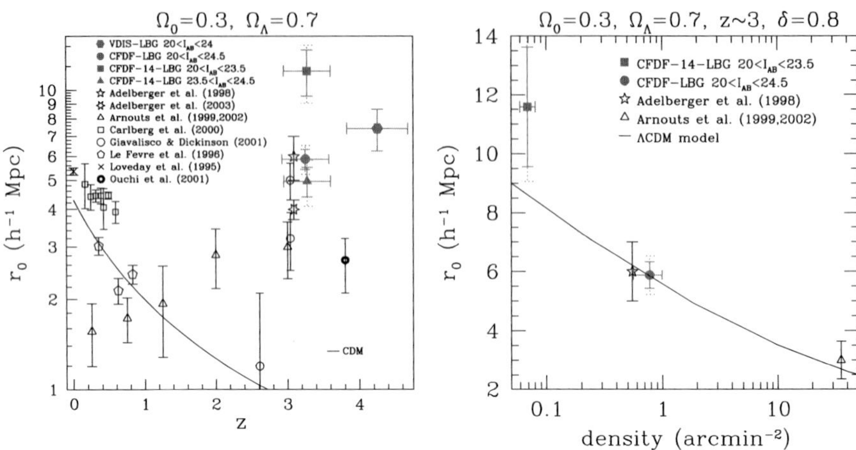

Figure 2. (a) The measured comoving correlation length r_0 for our three CFDF LBGs samples and our VDIS LBGs sample as a function of redshift for Λ-flat cosmological parameters. (b) The comoving correlation length r_0 for two LBGs samples at redshift $z \sim 3$ as a function of surface density for Λ-flat cosmological parameters.

Acknowledgments

This work has been partly performed under the framework of the VIRMOS consortium. SF's work has been supported by RTN Euro3D postdoctoral fellowship. SF wish to acknowledge the use of TERAPIX computer facilities at the Institut d'Astrophysique de Paris, where part of this work was carried out.

References

Adelberger, K. L., Steidel, C. C., Giavalisco, M., et al. 1998, ApJ, 505, 18

Adelberger, K. L., Steidel, C. C., Shapley, A. E., & Pettini, M. 2003, ApJ, 584, 45

Arnouts, S., Cristiani, S., Moscardini, L., et al. 1999, MNRAS, 310, 540

Arnouts, S., Moscardini, L., Vanzella, E., et al. 2002, MNRAS, 329, 355

Bardeen, J. M., Bond, J. R., Kaiser, N., & Szalay, A. S. 1986, ApJ, 304, 15

Carlberg, R. G., Yee, H. K. C., Morris, S. L., et al. 2000, ApJ, 542, 57

Foucaud, S., McCracken, H. J., Le Fèvre, O., et al. 2003, A&A *accepted*, astro-ph/0306585

Giavalisco, M. & Dickinson, M. 2001, ApJ, 550, 177

Landy, S. D. & Szalay, A. S. 1993, ApJ, 412, 64

Le Fèvre, O., Hudon, D., Lilly, S. J., et al. 1996, ApJ, 461, 534+

Le Fèvre, O., Mellier, Y., McCracken, H. J., et al. 2003, A&A *submitted*, astro-ph/0306252

Loveday, J., Maddox, S. J., Efstathiou, G., & Peterson, B. A. 1995, ApJ, 442, 457

McCracken, H. J., Le Fèvre, O., Brodwin, M., et al. 2001, A&A, 376, 756

McCracken, H. J., Radovich, M., Bertin, E., et al. 2003, A&A *accepted*, astro-ph/0306254

Moscardini, L., Coles, P., Lucchin, F., & Matarrese, S. 1998, MNRAS, 299, 95

Ouchi, M., Shimasaku, K., Okamura, S., et al. 2001, ApJ, 558, L83

Steidel, C. C., Adelberger, K. L., Giavalisco, M., et al. 1999, ApJ, 519, 1

ANGULAR CLUSTERING WITH PHOTOMETRIC REDSHIFTS IN THE SDSS: BIMODALITY IN THE CLUSTERING PROPERTIES OF GALAXIES

Tamás Budavári[1], Andrew J. Connolly[2], Alexander S. Szalay*[1], István Szapudi[3], István Csabai[4], Ryan Scranton[2]

[1] *Dept. of Physics and Astronomy, The Johns Hopkins University, Baltimore, MD 21218*

[2] *Department of Physics and Astronomy, University of Pittsburgh, Pittsburgh, PA 15260*

[3] *Institute for Astronomy, University of Hawaii, Honolulu, HI 96822*

[4] *Department of Physics, Eotvos University, Budapest, Pf. 32, Hungary, H-1518*

Abstract Understanding the clustering of galaxies has long been a goal of modern observational cosmology. Redshift surveys have been used to measure the correlation length as a function of luminosity and color. However, when subdividing the catalogs into multiple subsets, the errors increase rapidly.

Utilizing our photometric redshift technique a volume limited sample ($0.1 < z < 0.3$) containing more than 2 million galaxies is constructed from the SDSS galaxy catalog. In the largest such analysis to date, we study the angular clustering as a function of luminosity and spectral type. Using Limber's equation we calculate the clustering length for the full data set as $r_0 = 5.77 \pm 0.10\, h^{-1}\mathrm{Mpc}$. We find that r_0 increases with luminosity by a factor of 1.6 over the sampled luminosity range, in agreement with previous redshift surveys. We also find that both the clustering length and the slope of the correlation function depend on the galaxy type. In particular, by splitting the galaxies in four groups by their rest-frame type we find a bimodal behavior in their clustering properties.

Keywords: observational cosmology, galaxy clustering, photometric redshifts

1. Introduction

One of the primary tools for studying the evolution and formation of structure within the universe has been the angular correlation function. Angular clustering in magnitude-limited photometric surveys has the advantage of much larger catalogs, but suffers from a dilution of the clustering signal due to the broad

*Talk presented by Alexander S. Szalay

M. Plionis (ed.), Multiwavelength Cosmology, 55-58.

radial distribution of the sample. Also, up to now it has not been possible to select uniform subsamples based on physical parameters, like luminosity and rest-frame color.

We can overcome many of the disadvantages of the angular clustering if we utilize photometric redshifts (Connolly et al. 1995). Photometric redshifts provide a statistical estimate of the redshift, luminosity and type of a galaxy based on its broadband colors (Budavári et al. 2000, Csabai et al. 2003). As we can control the redshift interval from which we select the galaxies (and the distribution of galaxy types and luminosities) we can determine how the clustering signal evolves with redshift and invert it accurately to estimate the real space clustering length, r_0, for galaxies.

2. Rest-frame Selected Samples

We study a volume-limited sample of galaxies selected from the Sloan Digital Sky Survey (York et al. 2000) that extends out to a redshift $z_{\rm phot} = 0.3$ with limiting absolute magnitude $M_{r^*} = -19.97$. We further restrict the data set to those galaxies more distant than redshift of 0.1. The final catalog size is over 2 million galaxies. To analyze how angular clustering changes with luminosity, the volume limited sample is divided into three absolute magnitude bins. These subsamples represented by M_1, M_2 and M_3 have limiting absolute magnitudes $M_{r^*} > -21$, $-21 > M_{r^*} > -22$ and $-22 > M_{r^*} > -23$ respectively. The size of these subsets decrease by approximately a factor of two as a function of increasing luminosity.

We subdivide by spectral class breaking the luminosity classes into four subgroups (each with comparable numbers of galaxies). The first class consists of galaxies with SEDs similar to the Coleman, Wu & Weeman (1980) elliptical template (Ell), the second, third and fourth classes contain a broader distribution of galaxy types approximately corresponding to Sbc, Scd and Irr types, respectively.

3. The Analysis in a Nutshell

We utilize a novel approach to calculate the correlation function. Our implementation of the Landy–Szalay (1993) estimator is based on the Fast Fourier Transformation (Szapudi et al. 2001). Taking the masks into account, first we compute the 2D correlation function, then average over rings to derive $w(\theta)$. The method also proves to be an exceptionally sensitive diagnostic tool.

From the angular clustering we can derive the spatial correlation length, r_0, using Limber's equation. We apply Bayes' theorem to estimate the redshift distribution assuming a luminosity function and Gaussian uncertainty in the photometric redshifts.

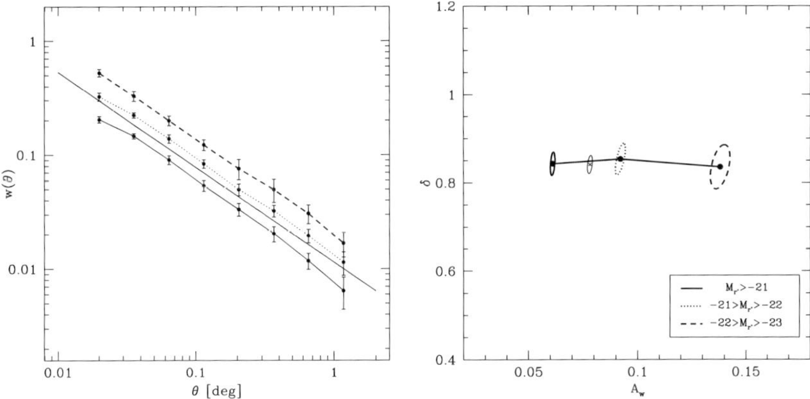

Figure 1. The clustering strength changes as a function of absolute magnitude without change in the slope. The correlation function and the parameter fits are shown in the left and right panels, respectively. The straight line represents the best fit to the fiducial sample.

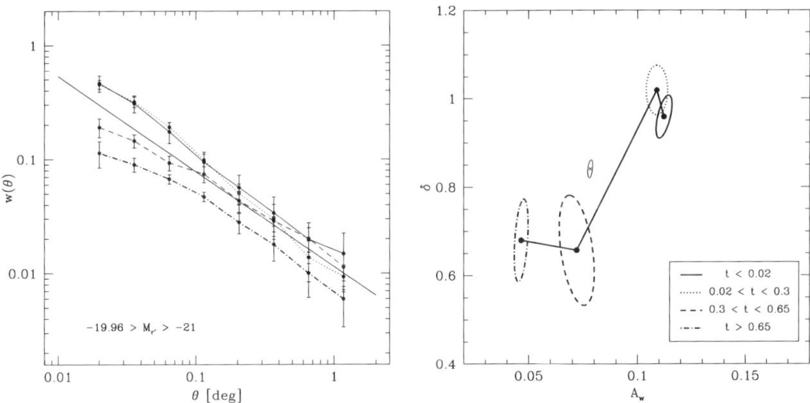

Figure 2. The slope of the angular correlation function changes significantly as a function of SED type. The figures illustrate the trend in the lower luminosity bin with $M_{r^*} > -21$. Note the bimodality in slope: the reddest two classes have the same slope, and the bluest two classes have the same slope, but the red slope differs distinctly from the blue slope.

4. Results

The amplitude and power of the correlation are calculated by fitting the usual formula, $w(\theta) = A_w (\theta/\theta_0)^{-\delta}$. Normalizing θ by $\theta_0 = 0.1^\circ$ enables the amplitude A_w be directly compared with the figures showing $w(\theta)$ measurements; A_w is essentially the value of the correlation function at θ_0, since $A_w \equiv w(\theta_0)$. For the full volume-limited sample the best fit to the data has a slope of $\delta = 0.84 \pm 0.02$ with an amplitude of $A_w = 0.078 \pm 0.001$.

The angular clustering of the galaxies in the three luminosity bins are compared in Figure 1.The left panel shows the correlation function, and the right panel gives the parameters of the power-law fits. As expected, the more luminous galaxies are clustered more strongly: the amplitudes are roughly larger by a factor of 1.5 from one sample to the next and are measured to be 0.061 ± 0.001, 0.092 ± 0.002 and 0.138 ± 0.004. The slope of the correlation functions are consistent for all luminosity bins and with the fiducial value derived for the entire volume (the estimated slope parameters scatter around $\delta = 0.84$).
We find that galaxies with spectral types similar to elliptical galaxies have a correlation length of $6.59 \pm 0.17\, h^{-1}\mathrm{Mpc}$ and a slope of the angular correlation function of 0.96 ± 0.05 while blue galaxies have a clustering length of $4.51 \pm 0.19\, h^{-1}\mathrm{Mpc}$ and a slope of 0.68 ± 0.09, see Figure 2. The two intermediate color groups behave like their more extreme 'siblings', rather than showing a gradual transition in slope (Budavári et al. 2003).

Acknowledgments

Funding for the creation and distribution of the SDSS Archive has been provided by the Alfred P. Sloan Foundation, the Participating Institutions, the National Aeronautics and Space Administration, the National Science Foundation, the U.S. Department of Energy, the Japanese Monbukagakusho, and the Max Planck Society. The SDSS Web site is http://www.sdss.org/.
The SDSS is managed by the Astrophysical Research Consortium (ARC) for the Participating Institutions. The Participating Institutions are The University of Chicago, Fermilab, the Institute for Advanced Study, the Japan Participation Group, The Johns Hopkins University, Los Alamos National Laboratory, the Max-Planck-Institute for Astronomy (MPIA), the Max-Planck-Institute for Astrophysics (MPA), New Mexico State University, University of Pittsburgh, Princeton University, the United States Naval Observatory, and the University of Washington.

References

Budavári, T., et al., 2000, Astronomical Journal, 120, 1588
Budavári, T., et al., 2003, Astrophysical Journal, 595, 59
Coleman, G.D., Wu., C.-C., & Weedman, D.W., 1980, Astrophysical Journal S., 43, 393
Connolly, A.J., et al., 1995a, Astronomical journal, 110, 2655
Csabai, I., et al., 2003, Astronomical Journal, 125, 580
Landy, S.D., & Szalay, A.S., 1993, Astrophysical Journal, 412, 64
Szapudi, I., Prunet, S., & Colombi, S., 2001, Astrophysical Journal, 561, 11
York, D.G., et al., 2000, Astronomical Journal, 120, 1579

OPTICALLY AND X-RAY SELECTED CLUSTERS OF GALAXIES IN THE XMM/2DF/SDSS SURVEY

S. Basilakos [1], M. Plionis [1,2], S. Georgakakis [1], I. Georgantopoulos [1]
T. Gaga [1,3], V. Kolokotronis [1], G. C. Stewart [4]
[1] *Institute of Astronomy & Astrophysics, National Observatory of Athens, I.Metaxa & B.Pavlou, Palaia Penteli, 152 36, Athens, Greece.*
[2] *Instituto Nacional de Astrofísica, Optica y Electrónica (INAOE) Apartado Postal 51 y 216, 72000, Puebla, Pue., Mexico.*
[3] *Physics Department, Univ. of Athens, Panepistimioupolis, Zografou, Athens, Greece.*
[4] *Department of Physics and Astronomy, University of Leicester, UK, LE1 7RH.*

Abstract In this work we present combined optical and X-ray cluster detection methods in an area near the North Galactic Pole area, previously covered by the SDSS and 2dF optical surveys. The same area has been covered by shallow (~ 1.8 deg^2) XMM-*Newton* observations. The optical cluster detection procedure is based on merging two independent selection methods - a smoothing+percolation technique, and a Matched Filter Algorithm. The X-ray cluster detection is based on a wavelet-based algorithm, incorporated in the SAS v.5.3 package. The final optical sample counts 9 candidate clusters with richness of more than 20 galaxies, corresponding roughly to APM richness class. Three, of our optically detected clusters are also detected in our X-ray survey.

1. Introduction

The cosmological significance of galaxy clusters has initiated a number of studies aiming to compile unbiased cluster samples to high redshifts, utilizing multiwavelength data (e.g. optical, X-ray, radio). From the optical point of view there are several available samples in the literature Abell, Corwin & Olowin 1989; Dalton et al. 1994; Olsen et al. 1999; Goto et al. 2002 which are playing a key role in astronomical research. Optical surveys suffer from projection effects Frenk et al. 1990 and thus, cluster detection in X-rays is a better approach, owing to the fact that the diffuse Intra-Cluster Medium (ICM) emits strongly in X-rays. The first such survey, was based on the Extended Einstein Medium Sensitivity Survey, containing 99 clusters Stocke et al. 1991. Recently, the *ROSAT* satellite allowed a leap forward in the X-ray cluster astronomy, producing large samples of both nearby and distant clusters Ebeling et al. 2000; Scharf et al. 1997.

M. Plionis (ed.), Multiwavelength Cosmology, 59-62.

However, even with the improved sensitivity of the XMM-*Newton*, optical surveys remain significantly more efficient and less expensive in telescope time for compiling cluster samples, albeit with some incompleteness and spurious detections. The aim of this work is to make a comparison of optical and X-ray cluster identification methods in order to quantify the selection biases introduced by these different techniques and to estimate the possible fraction of spurious optically selected clusters due to projection effects.

2. Observational Data

We analyze 9 *XMM-Newton* fields with nominal exposure time between 2 and 10 ksec, covering an area of 1.8 $\deg^2$. However, one of the fields, suffering from significantly elevated and flaring particle background, was excluded from the X-ray analysis. We apply our optical analysis to the the SDSS Early Data Release (EDR) Stoughton et al. 2002, that cover te above mentioned area. Note that Goto et al. (2002) also applied an objective cluster finding algorithm to the whole SDSS EDR and produced a list of 4638 galaxy clusters, with estimated photometric redshifts.

2.1 Optical & X-ray Cluster Finding Algorithms

One of our cluster detection algorithm, applied to the SDSS data, is based on smoothing the discrete distribution using a Gaussian smoothing kernel. We select all grid-cells with overdensities above a chosen critical threshold ($\delta \geq 1$) and then we use a friends-of-friends algorithm to form groups of connected cells, which we consider as our candidate clusters. Note that the grid cell size is such that at $z = 0.4$ it corresponds to 100 h^{-1}kpc ($\sim 19''$). The second optical cluster detection method is the matched filter algorithm (hereafter MFA) described by Postman et al. (1996). Finally, we construct our final cluster catalogue, in the $\sim$ 1.8 deg^2 area covered by our XMM survey, by adopting the conservative approach of considering as cluster candidates those identified by both independent selection methods. This sample, contains 9 clusters with SDSS richness of more than 20 galaxies, corresponding roughly to APM type clusters. Comparing our final list of 9 clusters with the Goto et al. (2002) clusters, we find 5 in common.

In order to detect candidate clusters in our XMM fields we use the soft 0.3-2 keV band since it maximizes the signal to noise ratio, especially in the case of relatively low temperature galaxy clusters. In particular, we utilize the *EWAVELET* detection algorithm of the *XMM-Newton* SAS v.5.2 analysis software package, which detects sources on the wavelet transformed images. We have detected 7 candidate clusters on the MOS mosaic, while 5 extended sources were detected on the PN images, out of which 3 overlap with the MOS candidates. After excluding obvious double point sources and MOS-edge ef-

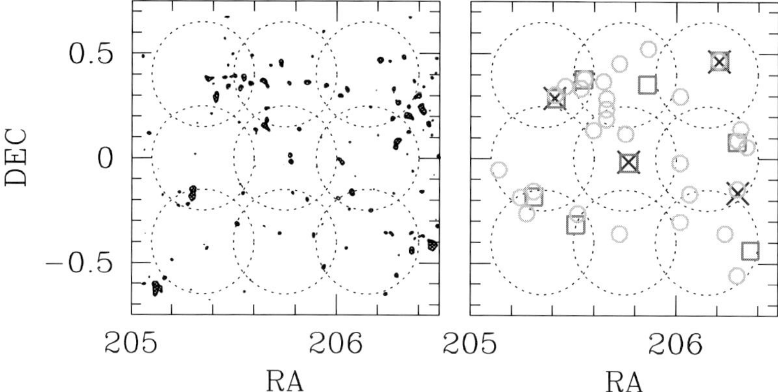

Figure 1. The smooth SDSS density field (left panel) in equatorial coordinates. The candidate cluster positions (right panel) within our shallow XMM survey. Open squares are the optically selected clusters using our technique, while open circles are the Goto et al. (2002) clusters. Crosses are our X-ray selected clusters. The large dotted circles represents the XMM 15 arcmin radius fields of view of our shallow XMM survey.

fects we are left with 4 X-ray candidate clusters. The faintest extended source has a flux of $\sim 2 \times 10^{-14}$erg cm^{-2} s^{-1}.

3. Results

In Fig. 1 (left panel) we plot the smoothed SDSS galaxy distribution, on the equatorial plane with contours delineating the $\delta_{\rm cr} = 1$ level. While in Fig. 1 (right panel), we plot the positions of (a) our detected cluster candidates, (b) the Goto et al. (2002) clusters and (c) the X-ray detected XMM clusters. All four X-ray detections coincide with optical cluster candidates from different methods, with the largest coincidence rate (3 out of 9) being with our methods. The most distant cluster, $z = 0.67$ Couch et al. 1991 in our optical sample, located at RA =13h 43min 4.84sec and $DEC = 00^o\ 00^{'}\ 56.26^{''}$, found also in X-rays, is missed by Goto et al. (2002). However, there are still 6 optical clusters that do not appear to have X-ray counterparts. This could be a hint that these clusters are either the results of projection effects, or that our XMM survey is too shallow to reveal the probably weak X-ray emission from these clusters.

In order to address this final issue and to study the relation between the limiting flux of our X–ray survey with respect to exposure time, we have carried out the following experiment. We have analyzed observations taken from 15

XMM public fields with mean exposure times ~ 21 ksec, after filtering to correct for the high particle background in the soft 0.3-2 keV band. Using the parameters of the SAS software as described previously, we have detected 31 candidate clusters. The faintest cluster detected, with a flux of $\sim 5 \times 10^{-15}$ erg cm^{-2} s^{-1}, was found in the deepest field with an exposure time of 37 ksec. We then reduce the exposure times to a new mean value of ~ 5 ksec, similar to our shallow survey, and find only 9 out of the 31 previously identified candidate clusters (29%), having a limiting flux of $\sim 2.3 \times 10^{-14}$ erg cm^{-2} s^{-1}. Therefore, had we had deeper XMM observations (by an average factor of ~ 5 in exposure time) we would have detected ~ 13 X-ray candidate clusters in the region covered by our shallow XMM survey, which is consistent (within 1σ) with the number of our optical cluster candidates.

4. Conclusions

We have made a direct comparison between optical and X-ray based techniques used to identify clusters. We have searched for extended emission in our shallow XMM-*Newton* Survey, which covers a $\sim 1.6 \ \deg^2$ area (8 out of 9 original XMM pointings) near the North Galactic Pole region and we have detected 4 candidate X-ray clusters. Out of the 4 X-ray candidate clusters 3 are common with our optical cluster list .This relatively, small number of optical cluster candidates observed in X-rays suggest that some of the optical cluster candidates are either projection effects or poor clusters and hence they are fainter in X-rays than the limit of our shallow survey $f_x(0.3-2keV) \simeq 2\times10^{-14}$erg cm^{-2} s^{-1}. This latter explanation seems to be supported from an analysis of public XMM fields with larger exposure times.

References

Abell, G.O., Corwin, H.G., Olowin, R.P., 1989, ApJS, 70, 1

Couch, W.J., Ellis, R. S., MacLaren, I., Malin, D. F., 1991, MNRAS, 249, 606

Dalton, G.B., Efstathiou, G., Maddox, S. J., Sutherland, W. J., 1994, MNRAS, 269, 151

Ebeling, H., et al., 2000, ApJ, 534, 133

Frenk, C. S., White, S. D. M., Efstathiou, G., Davis, M., 1990, ApJ, 351, 10

Goto, T., et al., 2002, AJ, 123, 1807

Olsen, L. F., et al., 1999, A&A, 345, 681

Postman, M., Lubin, L. M., Gunn, J. E., Oke, J. B., Hoessel, J. G., Schneider, D. P.,Christensen, J. A., 1996, AJ, 111, 615

Scharf, C. A., Jones, L. R., Ebeling, H., Perlman, E., Malkan, M., Wegner, G., 1997, ApJ, 477, 79

Stocke, J. T., Morris, S. L., Gioia, I. M., Maccacaro, T., Schild, R., Wolter, A., Fleming, T. A., Henry, J. P., 1991, ApJS,76,813

Stoughton, C., et al., 2002, AJ, 123, 485

STRUCTURE FORMATION AND GALAXY EVOLUTION AT $Z = 3 - 7$ PROBED BY 2,600 GALAXIES IN THE SUBARU DEEP FIELDS

Masami Ouchi,[1] Kazuhiro Shimasaku,[1] Sadanori Okamura,[1], and SDS team

[1] *University of Tokyo*

ouchi@astron.s.u-tokyo.ac.jp

Abstract We investigate photometric and clustering properties of galaxies at $z = 3.5 - 5.2$ based on large samples of 2,600 Lyman Break Galaxies (LBGs) and Lyman α Emitters (LAEs) detected in deep ($i' \sim 27$) and wide-field (1,200 arcmin^2) images taken in the Subaru Deep Field (SDF) and the Subaru/XMM Deep Field (SXDF). We find in the luminosity functions of LBGs that the number density of bright galaxies ($M_{1700} < -22$; corresponding to $SFR_{corr} > 100 M_{\odot}\mathrm{yr}^{-1}$) decreases significantly from $z = 4$ to 5. We investigate the evolution of UV-luminosity density at 1700A, ρ_{UV}, and find that ρ_{UV} does not significantly change from $z = 3$ to $z = 5$, i.e., $\rho_{\mathrm{UV}}(z = 4)/\rho_{\mathrm{UV}}(z = 3) = 1.0 \pm 0.2$ and $\rho_{\mathrm{UV}}(z = 5)/\rho_{\mathrm{UV}}(z = 3) = 0.8 \pm 0.4$. We find clear clustering signals for both LBGs and LAEs, and estimate the correlation length (and the galaxy-dark matter bias) to be $r_0 = 5.1^{+1.0}_{-1.1}$ and $5.9^{+1.3}_{-1.7} h^{-1}$ Mpc ($b_g = 3.5^{+0.6}_{-0.7}$ and $4.6^{+0.9}_{-1.2}$ for $L > L^*$ LBGs at $z = 4$ and 5, respectively. We find that the correlation length of $L > L^*$ LBGs is almost constant, $\sim 5h^{-1}$ Mpc, at $z > 3$, while the bias monotonically increases with redshift at $z > 3$. These findings support the biased galaxy formation picture.

1. Introduction

Formation history of galaxies is basically understood as a combination of two fundamental evolutionary processes, i.e., production of stars and accumulation of dark matter, in the standard framework of galaxy formation which is the Cold Dark Matter (CDM) models. In order to investigate the formation history of stars, many efforts have been made to search for high-z galaxies up to $z \simeq 7$ (e.g., Kodaira et al. 2003). On the other hand, clustering properties of galaxies are closely related to the distribution and amount of underlying dark matter (e.g., Verde et al. 2002). Thus, measuring clustering properties of galaxies is very useful for understanding the dark side of galaxy properties (different from stellar properties). We have launched surveys of two blank fields (SDF

M. Plionis (ed.), Multiwavelength Cosmology, 63-66.

and SXDF) with the Subaru Prime Focus Camera (Suprime-Cam; Miyazaki et al. 2002), which is the optical wide-field camera mounted on 8m-Subaru. These surveys cover wide areas with deep images which are appropriate for deriving two basic fundamental measures of high-z galaxies, i.e., luminosity functions and correlation functions. In this paper, we present the photometric and clustering properties of Lyman Break Galaxies (LBGs) and Lyman α Emitters (LAEs) at $z = 3.5 - 5.2$. Throughout this paper, we use a set of the cosmological parameters of $(\Omega_m, \Omega_\Lambda, n, \sigma_8) = (0.3, 0.7, 1.0, 0.9)$.

2. Galaxy Samples

We make three photometric samples of LBGs at $z = 4 - 5$ and one sample of LAEs at $z = 4.9$ from multi-band (B,V,R,i',z', and NB_{711}), deep ($i' \sim 27$) and wide-field (1,200 arcmin^2 in total) optical imaging with Suprime-Cam in two blank fields. One is the Subaru Deep Field (SDF: $13^h24^m21.4^s$,$+27°29'23''$ [J2000]) covering a 616 arcmin^2 area, and the other is the Subaru/XMM Deep Field (SXDF: $2^h18^m00^s$,$-5°12'00''$[J2000]) covering a 653 arcmin^2 area. The selection criteria for LBGs and LAEs we adopt are:

$B - R > 1.2,\ R - i' < 0.7,\ B - R > 1.6(R - i') + 1.9 \quad (BRi - \text{LBGs}),$
$V - i' > 1.2,\ i' - z' < 0.7,\ V - i' > 1.8(i' - z') + 1.7 \quad (Viz - \text{LBGs}),$
$R - i' > 1.2,\ i' - z' < 0.7,\ R - i' > 1.0(i' - z') + 1.0 \quad (Riz - \text{LBGs}),$
$Ri - NB_{711} > 0.8, R - i' > 0.5, i' - NB_{711} > 0,\ Ri \equiv (R + i')/2 \quad (\text{LAEs}).$

We apply these selection criteria to our photometric catalogs, and we find 1,438 (732), 246(34), and 68 (38) objects for BRi-LBGs at $z = 4.0 \pm 0.5$, Viz-LBGs at $z = 4.7 \pm 0.5$, and Riz-LBGs at $z = 4.9 \pm 0.3$ in the SDF (SXDF), and 87 LAEs at $z = 4.86 \pm 0.03$. In order to examine how well these selection criteria isolate true LBGs and LAEs, we use 85 spectroscopically identified objects at $0 < z < 5$ in the SDF obtained by Kashikawa et al. (2003), Shimasaku et al. (2003), and Ouchi et al. (2003b). We find that 6 (4) out of the 85 objects are LBGs (LAEs) at $z > 3.5$, and that no (one) spectroscopically identified low-z interloper is included in our LBG (LAE) samples. Thus, our photometric samples are thought to be reliable.

3. Luminosity Functions

The UV-luminosity functions (LFs) of LBGs and LAEs are derived from the samples defined in section 2. We show LFs of LBGs in Figure 1a, together with those at $z = 0$ (Sullivan et al. 2000) and $z \sim 3$ (Steidel et al. 1999). The LF of our $z \sim 4$ LBGs is consistent with the one derived by Steidel et al. (1999). In Figure 1a, we find that the number density of bright LBGs with $M_{1700} \sim -22$ decreases from $z = 4$ to $z = 5$. The difference between $z = 4$ and 5 is about one order in number density at $M_{1700} \sim -22$. We calculate UV-luminosity densities of our LBGs by integrating the LFs (Figure

1a) down to $0.1L^*$. The luminosity density of LBGs is $\rho_{\rm UV} = 1.9 \pm 0.2 \times 10^{26}$, $2.0 \pm 0.2 \times 10^{26}$, and $1.6 \pm 0.7 \times 10^{26}$ erg s^{-1} Hz^{-1} Mpc^{-3} for $z = 3$, $z = 4$, and $z = 5$, where the value at $z = 4$ is the mean of our measurement and Steidel et al.'s (1999). Thus the ratio of the UV-luminosity density at $z = 4$ and 5 to that at $z = 3$ is $\rho_{\rm UV}(z = 4)/\rho_{\rm UV}(z = 3) = 1.0 \pm 0.2$ and $\rho_{\rm UV}(z = 5)/\rho_{\rm UV}(z = 3) = 0.8 \pm 0.4$. The total UV-luminosity density may slightly decrease toward $z = 5$, but the amount of the decrease is 20% at most. Although the number density of bright LBGs ($M_{1700} < -22$) significantly decreases toward $z = 5$ (Figure 1a), the total UV-luminosity density does not change largely. This is because the total UV-luminosity density is mainly contributed by LBGs fainter than $M_{1700} \sim -22$.

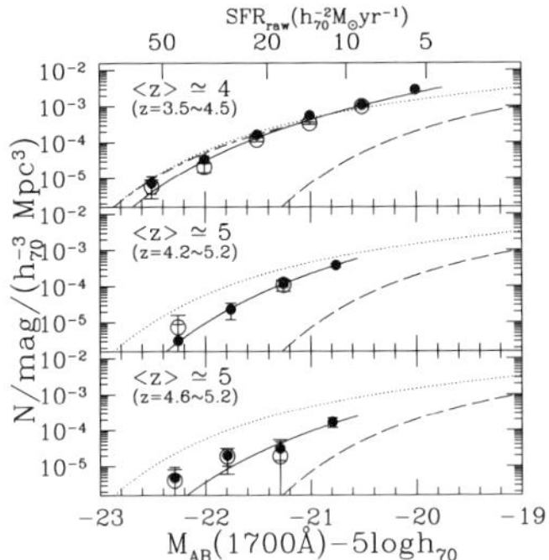

Figure 1a. Luminosity functions (LFs) of LBGs at $z = 4 - 5$. LFs of BRi-, Viz-, and Riz-LBGs are given in the top panel, middle panel, and bottom panel, respectively. In each panel, filled circles (open circles) are the LFs derived from the SDF (SXDF) data. The solid lines are the best-fit Schechter functions. The dashed lines denote the LF of UV-selected galaxies in the local universe (Sullivan et al. 2000), while the dotted lines are the LF of LBGs at $z \simeq 3$ derived by Steidel et al. (1999). In the top panel, the best-fit Schechter function of LBGs at $z = 4$ (Steidel et al. 1999) is shown by the dash-long dashed line down to ~ -21 mag. The upper abscissa axis, $SFR_{\rm raw}$, indicates the star-formation rate without extinction correction. The true (extinction-corrected) star-formation rate is about a factor of 4 larger than the raw rate (Ouchi et al. 2003b).

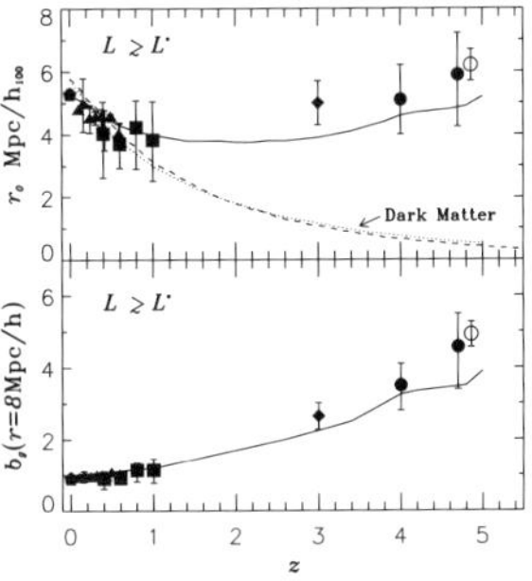

Figure 1b. Correlation length r_0 (top panel) and galaxy-dark matter bias b_g (bottom panel) as a function of redshift. The filled circles indicate our LBGs at $z = 4 - 5$, and the filled diamond is for LBGs at $z = 3$ (Giavalisco & Dickinson 2001); these LBGs at $z = 3 - 5$ have a similar luminosity ($> L^*$). The open circle indicates LAEs at $z = 4.9$ obtained by Ouchi et al. (2003a). The filled pentagons, triangles, and squares are r_0 of galaxies at $z = 0 - 1$ shown in the literature (see Ouchi et al. 2003c and references therein). The dotted line shows the correlation length of the underlying dark matter obtained with a non-linear theory (Peacock & Dodds 1996). The solid lines indicate r_0 and b_g of galaxies with $M_B < -19 + 5\log h$ predicted by the semi analytic model of Kauffmann et al. (1999).

4. Angular Correlation Functions

We derive angular correlation functions of these sample galaxies, and find a clear clustering signal with a significant level in each galaxy sample. We fit a power law, $\omega(\theta) = A_\omega \theta^{-\beta}$, to the data points, and we calculate the correlation lengths, r_0, from the best-fit A_ω by Limber de-projection. Since the clustering amplitude (i.e., correlation length) depends on the brightness of galaxies , we compare the correlation lengths of LBGs whose luminosities are $L > L^*$. We make subsamples composed of $L > L^*$ LBGs at $z = 4$ and 5 with $i' < 25.3$ and $z' < 25.8$ corresponding to $M < -20.8$ and $M < -20.5$ ($\simeq M^*$; Ouchi et al. 2003b), and calculate the correlation length to be $r_0 = 5.1^{+1.0}_{-1.1} h^{-1}_{100}$ Mpc and $5.9^{+1.3}_{-1.7} h^{-1}_{100}$ Mpc, respectively. The top panel of Figure 1b plots r_0 as a function of redshift, together with r_0 of the underlying dark matter calculated by the non-linear model of Peacock & Dodds (1996). We also show the correlation lengths of LBGs with $L > L^*$ at $z = 3$ given by Giavalisco & Dickinson (2001) and of all LAEs at $z = 4.9$ obtained by Ouchi et al. (2003a). The correlation length of LBGs with $L > L^*$ is almost constant around $r_0 \sim 5h^{-1}_{100}$Mpc at $z = 3 - 5$, although it might increase slightly with redshift. It is also found that the correlation amplitudes of LBGs (and LAEs) are much higher than that of underlying dark matter. We plot values of the bias against underlying dark matter, $b_g \equiv \sqrt{\xi_g/\xi_{\rm DM}}$ at $8h^{-1}_{100}$Mpc, in the bottom panel of Figure 1b. LBGs and LAEs found to be biased against dark matter by $b_g \sim 3 - 5$, and the bias becomes stronger at higher redshifts. This increase in LBGs' bias can be regarded as a piece of evidence supporting the biased-galaxy formation scenario (e.g., Baugh et al. 1999).

References

Bruzual A., G. & Charlot, S. 1993, ApJ, 405, 538

Baugh, C. M., Benson, A. J., Cole, S., Frenk, C. S., & Lacey, C. G. 1999, MNRAS, 305, L21

Giavalisco, M. & Dickinson, M. 2001, ApJ, 550, 177

Kashikawa, N. et al. 2003, AJ, 125, 53

Kauffmann, G., Colberg, J. M., Diaferio, A., & White, S. D. M. 1999, MNRAS, 307, 529

Kodaira, K. et al. 2003, PASJ, 55, L17

Miyazaki, S. et al. 2002, PASJ, 54, 833

Ouchi, M. et al. 2003a, ApJ, 582, 60

Ouchi, M. et al. 2003b, submitted to ApJ (astro-ph/0309655)

Ouchi, M. 2003c, submitted to ApJ (astro-ph/0309657)

Peacock, J. A. & Dodds, S. J. 1996, MNRAS, 280, L19

Shimasaku, K. et al. 2003, ApJ, 586, L111

Steidel, C.C., Adelberger, K.L., Giavalisco, M., Dickinson, M., & Pettini, M. 1999, ApJ, 519, 1

Sullivan, M., Treyer, M. A., Ellis, R. S., Bridges, T. J., Milliard, B., & Donas, J. ; 2000, MNRAS, 312, 442

Verde, L. et al. 2002, MNRAS, 335, 432

ILLUMINATING PROTOGALAXIES? DISCOVERY OF EXTENDED LYMAN-α EMISSION AROUND A $Z = 4.5$ RADIO-QUIET QSO

Andrew Bunker
Institute of Astronomy, University of Cambridge, Cambridge, CB3 0HA, UK
bunker@ast.cam.ac.uk

Abstract We have discovered extended Lyman-α emission around a $z = 4.5$ QSO in a deep long-slit spectrum with Keck/LRIS at moderate spectral resolution ($R \approx 1000$). The line emission extends 5 arcsec beyond the continuum of the QSO and is spatially asymmetric. This extended line emission has a spectral extent of 1000km/s, much narrower in velocity spread than the broad Lyman-α from the QSO itself and slightly offset in redshift. No evidence of continuum is seen for the extended emission line region, suggesting that this recombination line is powered by reprocessed QSO Lyman continuum flux rather than by local star formation. This phenomenon is rare in QSOs which are not radio loud, and this is the first time it has been observed at $z > 4$. It seems likely that the QSO is illuminating the surrounding cold gas of the host galaxy, with the ionizing photons producing Lyman-α fluorescence. As suggested by Haiman & Rees (2001), this "fuzz" around a distant quasar may place strong constraints on galaxy formation and the extended distribution of cold, neutral gas.

Keywords: line: profiles, formation; techniques: spectroscopic; galaxies: formation, evolution; quasars: emission lines, high-redshift, individual - PC0953+4749

1. Introduction

Although it is by now well-established that QSOs are hosted by galaxies, QSOs themselves are difficult to interpret as probes of galaxy evolution. The evolution of the QSO and galaxy populations is expected to be linked, but in fact they exhibit very different behaviour: the extinction-corrected Madau-Lilly diagram of volume-averaged star formation is flat at redshifts beyond $z = 1$ (Hopkins, Connolly & Szalay 2000 AJ 120, 2843), yet the QSO luminosity function peaks sharply at $z = 2$ (Schmidt, Schneider & Gunn 1995 AJ, 110, 68).

A quasar phase may be a natural (but brief) evolutionary stage in the life of all massive galaxies, and theory predicts the proto-galaxy to be enveloped by

M. Plionis (ed.), Multiwavelength Cosmology, 67-70.

spatially-extended cold gas of temperature $\sim$ 10,000 K, with a radiative cooling time shorter than the dynamical time. Interesting questions include: What is the physical effect of a QSO turning on within an assembling galaxy? And is this observable?

Recent theoretical work predicts a Lyman-α halo with a surface brightness of $\sim 10^{-17}\,\mathrm{ergs\,s^{-1}\,cm^{-2}\,arcsec^{-2}}$ (Haiman & Rees 2001 ApJ 556, 87). Such haloes have been seen in radio galaxies and radio-loud QSOs (*e.g.*, Bremer et al. 1992 MNRAS 258, 3) but in these cases it is probably related to outflows. Until now, this phenomenon has yet to be seen for quasars which are not radio loud (see Hu & Cowie 1987 ApJLett 317, 7). If no extended emission can be found, this may imply that QSOs only turn on when the gas has settled into a thin disk or the cold gas has already been consumed in star formation.

The ionizing photons from the QSO will generate recombination line emission from optically-thick neutral hydrogen clouds around the QSO. There will be low surface-brightness Lyman-α "fuzz" anyway from line cooling of gas in the halo potential, and external photoionization by UV background (*e.g.*, Bunker, Marleau & Graham 1998 AJ 116, 2086), but the presence of the QSO will greatly enhance this. Haiman & Rees (2001) predict a Lyman-α halo extending out to a significant fraction of virial radius ($10 - 100\,\mathrm{kpc}$), corresponding to an angular size of $\sim 3''$. The predicted surface brightness of $\sim 10^{-17}\,\mathrm{ergs\,s^{-1}\,cm^{-2}\,arcsec^{-2}}$ is accessible to large telescopes with spectroscopy or deep narrow-band imaging. Another recent theory paper by Alam & Miralda-Escudè (2002 ApJ 568, 576) claims 100 times fainter surface brightness and very small extent for the extended Lyman-α emission ($0.4''$). The large discrepancy in the predictions arises from different assumptions about the clumping factor of cold gas and central concentration of the line-emitting region.

2. Our Observations

We have been undertaking an extensive study of the radio-quiet quasar PC0953+4749 at $z = 4.46$ (Schneider, Schmidt & Gunn 1991 AJ 101, 2004) which has 3 damped Lyman-α systems at $z > 3$ (Figure 1). The long-slit spectroscopy was obtained with Keck/LRIS (Oke et al. 1995 PASP 107, 375) for 1 hour at spectral resolution of 300km/s. Inspection of the 2D spectrum reveals Lyman-α at the QSO redshift but extended spatially beyond the continuum of the QSO (Figure 2). This is the first time this phenomenon has been seen at $z > 4$ in a QSO which is not radio-loud.

This line emission extends over $\sim 5''$ beyond the QSO point spread function. The emission is asymmetric, which implies either that gas is clumpy, or that the radiation is beamed anisotropically.

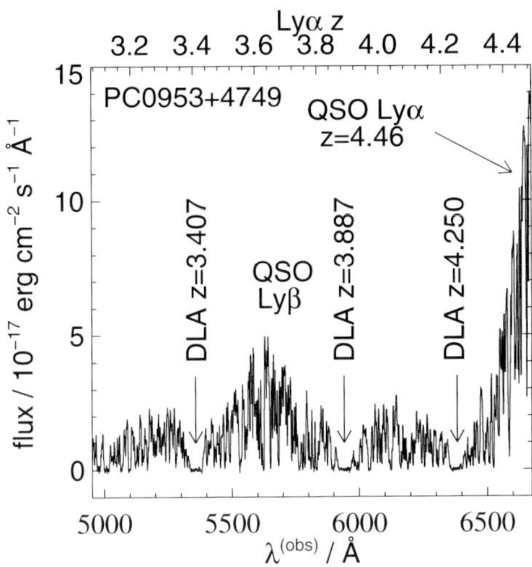

Figure 1. Our Keck/LRIS spectrum of the radio-quiet QSO PC0953+4749 at $z = 4.46$. We confirm 3 damped Lyman-α systems at $z > 3$.

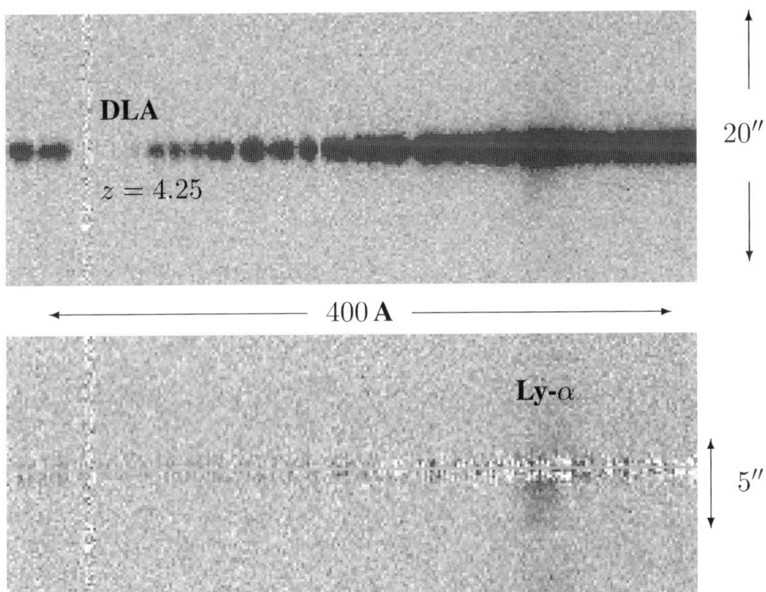

Figure 2. The upper panel shows the 2D longslit Keck/LRIS spectrum of the QSO PC0953+4749. Wavelength increases left-to-right, and the slit axis is vertical. The region around Lyman-α from the QSO at $z = 4.46$ is shown, and the damped Lyman-α absorption system at $z = 4.25$ can also be seen near the 6363 A sky line. The spectrum of the QSO point source has been subtracted in the lower panel to reveal residual Lyman-α emission at 6670 A ($z = 4.49$) extended over ≈ 5 arcsec.

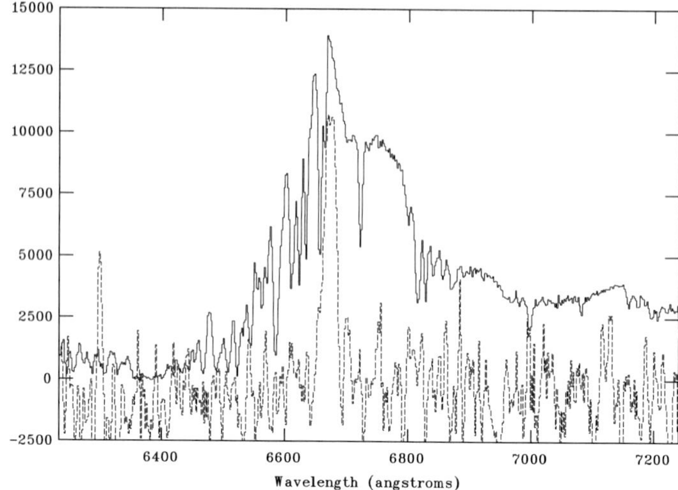

Figure 3. The extracted 1D spectrum of the QSO (solid line) compared with that of the spatially-extended Lyman-α emission (dashed line, flux scaled up by $\times$ 30).

The extended line emission (the dashed line in Figure 3) covers a spectral extent of $\sim$ 1000 km/s FWHM. This is not a good measure of the velocity dispersion of the gas, as this line is resonantly broadened. The spatially extended line emission is much narrower than Lyman-α from the QSO (solid line). No evidence of continuum is seen for the extended emission line region. This indicates that the recombination line is probably powered by reprocessed QSO UV flux rather than by local star formation.
The H I cloud of this host galaxy is $> 35\, h_{70}^{-1}\, kpc$ ($\Omega = 0.3$). The size and surface brightness agree more closely with the theoretical prediction of Haiman & Rees (2001) than with that of Alam & Miralda-Escudè (2002). However, we stress that this is only one example: other deep longslit spectra of high-redshift QSOs need to be studied to see if this extended emission is a generic feature of QSOs in young galaxies.

Acknowledgments

This work was done in collaboration with Joanna Smith, Hyron Spinrad, Daniel Stern and Stephen Warren.

SPECTRO-MORPHOLOGY OF GALAXIES

Sebastien Lauger, Denis Burgarella and Veronique Buat
Laboratoire d'Astrophysique de Marseille
Traverse du Siphon B.P.8, F-13376 Marseille Cedex 12
sebastien.lauger@oamp.fr

Abstract We present a quantitative method of classification of the multi-wavelength morphology of galaxies, elaborated from the nearby galaxies and usable to high redshift objects. We use the morphological parameters of concentration of light at the galaxy center and 180°-rotational asymmetry, computed in several wavelengths, from ultraviolet (UV) to R band. The variation of these indices with λ reflects the proportion of young and old stellar populations in galaxies, that is characteristic of every of morphological types : elliptical/lenticular, spiral and irregular. We need to elaborate some templates of these variation spectra from the study of nearby objects, for an application to distant galaxies.

Keywords: galaxies,morphology, high redshift, ultraviolet, multi-wavelength, concentration, asymmetry.

1. Introduction

⋆ The number of unclassified galaxies with the Hubble's classification scheme increases with the distance: 7% in local universe, 39% in HDF, according to van den Bergh et al.(1996). These galaxies may be : low surface-brightness galaxies, small angular-size galaxies or **series interacting objects**.
⋆ Hubble's classification was elaborated in optical wavelength, but galaxy morphology changes with λ.
⋆ We need a new system of morphological classification, that accounts for the physical processes of galaxies formation. Such a system must be successful for nearby objects and high-redshift galaxies.

2. The Data

We work on UV images from UIT (1500Aand 2500A) and FOCA telescope (2000A) associated to optical data in U, B, V, R bands from Marcum *et al.*(2001) and Larsen & Richtler(1999) samples. The galaxies we used are the following : M 81, M 82, M 100, NGC 1156, NGC 1316, NGC 1317, NGC 1399, NGC 2835, NGC 5204.

M. Plionis (ed.), Multiwavelength Cosmology, 71-74.

3. The Method

Concentration of Kent: Defined as: $C_K = 5\log(\frac{r_{80\%}}{r_{20\%}})$ by Kent(1985) and based on the ratio of two radii including 80% and 20% of total flux. Bershady *et al.* (2000) showed that C_K is very steady against a spatial resolution degradation and allows the study of high redshift galaxies.
Asymmetry index: A is a measure of the asymmetry of galaxies by a 180° rotation. It consists in computing the difference pixel by pixel between the original image and its 180° rotation, in absolute value, and in normalizing the result by the sum of the original pixels. A correction term must be added to A, because of the subtraction in absolute value which systematically introduces a positive value.

$$A = \frac{1}{2}[\frac{\sum_{I>n\sigma_{sky}} |I - I_0|}{\sum I_0} - \frac{k_{scale} \sum |B - B_0|}{\sum I_0}]$$

That definition is inspired from Kuchinski et al.(2000), that uses in the subtraction only the pixels above the $n\sigma_{sky}$-level ; k_{scale} is scale factor on the number of pixel used. The correction term $\sum |B - B_0|$ is statistically computed with the relation: $\sum |B - B_0| = \frac{2}{\sqrt{(\pi)}}\sigma_{sky}N_{pix}$. That definition minimizes the influence of signal-to-noise ratio and is useful until $S/N \sim 1$, in UV and optical bands.
Bandshifting: Appearance of galaxies changes with the wavelength λ: in UV, spiral galaxies present a faint or an inexistent bulge; star-forming regions, scattered on the spiral arms, give a patchy aspect. At longer wavelengths, like in R bands, the central bulge is prominent; and the red old-stellar population gives to the disk a uniform appearance. Thus, generally speaking, spiral galaxies have a later type in UV than in visible (Figure 1 on top left) As expected, we obtained the mean profiles (Figure 1 on bottom left) of C_K and $A3$ with λ ("*Morpho-spectra*") : the highest variations are those of spiral galaxies.
To quantify the variations of asymmetry and concentration as a function of the wavelength, we define the spectro-morphological shift (Figure 1):

$$S_{SpM} = \sqrt{(A_{UV} - A_{Vis.})^2 - (C_{UV} - C_{Vis.})^2}$$

That new morphological parameter accounts for physics of galaxies by separating one and two-stellar-component galaxies: A correlation with asymmetry

One-stellar-component galaxies	E/SO and Irregular types	$S_{SpM} \leq 0.45$
Two-stellar-component galaxies	Late Spiral type	$0.45 \leq S_{SpM} \leq 0.95$
	Early Spiral type	$S_{SpM} \geq 0.95$

or color index makes it possible to discriminate irregular and elliptical types (Figure 2 on top right)

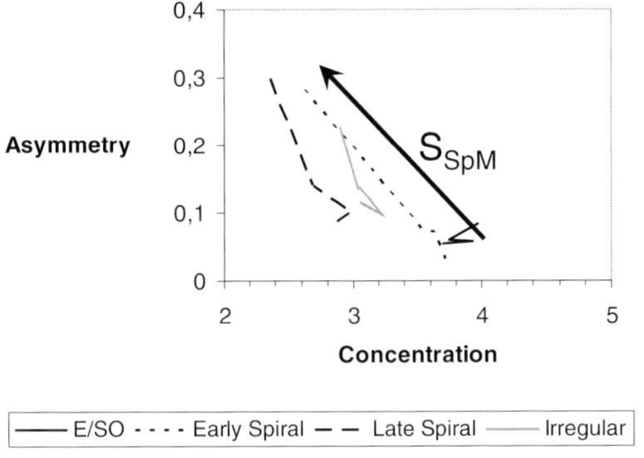

Figure 1. Spectro-Morphological Shift S_{SpM}

4. An example of application at high z

We have used that method on 3 galaxies of the HDF: HDF 4-378, 4-474 and 4-550 (see Bunker *et al.*(2000)). Their morpho-spectra clearly show that they are 3 spirals (Figure 2 on bottom right). Moreover the positions of those 3 galaxies in (S_{SpM}, A_{UV}) and $(S_{SpM}, B - V)$ diagrams confirm that result (Figure 2 on top right).

5. Conclusions and Perspectives

A larger sample of galaxies will be helpful to improve our template at low z. UV is the most constraining band to go on and the UV survey GALEX launched on the 28th April 2003 will certainly be a key point. We must also understand the behavior of active nuclei galaxies in such a system of classification. Galaxies of the Hubble Deep Field are under study, in collaboration with C. Conselice

References

Bershady, M. A., Jangren, A., & Conselice, C. J. 2000, AJ, 119, 2645.

Bunker, A. et al. 2000, astro-ph/0004348

Kent, S. M. 1985, apjs, 59, 115

Kuchinski, L. E. et al. 2000, ApJS, 131, 441

Larsen, S. S. & Richtler, T. 1999, a, 345, 59

Marcum, P. M. et al. 2001, ApJS, 132, 129.

van den Bergh, S., Abraham, R. G., Ellis, R. S., Tanvir, N. R., Santiago, B. X., & Glazebrook, K. G. 1996, AJ, 112, 359

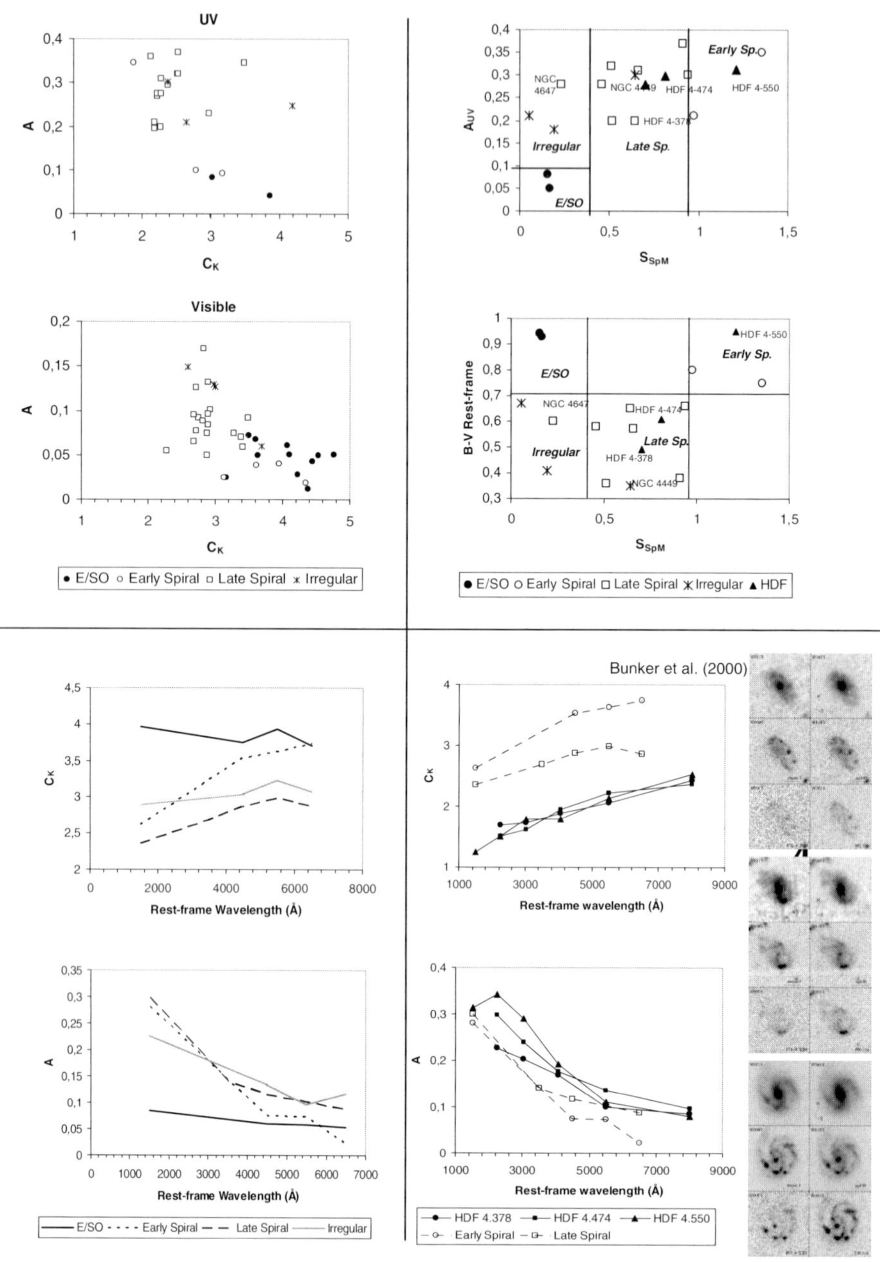

Figure 2. On top left: (A, C) diagrams in UV and visible. There is a global shift toward the late type from visible to UV. On bottom left: templates of morpho-spectra elaborated from a sample of nearby galaxies. On top right: (S_{SpM}, A_{UV}) and $(S_{SpM}, B-V)$ diagrams allow to distinguish 4 morphological types. On bottom right: morpho-spectra found for 3 high-z galaxies.

MICROWAVE WAVELENGTHS

RECENT CMB OBSERVATIONS (*INVITED*)

Edward L. Wright
UCLA Astronomy
PO Box 951562, Los Angeles CA 90095-1562
USA[*]
wright@astro.ucla.edu

Abstract The past year has seen very dramatic progress in the observational study of the CMB anisotropy. New results include the first detection of polarization anisotropy of the CMB by DASIPOL; extensive sky coverage from the ARCHEOPS balloon experiment; and very high angular resolution data from CBI and ACBAR. But the most significant new data are the first year results from the WMAP satellite, which cover 100% of the sky in 5 bands from 23 to 94 GHz with angular resolutions up to $13'$.

Keywords: cosmic microwave background, anisotropy

1. Introduction

Shortly after the Cosmic Microwave Background (CMB) was discovered (Penzias and Wilson, 1965), the first anisotropy in the CMB was seen: the dipole pattern due to the motion of the observer relative to the rest of the Universe (Conklin, 1969). After confirmation by Henry, 1971 and by Corey and Wilkinson, 1976 the fourth "discovery" of the dipole (Smoot et al., 1977) showed a very definite cosine pattern as expected for a Doppler effect, and placed an upper limit on any further variations in T_{CMB}.
Polarization of the CMB was shown to be < 300 μK (Lubin and Smoot, 1981). At the time of the launch of COBE in 1989 (Boggess et al., 1992), the CMB was known to have a blackbody spectrum to within 10%, although the Berkeley-Nagoya rocket experiment had suggested deviations from a Planck spectrum at that level. Furthermore, the anisotropy with spherical harmonic index $\ell > 1$ (to exclude the isotropic $\ell = 0$ and the dipole $\ell = 1$ terms) was known to be < 56 μK (Klypin et al., 1987) from the results of the Russian Relikt satellite.

[*] *WMAP* is the result of a partnership between Princeton University and NASA's Goddard Space Flight Center. Scientific guidance is provided by the *WMAP* Science Team.

M. Plionis (ed.), Multiwavelength Cosmology, 77-84.

COBE showed that the CMB spectrum was a 2.725 K blackbody to within 50 parts per million (Fixsen et al., 1996, Mather et al., 1999) and that there was an intrinsic ($\ell > 1$) anisotropy with $\Delta T_2 = \sqrt{2.4} Q_{rms-ps} = 27\ \mu$K (Bennett et al., 1996). The polarization of the anisotropy was found to be less than 15 μK.

Theoretically the first predictions of $\Delta T/T = 10^{-2}$ (Sachs and Wolfe, 1967) and $\Delta T/T = 10^{-3.5}$ (Silk, 1968) had been superseded by predictions based on cold dark matter (Peebles, 1982, Bond and Efstathiou, 1987). These predictions were consistent with the small anisotropy seen by COBE and furthermore predicted a large peak at a particular angular scale due to acoustic oscillations in the baryon/photon fluid prior to recombination. The position of this big peak and other peaks in the angular power spectrum of the CMB anisotropy depends on a combination of the density parameter Ω_m and the vacuum energy density Ω_V. Thus measuring this peak provides a means to determine the density of the Universe (Jungman et al., 1996). By 1994 a tentative detection of the big peak at the position predicted for a flat Universe had been made (Scott et al., 1995). By the beginning of 2000 the peak had been localized to $\ell_{pk} = 229 \pm 8.5$ (Knox and Page, 2000). Later that year the BOOMERanG group claimed to have made a dramatic improvement in this datum to $\ell_{pk} = 197 \pm 6$ (de Bernardis et al., 2000). This smaller value for ℓ_{pk} favored a slightly closed model for the Universe.

The predicted polarization level was an order of magnitude lower than the anisotropy. This predicted polarization was caused by electron scattering during the late stages of recombination. The magnitude of the polarization depends on the anisotropy being in place at recombination, as is the case for primordial adiabatic perturbations but not for topological defects; the electron scattering cross-section; and the recombination coefficient of hydrogen. A detection of this polarization would be a very strong confirmation of the standard model for CMB anisotropy, but due to the small signal no detections were made until late in 2002.

Because polarization is a vector field, two distinct modes or patterns can arise (Kamionkowski et al., 1997, Seljak and Zaldarriaga, 1997): the gradient of a scalar field (the “E” mode) or the curl of a vector field (the “B” mode). Electron scattering only produces the E mode. Electron scattering gives a polarization pattern that is correlated with the temperature anisotropy, so the E modes can be detected by cross-correlating the polarization with the temperature. The B modes cannot be detected this way, and the predicted level of the B modes is at least another order of magnitude below the E modes, or two orders of magnitude below the temperature anisotropy.

In this paper I will discuss the new observations that have been released in the year prior to this meeting: June 2002 to June 2003. I will discuss these results in time order.

2. DASIPOL

The Degree Angular Scale Interferometer (DASI) is a very small array that operates at 26-36 GHz and the South Pole. After measuring the angular power spectrum of the anisotropy (Halverson et al., 2002) the instrument was converted into a polarization sensitive interferometer which detected the E mode polarization at 5.5σ by looking at a small patch of sky for most of a year of integration time (Kovac et al., 2002). The level agreed well with the solid predictions for adiabatic primordial perturbations. Since the measured quantity was the EE autocorrelation, the 5.5σ corresponds to a 9% accuracy in the polarization amplitude.
The TE cross-correlation was also seen, but with only 50% accuracy. As expected, the B modes were not seen.

3. ARCHEOPS

ARCHEOPS is a balloon-borne experiment built to test the detectors and the cryogenic system planned for the ESA Planck Explorer mission High Frequency Instrument. With bolometers cooled below 0.1 K, this experiment achieves a very high instantaneous sensitivity and was able to map a substantial fraction of the sky with good SNR in only 1 night of observing. With this large sky coverage came a lower level of cosmic or sample variance, so the ARCHEOPS data gave much better results on the low-ℓ side of ℓ_{pk}. This gave a peak location of $\ell_{pk} = 220 \pm 6$ (Beno^t et al., 2003).

4. ACBAR

ACBAR is a large format bolometer camera array. It operates at several frequencies on both sides of the peak of the spectrum of the CMB, and thus can be used to make a sensitive test for the Sunyaev-Zeldovich effect. However, the data released to date are only at one frequency. ACBAR is placed on the VIPER telescope, which has a 2 meter diameter and is located at the South Pole. ACBAR was able to measure the CMB angular power spectrum up to $\ell > 2000$ (Kuo et al., 2002), but the size of the surveyed regions was so small that ACBAR's ℓ resolution was too limited to provide much information about ℓ_{pk}. What ACBAR was able to do with its high observing frequency was show that the excess at $\ell > 2000$ seen by CBI (Mason et al., 2003) was not a primary CMB anisotropy. It could be due to point sources, or it could be due to the S-Z effect, both of which would be much weaker in the ACBAR band than in the CBI 26-36 GHz band.

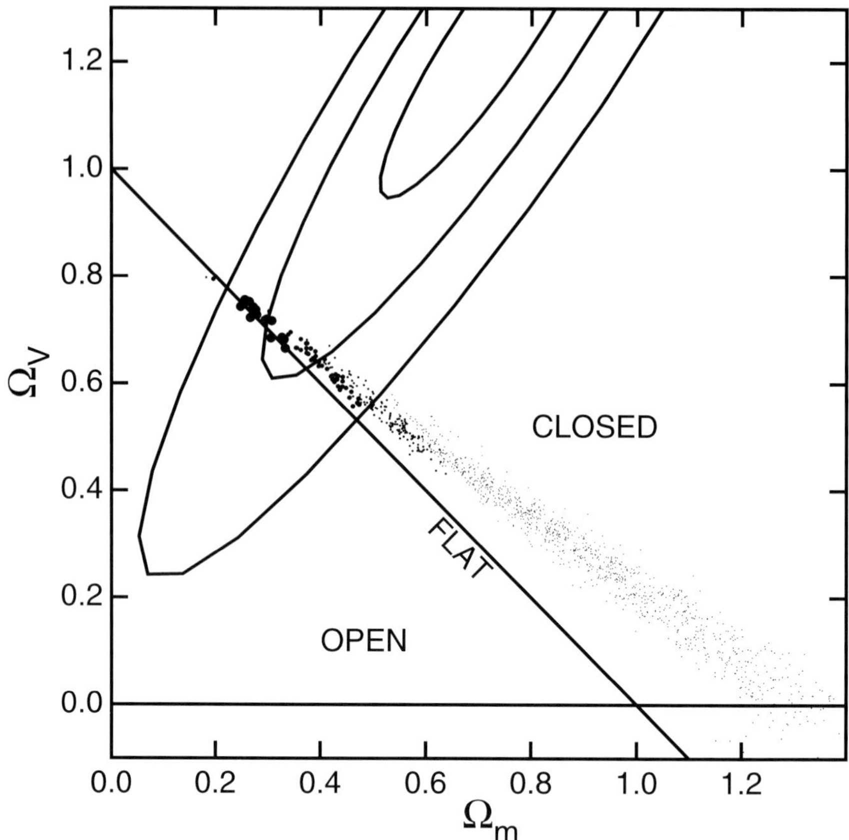

Figure 1. Cloud of points from a Monte Carlo Markov chain sampling of the likelihood of models fit to the *WMAP* plus other CMB datasets. The size of the points indicates how consistent the model is with the HST Key Project on the Distance Scale value for the Hubble constant. The contours show the likelihood computed for 230 Type Ia supernovae (Tonry et al., 2003).

5. WMAP

The *WMAP* satellite was launched on 30 June 2001. On 11 Feb 2003 the results from the first year of data were released. At the same time, the mission was renamed the Wilkinson Microwave Anisotropy Probe to honor the late David T. Wilkinson who was a key member of both the COBE and the *WMAP* teams until his death in September 2002. The results from *WMAP* and their cosmological significance are described in 13 papers and will not be repeated here. Bennett et al., 2003a gives a description of the *WMAP* mission. Bennett et al., 2003b summarizes the results from first year of *WMAP* observations. Bennett et al., 2003c describes the observations of galactic and extragalactic foreground sources. Hinshaw et al., 2003b gives the angular power spectrum derived from the the *WMAP* maps. Hinshaw et al., 2003a describes the *WMAP* data processing and systematic error limits. Page et al., 2003a discusses the beam sizes and window functions for the *WMAP* experiment. Page et al., 2003b discusses results that can be derived simply from the positions and heights of the peaks and valleys in the angular power spectrum. Spergel et al., 2003 describes the cosmological parameters derived by fitting the *WMAP* data and other datasets. Verde et al., 2003 describes the fitting methods used. Peiris et al., 2003 describes the consequences of the *WMAP* results for inflationary models. Jarosik et al., 2003 describes the on-orbit performance of the *WMAP* radiometers. Kogut et al., 2003 describes the *WMAP* observations of polarization in the CMB. Barnes et al., 2003 describes the large angle sidelobes of the *WMAP* telescopes. Komatsu et al., 2003 addresses the limits on non-Gaussianity that can be derived from the *WMAP* data.

From the *WMAP* data plus CBI and ACBAR, the position of the big peak in the angular power spectrum is found to be $\ell_{pk} = 220.1 \pm 0.8$. The position of the big peak defines a track in the Ω_m-Ω_V plane shown in Figure 1.

The ratios of the anisotropy powers below the peak at $\ell \approx 50$, at the big peak at $\ell \approx 220$, in the trough at $\ell \approx 412$, and at the second peak at $\ell \approx 546$ can be precisely determined using the *WMAP* data which has a single consistent calibration for all ℓ's. Previously, these ℓ ranges had been measured by different experiments having different calibrations so the ratios were poorly determined. Knowing these ratios determines the photon:baryon:CDM density ratios, and since the photon density was precisely determined by FIRAS on COBE, one obtains accurate values for the baryon density and the dark matter density. These values are $\Omega_b h^2 = 0.0224 \pm 4\%$, and $\Omega_m h^2 = 0.135 \pm 7\%$. The ratio of CDM to baryon densities is 5.0:1.

Since the matter density $\Omega_m h^2$ is fairly well constrained by the amplitudes, the positions of a point in Figure 1 serves to define a value of the Hubble constant. The size of the points in Figure 1 indicates how well this derived Hubble constant agrees with the $H_\circ = 72 \pm 8$ from the HST Key Project (Freedman et al., 2001). Also shown as contours are the $\Delta\chi^2 = 1,\ 4,\ \&\ 9$ contours from fits to 230 SNe Ia (Tonry et al., 2003). Clearly the CMB data, the HST data, and the SNe data are all consistent at a three-way crossing that is very close to the flat Universe line.

The *WMAP* data show a TE (temperature-polarization) cross-correlation. At small angles the TE amplitude is perfectly consistent with the standard picture of the recombination era. But there is also a large angle TE signal that gives an estimate for the electron scattering optical depth since reionization: $\tau = 0.17 \pm 0.04$.

6. Discussion

The CMB data released in the past year has moved cosmology into a new precision era, with the baryon, dark matter and vacuum energy densities all known along with the Hubble constant.

References

Barnes, C., Hill, R. S., Hinshaw, G., Page, L., Bennett, C. L., Halpern, M., Jarosik, N., Kogut, A., Meyer, S. S., Tucker, G. S., Wollack, E., and Wright, E. L. (2003). First Year Wilkinson Microwave Anisotropy Probe (*WMAP*) Observations: Galactic Signal Contamination from Sidelobe Pickup. *Astrophys. J., Suppl. Ser.*, 148:51.

Bennett, C. L., Banday, A. J., G®rski, K. M., Hinshaw, G., Jackson, P., Keegstra, P., Kogut, A., Smoot, G. F., Wilkinson, D. T., and Wright, E. L. (1996). Four-Year COBE DMR Cosmic Microwave Background Observations: Maps and Basic Results. *ApJ*, 464:L1.

Bennett, C. L., Bay, M., Halpern, M., Hinshaw, G., Jackson, C., Jarosik, N., Kogut, A., Limon, M., Meyer, S. S., Page, L., Spergel, D. N., Tucker, G. S., Wilkinson, D. T., Wollack, E., and Wright, E. L. (2003a). The Microwave Anisotropy Probe Mission. *ApJ*, 583:1–23.

Bennett, C. L., Halpern, M., Hinshaw, G., Jarosik, N., Kogut, A., Limon, M., Meyer, S. S., Page, L., Spergel, D. N., Tucker, G. S., Wollack, E., Wright, E. L., Barnes, C., Greason, M. R., Hill, R. S., Komatsu, E., Nolta, M. R., Odegard, N., Peiris, H., Verde, L., and Weiland, J. L. (2003b). First Year Wilkinson Microwave Anisotropy Probe (*WMAP*) Observations: Preliminary Maps and Basic Results. *Astrophys. J., Supp.*, 148:1.

Bennett, C. L., Hill, R. S., Hinshaw, G., Nolta, M. R., Odegard, N., Page, L., Spergel, D. N., Weiland, J. L., Wright, E. L., Halpern, M., Jarosik, N., Kogut, A., Limon, M., Meyer, S. S., Tucker, G. S., and Wollack, E. (2003c). First Year Wilkinson Microwave Anisotropy Probe (*WMAP*) Observations: Foreground Emission. *Astrophys. J., Supp.*, 148:97.

Beno^t, A., et al. (2003). The cosmic microwave background anisotropy power spectrum measured by Archeops. *A&A*, 399:L19–L23.

Boggess, N. W., Mather, J. C., Weiss, R., Bennett, C. L., Cheng, E. S., Dwek, E., Gulkis, S., Hauser, M. G., Janssen, M. A., Kelsall, T., Meyer, S. S., Moseley, S. H., Murdock, T. L., Shafer, R. A., Silverberg, R. F., Smoot, G. F., Wilkinson, D. T., and Wright, E. L. (1992). The COBE mission - Its design and performance two years after launch. *ApJ*, 397:420–429.

Bond, J. R. and Efstathiou, G. (1987). The statistics of cosmic background radiation fluctuations. *MNRAS*, 226:655–687.

Conklin, E. K. (1969). Velocity of the Earth with Respect to the Cosmic Background Radiation. *Nature*, 222:971–972.

Corey, B. E. and Wilkinson, D. T. (1976). A Measurement of the Cosmic Microwave Background Anisotropy at 19 GHz. *BAAS*, 8:351–351.

de Bernardis, P., et al. (2000). A flat Universe from high-resolution maps of the cosmic microwave background radiation. *Nature*, 404:955–959.

Fixsen, D. J., Cheng, E. S., Gales, J. M., Mather, J. C., Shafer, R. A., and Wright, E. L. (1996). The Cosmic Microwave Background Spectrum from the Full COBE FIRAS Data Set. *ApJ*, 473:576.

Freedman, W. L., Madore, B. F., Gibson, B. K., Ferrarese, L., Kelson, D. D., Sakai, S., Mould, J. R., Kennicutt, R. C., Ford, H. C., Graham, J. A., Huchra, J. P., Hughes, S. M. G., Illingworth, G. D., Macri, L. M., and Stetson, P. B. (2001). Final Results from the Hubble Space Telescope Key Project to Measure the Hubble Constant. *ApJ*, 553:47–72.

Halverson, N. W., Leitch, E. M., Pryke, C., Kovac, J., Carlstrom, J. E., Holzapfel, W. L., Dragovan, M., Cartwright, J. K., Mason, B. S., Padin, S., Pearson, T. J., Readhead, A. C. S., and Shepherd, M. C. (2002). Degree Angular Scale Interferometer First Results: A Measurement of the Cosmic Microwave Background Angular Power Spectrum. *ApJ*, 568:38–45.

Henry, P. S. (1971). Isotropy of the 3K Background. *Nature*, 231:516–518.

Hinshaw, G. F., Barnes, C., Bennett, C. L., Greason, M. R., Halpern, M., Hill, R. S., Kogut, N. Jarosik A., Limon, M., Meyer, S. S., Odegard, N., Page, L., Spergel, D. N., Tucker, G. S., Wollack, J. L. Weiland E., and Wright, E. L. (2003a). First Year Wilkinson Microwave Anisotropy Probe (*WMAP*) Observations: Data Processing Methods and Systematic Error Limits. *Astrophys. J., Supp.*, 148:63.

Hinshaw, G. F., Spergel, D. N., Verde, L., Hill, R. S., Meyer, S. S., Barnes, C., Bennett, C. L., Halpern, M., Jarosik, N., Kogut, A., Komatsu, E., Limon, M., Page, L., Tucker, G. S., Weiland, J. L., Wollack, E., and Wright, E. L. (2003b). First Year Wilkinson Microwave Anisotropy Probe (*WMAP*) Observations: The Angular Power Spectrum. *Astrophys. J., Supp.*, 148:135.

Jarosik, N., Barnes, C., L.Bennett, C., M.Halpern, G.Hinshaw, A.Kogut, Limon, Meyer, S. S., Page, L., Spergel, D. N., Tucker, G. S., Weiland, J. L., Wollack, E., and Wright, E. L. (2003). First Year Wilkinson Microwave Anisotropy Probe (*WMAP*) Observations: In-Orbit Radiometer Characterization. *Astrophys. J., Supp.*, 148:29.

Jungman, G., Kamionkowski, M., Kosowsky, A., and Spergel, D. N. (1996). Weighing the Universe with the Cosmic Microwave Background. *Physical Review Letters*, 76:1007–1010.

Kamionkowski, M., Kosowsky, A., and Stebbins, A. (1997). Statistics of cosmic microwave background polarization. *Phys. Rev. D*, 55:7368–7388.

Klypin, A. A., Sazhin, M. V., Strukov, I. A., and Skulachev, D. P. (1987). Anisotropy of the relic radiation according to the Relikt experiment. *Pis ma Astronomicheskii Zhurnal*, 13:259–267.

Knox, L. and Page, L. (2000). Characterizing the Peak in the Cosmic Microwave Background Angular Power Spectrum. *Phys. Rev. Lett.*, 85:1366–1369.

Kogut, A., Spergel, D. N., Barnes, C., Bennett, C. L., Halpern, M., Hinshaw, G., Jarosik, N., Limon, M., Meyer, S. S., Page, L., Tucker, G. S., Wollack, E., and Wright, E. L. (2003). First Year Wilkinson Microwave Anisotropy Probe (*WMAP*) Observations: Temperature-Polarization Correlation. *Astrophys. J., Supp.*, 148:161.

Komatsu, E., Kogut, A., Nolta, M. R., Bennett, C. L., Halpern, M., Hinshaw, G., Jarosik, N., Limon, M., Meyer, S. S., Page, L., Spergel, D. N., Tucker, G. S., Verde, L., Wollack, E., and Wright, E. L. (2003). First Year Wilkinson Microwave Anisotropy Probe (*WMAP*) Observations: Test of Gaussianity. *Astrophys. J., Supp.*, 148:119.

Kovac, J. M., Leitch, E. M., Pryke, C., Carlstrom, J. E., Halverson, N. W., and Holzapfel, W. L. (2002). Detection of polarization in the cosmic microwave background using DASI. *Nature*, 420:772–787.

Kuo, C. L. et al. (2002). High resolution observations of the cmb power spectrum with acbar. *Astrophys. J.* astro-ph/0212289.

Lubin, P. M. and Smoot, G. F. (1981). Polarization of the cosmic background radiation. *ApJ*, 245:1–17.

Mason, B. S., Pearson, T. J., Readhead, A. C. S., Shepherd, M. C., Sievers, J., Udomprasert, P. S., Cartwright, J. K., Farmer, A. J., Padin, S., Myers, S. T., Bond, J. R., Contaldi, C. R., Pen, U., Prunet, S., Pogosyan, D., Carlstrom, J. E., Kovac, J., Leitch, E. M., Pryke, C., Halverson, N. W., Holzapfel, W. L., Altamirano, P., Bronfman, L., Casassus, S., May, J., and Joy, M. (2003). The Anisotropy of the Microwave Background to l = 3500: Deep Field Observations with the Cosmic Background Imager. *ApJ*, 591:540–555.

Mather, J. C., Fixsen, D. J., Shafer, R. A., Mosier, C., and Wilkinson, D. T. (1999). Calibrator Design for the COBE Far-Infrared Absolute Spectrophotometer (FIRAS). *ApJ*, 512:511–520.

Page, L., Barnes, C., Hinshaw, G., Spergel, D. N., Weiland, J. L., Wollack, E., Bennett, C. L., Halpern, M., Jarosik, N., Kogut, A., Limon, M., Meyer, S. S., Tucker, G. S., and Wright, E. L. (2003a). First Year Wilkinson Microwave Anisotropy Probe (*WMAP*) Observations: Beam Profiles and Window Functions. *Astrophys. J., Supp.*, 148:39.

Page, L., Nolta, M. R., Barnes, C., Bennett, C. L., Halpern, M., Hinshaw, G., Jarosik, N., Kogut, A., Limon, M., Meyer, S. S., Peiris, H., Spergel, D. N., Tucker, G. S., Wollack, E., and Wright, E. L. (2003b). First Year Wilkinson Microwave Anisotropy Probe (*WMAP*) Observations: Interpretation of the TT and TE Angular Spectrum Peaks. *Astrophys. J., Supp.*, 148:233.

Peebles, P. J. E. (1982). Large-scale background temperature and mass fluctuations due to scale-invariant primeval perturbations. *ApJ*, 263:L1–L5.

Peiris, H., Komatsu, E., Verde, L., Bennett, D. N. SpergelC. L., Halpern, M., Hinshaw, G., Jarosik, N., Kogut, A., Limon, M., Meyer, S. S., Page, L., Tucker, G. S., Wollack, E., and Wright, E. L. (2003). First Year Wilkinson Microwave Anisotropy Probe (*WMAP*) Observations: Implications for Inflation. *Astrophys. J., Supp.*, 148:213.

Penzias, A. A. and Wilson, R. W. (1965). A Measurement of Excess Antenna Temperature at 4080 Mc/s. *ApJ*, 142:419–421.

Sachs, R. K. and Wolfe, A. M. (1967). Perturbations of a Cosmological Model and Angular Variations of the Microwave Background. *Astrophys. J.*, 147:73.

Scott, D., Silk, J., and White, M. (1995). From Microwave Anisotropies to Cosmology. *Science*, 268:829–835.

Seljak, U. and Zaldarriaga, M. (1997). Signature of Gravity Waves in the Polarization of the Microwave Background. *Physical Review Letters*, 78:2054–2057.

Silk, J. (1968). Cosmic Black-Body Radiation and Galaxy Formation. *Astrophys. J.*, 151:459.

Smoot, G. F., Goernstein, M. V., and Muller, R. A. (1977). Detection of Anisotropy in the Cosmic Blackbody Radiation. *prl*, 39:898–901.

Spergel, D. N., Verde, L., Peiris, H., Komatsu, E., Nolta, M. R., Bennett, C. L., Halpern, M., Hinshaw, G., Jarosik, N., Kogut, A., Limon, M., Meyer, S. S., Page, L., Tucker, G. S., Weiland, J. L., Wollack, E., and Wright, E. L. (2003). First Year Wilkinson Microwave Anisotropy Probe (*WMAP*) Observations: Determination of Cosmological Parameters. *Astrophys. J., Supp.*, 148:175.

Tonry, J., Schmidt, B. P., Barris, B., Candia, P., Challis, P., Clocciatti, A., L., Coil A., Filipenko, A. V., Garnavich, P., and many others (2003). Cosmological Results from High-z Supernovae. ApJ, in press, astro-ph/0305008.

Verde, L., Peiris, H., Spergel, D. N., Nolta, M. R., Bennett, C. L., Halpern, M., Hinshaw, G., Jarosik, N., Kogut, A., Limon, M., Meyer, S. S., Page, L., Tucker, G. S., Wollack, E., and Wright, E. L. (2003). First Year Wilkinson Microwave Anisotropy Probe (*WMAP*) Observations: Parameter Estimation Methodology. *Astrophys. J., Supp.*, 148:195.

NEW RESULTS & CURRENT WORK WITH THE CBI

Brian S. Mason*
National Radio Astronomy Observatory Green Bank, WV 24944
bmason@gb.nrao.edu

Keywords: Cosmic Microwave Background, Cosmology, Radio Astronomy

1. Introduction

Anisotropies in the Cosmic Microwave Background (CMB) contain a wealth of information about fundamental cosmological parameters (White et al. 1994), as well as providing a direct link to theories of high-energy physics and structure formation. The most straightforward models for anisotropies in the CMB (Bond & Efstathiou, 1987) feature two key physical scales: a length scale associated with acoustic oscillations of density fluctuations in the primordial plasma, and an exponential damping of these fluctuations caused by photon diffusion. Both of these physical scales depend upon the values of the cosmological parameters at and before the time of recombination, and the relation of these scales to observable angular scales on the sky is determined by the angular diameter distance between the present and $z = 1100$. Detections of acoustic peaks have been reported (e.g. Miller et al. 1999, de Bernardis et al. 2000, Halverson et al. 2002). The detailed signature of causal processes on the spectrum also conveys information about cosmological parameters.

In addition, the *polarization* of the CMB can break degeneracies between the primordial spectrum and later processes such as acoustic oscillations, and provides a fundamental test of our theories. CMB polarization anisotropies have been detected (Kovac et al. 2002), although the shape of the power spectrum of polarized anisotropies is not yet well characterized. Secondary anisotropies such as the Sunyaev-Zel'dovich Effect (SZE) probe more recent epochs of structure formation.

The Cosmic Background Imager (CBI) is an interferometer designed to measure CMB anisotropies on angular scales of $400 < \ell < 4250$. The CBI consists of an array of 13 0.9-meter diameter antennas mounted on a 6 meter

*on behalf of the CBI collaboration

M. Plionis (ed.), Multiwavelength Cosmology, 85-88.

steerable platform, and operating in 10 1-GHz bands between 26 and 36 GHz ($\sim 1\,\mathrm{cm}$); it is sited high in the Chilean Andes. The CBI is well suited to measure intrinsic anisotropies in both intensity and polarization, and secondary SZE anisotropies. In these proceedings we report on the status of CBI efforts in these directions.

2. Previous Results from the CBI

Previous intensity anisotropy results from the CBI have been reported by Padin et al. 2001, Pearson et al. 2003, and Mason et al. 2003. These results were based on 240 nights of observing divided between 3 deep fields (individual pairs of CBI pointings) and 3 mosaic fields. The mosaic fields provided a total sky coverage of $40\,\mathrm{deg}^2$. The deep fields allowed extensive data consistency checks and had very low thermal noise, and thus formed the basis for results at high ℓ. These data indicate a drop in the anisotropy power spectrum from $\ell \sim 900$ and out to $\ell \sim 2000$. The data points in this region are consistent with the theoretical curve obtained by fitting to the low-ℓ data. This strongly supports the physics underlying these models, *i.e.*, an approximately scale-invariant spectrum of primordial density fluctuations modified primarily by acoustic oscillations and Silk damping (Silk 1968) before recombination. See Sievers et al. 2003 for a cosmological parameter analysis of the CBI mosaic data.

Beyond $\ell = 2000$ the CBI deep field data show a significant detection of power. It is unlikely that a significant fraction of this excess power is due to residual radio source contamination, as the combination of OVRO data and 30 GHz source count information from the CBI maps— together with the conservative approach of projecting all NVSS source modes out of the dataset— have enabled us to place strong limits on foreground contamination. For σ_8 values of ~ 1 the observed signal is consistent with a secondary SZE anisotropy due to high-redshift galaxy clusters. The small sky coverage in the CBI deep fields is a limiting factor, since the SZE signal is highly non-Gaussian and has a large "Cosmic Variance". This point, and plans to improve point source constraints, will be discussed below.

3. New Results & Current Work

3.1 Intensity:

Since the results of Pearson et al. 2003 and Mason et al. 2003, the CBI has continued to observe. Sky coverage and total integration time have each roughly doubled, resulting in $\sim \sqrt{2}$ better in ℓ resolution at large angular scales. There are a number of further improvements to our analysis:

- WMAP has made a more accurate 30 GHz calibration feasible. We have used WMAP's measurement of Jupiter's brightness temperature to reduce our calibration uncertainty to 2% (4% in bandpower).

- We have improved our understanding of the thermal noise in the CBI data. This permits us to combine the mosaic and deep data for a single unified CBI power spectrum. The high ℓ results then also come from a larger sky area, reducing the non-Gaussian Cosmic Variance uncertainty.

- For the unified spectrum we have adopted a variable binning scheme with a single wide bin at high-ℓ breaking down into narrow ($\delta\ell = 100$) bins at the largest scales.

Preliminary results from this analysis are shown in Figure 1. The improved resolution at low-ℓ is notable. Results at high-ℓ are consistent with those reported in Mason et al.

We are working in collaboration with NRAO to field a sensitive 30 GHz receiver and wideband continuum backend on the GBT. When completed this instrument will, by reducing the uncertainty in the residual source level and allowing many NVSS sources to be vetoed, further improve results on small angular scales.

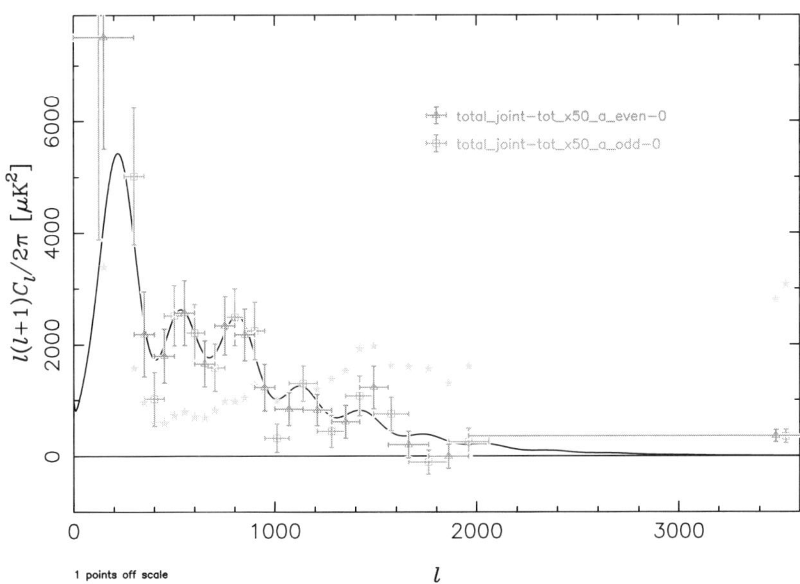

Figure 1. CMB power spectrum from CBI 2000 and 2001 data. Blue and red show alternate binnings of the data; grey stars show the sum of the thermal noise power spectrum and the residual point source spectrum (which is sub-dominant at all ℓ).

3.2 Polarization:

Polarization anisotropies in the CMB are expected to peak at higher ℓ (~ 900) than intensity anisotropies. This is near the peak of the CBI's sensitivity. During years 2000 and 2001 one antenna was kept in a cross-polarized state; this data has been used to place stringent limits on polarized anisotropy from $400 < \ell < 2000$ (Cartwright et al. in preparation).
The CBI has been reconfigured into a compact, highly redundant configuration to make these measurements; it has also been outfitted with upgraded HEMT amplifiers which will save 25% in observing time, and switchable broadband circular polarizers. Polarization observations with the upgraded system have been underway since Fall 2002, and the analysis of these data are underway.

4. Summary

We have summarized previous results from the Cosmic Background Imager, as well as current results and prospects for current CBI observing campaigns focused on higher ℓ resolution, small angular scales, and CMB polarization.

Acknowledgments

The CBI is an international collaboration based at the California Institute of Technology (A.C.S. Readhead PI), with collaborators at the Canadian Institute for Theoretical Astrophysics, the National Radio Astronomy Observatory, the University of Alberta, the University of Chicago, UC Berkeley, the University of Chile, and Marshall Space Flight Center. This project has been supported by funds from the NSF, Caltech, the Canadian Institute for Advanced Research, and private donors.

References

J. R. Bond, & G. Efstathiou, MNRAS **226**, 655 (1987)
J. R. Bond, A. H. Jaffe, & L. Knox, Phys. Rev. D **57**, 2117 (1998)
J. R. Bond et al., submitted to Ap.J. (astro-ph/0205386)
P. de Bernardis et al., Nature **404**, 955 (2000)
N. Halverson et al., Ap.J. **568**, 38 (2002)
J. Kovac et al., Nature **420**, 772 (2002)
B. Mason et al., Ap. J. **591**, 540 (2003)
A. D. Miller et al., Ap.J. **524**, L1 (1999)
Padin, S. et al., Ap.J. **549**, L1 (2001)
T. Pearson et al., Ap. J. **591**, 556 (2003)
J. Sievers et al., Ap. J. **591**, 599 (2003)
J. Silk, Ap.J. **151**, 459 (1968)
M. White, D. Scott, & J. Silk, Ann. Rev. A. & A. **32**, 319 (1994).

THE GALACTIC DUST AS A FOREGROUND TO COSMIC MICROWAVE BACKGROUND MAPS

X. Dupac (ESA-ESTEC, Noordwijk, the Netherlands, xdupac@rssd.esa.int), J.-P. Bernard, N. Boudet, M. Giard (CESR Toulouse), J.-M. Lamarre (LERMA Paris), C. Meny (CESR), F. Pajot (IAS Orsay), I. Ristorcelli (CESR)

Abstract We present results obtained with the PRONAOS balloon-borne experiment on interstellar dust. In particular, the submillimeter / millimeter spectral index is found to vary between roughly 1 and 2.5 on small scales (3.5′ resolution). This could have implications for component separation in Cosmic Microwave Background maps.

Keywords: dust, extinction — infrared: ISM — submillimeter — cosmology: Cosmic Microwave Background

1. Introduction

To accurately characterize dust emissivity properties represents a major challenge of nowadays astronomy. It is crucial for deriving very accurate maps of the Cosmic Microwave Background fluctuations, as well as for understanding the physics of the interstellar medium. In the submillimeter domain, large grains at thermal equilibrium (e.g. Desert et al. 1990) dominate the dust emission. This thermal dust is characterized by a temperature and a spectral dependence of the emissivity which is usually simply modeled by a spectral index. The temperature, density and opacity of a molecular cloud are key parameters which control the structure and evolution of the clumps, and therefore, star formation. The spectral index (β) of a given dust grain population is directly linked to the internal physical mechanisms and the chemical nature of the grains.

It is generally admitted from Kramers-König relations that 1 is a lower limit for the spectral index. $\beta = 2$ is particularly invoked for isotropic crystalline grains, amorphous silicates or graphite grains. However, it is not the case for amorphous carbon, which is thought to have a spectral index equal to 1. Spectral indices above 2 may exist, according to several laboratory measurements on grain analogs. Observations of the diffuse interstellar medium at large scales

M. Plionis (ed.), Multiwavelength Cosmology, 89-92.

favour β around 2 (e.g. Boulanger et al. 1996, Dunne & Eales 2001). In the case of molecular clouds, spectral indices are usually found to be between 1.5 and 2. However, low values (0.2-1.4) of the spectral index have been observed in circumstellar environments, as well as in molecular cloud cores.

2. PRONAOS observations

PRONAOS (PROgramme NAtional d'Observations Submillimetriques) is a French balloon-borne submillimeter experiment (Ristorcelli et al. 1998). Its effective wavelengths are 200, 260, 360 and 580 μm, and the angular resolutions are $2'$ in bands 1 and 2, $2.5'$ in band 3 and $3.5'$ in band 4. The data analyzed here were obtained during the second flight of PRONAOS in September 1996, at Fort Sumner, New Mexico. The data processing method, including deconvolution from chopped data, is described in Dupac et al. (2001). This experiment has observed various phases of the interstellar medium, from diffuse clouds in Polaris (Bernard et al. 1999) and Taurus (Stepnik et al. 2003) to massive star-forming regions in Orion (Ristorcelli et al. 1998, Dupac et al. 2001), Messier 17 (Dupac et al. 2002), Cygnus B, and the dusty envelope surrounding the young massive star GH2O 092.67+03.07 in NCS. The ρ Ophiuchi low-mass star-forming region has also been observed, as well as the edge-on spiral galaxy NGC 891 (Dupac et al. 2003b).

3. Analysis of the Galactic dust emission

We fit a modified black body law to the spectra: $I_\nu = \epsilon_0 \; B_\nu(\lambda, T) \; (\lambda/\lambda_0)^{-\beta}$, where I_ν is the spectral intensity (MJy/sr), ϵ_0 is the emissivity at λ_0 of the observed dust column density, B_ν is the Planck function, T is the temperature and β is the spectral index. In most of the areas, we use either only PRONAOS data or PRONAOS + IRAS 100 μm data. We restrain the analysis to all fully independent ($3.5'$ side) pixels for which both relative errors on the temperature and the spectral index are less than 20%. This procedure allows to reduce the degeneracy effect between the temperature and the spectral index. Dupac et al. (2001) and Dupac et al. (2002) have shown by fitting simulated data that this artificial anticorrelation effect was small compared to the effect observed in the data.

We present in Fig. 1 the spectral index - temperature relation observed. The temperature in this data set ranges from 11 to 80 K, and the spectral index also exhibits large variations from 0.8 to 2.4. One can observe an anticorrelation on these plots between the temperature and the spectral index, in the sense that the cold regions have high spectral indices around 2, and warmer regions have spectral indices below 1.5. In particular, no data points with $T > 35$ K and $\beta > 1.6$ can be found, nor points with $T < 20$ K and $\beta < 1.5$. This anticorrelation effect is present for all objects in which we observe a large

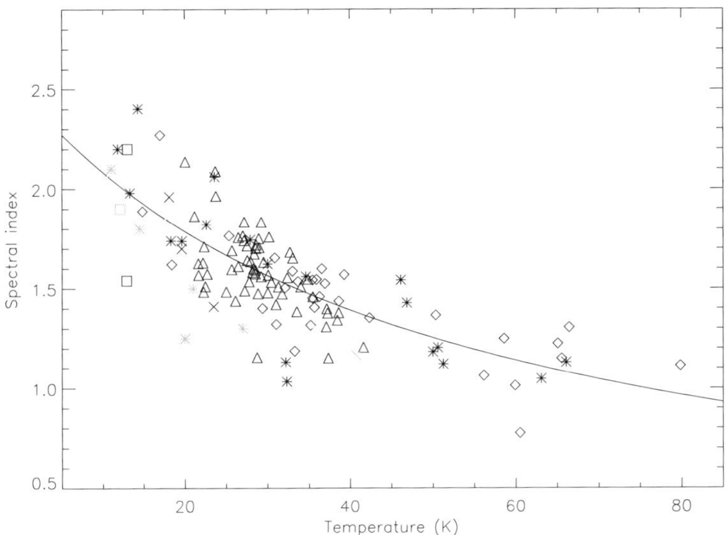

Figure 1. Spectral index versus temperature, for fully independent pixels in Orion (black asterisks), M17 (diamonds), Cygnus (triangles), ρ Ophiuchi (grey asterisks), Polaris (black squares), Taurus (grey square), NCS (grey cross) and NGC 891 (black crosses). The full line is the result of the best hyperbolic fit: $\beta = \frac{1}{0.4+0.008T}$

range of temperatures, namely Orion, M17, Cygnus and ρ Ophiuchi. It is also remarkable that the few points from other regions are well compatible with this general anticorrelation trend. The temperature dependence of the emissivity spectral index is well fitted by a hyperbolic approximating function.

Several interpretations are possible for this effect: one is that the grain sizes change in dense environments, another is that the chemical composition of the grains is not the same in different environments and that this correlates to the temperature, a third one is that there is an intrinsic dependence of the spectral index on the temperature, due to quantum processes such as two-level tunneling effects. Additional modeling, as well as additional laboratory measurements and astrophysical observations, are required in order to discriminate between these different interpretations. More details about this analysis and the possible interpretations can be found in Dupac et al. (2003a).

4. Implications for high-redshift far-infrared observations

A recent paper from Eales et al. (2003) showed that the 850 / 1200 μm ratio of their sample of extragalactic millimeter sources was very low, which could be explained by spectral indices of the dust around 1 in high-redshift galaxies. Though this result is still uncertain, it might confirm our measurements be-

cause high-redshift galaxies are likely to be warmer than low-redshift galaxies (because of cosmic expansion).
The anticorrelation between the temperature and the emissivity spectral index can indeed have major implications for deriving the redshifts, masses, temperatures, luminosities, etc, of extragalactic objects.

5. Implications for Cosmic Microwave Background measurements

Even high Galactic latitude clouds can have a non-uniform spectral index (e.g. Bernard et al. 1999). Since the Galactic dust is the major contributor to CMB foregrounds in the submillimeter and millimeter domains, even small variations of the dust spectral energy distribution can harm the CMB measurements if the component-separation methods assume a uniform dust spectral index.

References

Bernard, J.-P., Abergel, A., Ristorcelli, I., et al.: 1999, A&A, 347, 640

Boulanger, F., Abergel, A., Bernard, J.-P., et al.: 1996, A&A, 312, 256

Desert, F.-X., Boulanger, F., Puget, J.-L.: 1990, A&A, 237, 215

Dunne, L., Eales, S.A.: 2001, MNRAS, 327, 697

Dupac, X., Giard, M., Bernard, J.-P., et al.: 2001, ApJ, 553, 604

Dupac, X., Giard, M., Bernard, J.-P., et al.: 2002, A&A, 392, 691

Dupac, X., Bernard, J.-P., Boudet, N., et al.: 2003a, A&A Lett., 404, L11

Dupac, X., del Burgo, C., Bernard, J.-P., et al.: 2003b, MNRAS, in press, astro-ph/0305230

Eales, S., Bertoldi, F., Ivison, R., Carilli, C., Dunne, L., Owen, F.: 2003, MNRAS, in press, astro-ph/0305218

Ristorcelli, I., Serra, G., Lamarre, J.-M., et al.: 1998, ApJ, 496, 267

Stepnik, B., Abergel, A., Bernard, J.-P., et al.: 2003, A&A, 398, 551

NEUTRINO PHYSICS IN THE LIGHT OF WMAP

Elena Pierpaoli
Physics Department and Astronomy Department
Princeton University, Princeton, NJ, 08544, USA *
epierpao@princeton.edu

Abstract We investigate the implications of recent cosmic microwave background radiation (CMB) and large scale structure (LSS) results the neutrino background and mass. We use the recent WMAP data, together with previous CBI data and 2dF matter power spectrum, to constrain the effective number of neutrino species N_{eff} and the neutrino mass in a general Cosmology. We find that $N_{eff} = 4.08$ with a 95 per cent C.L. $1.9 \leq N_{eff} \leq 6.62$. The curvature we derive is still consistent with flat, but assuming a flat Universe from the beginning implies a bias toward lower N_{eff}, as well as artificially smaller error bars. We analyze and discuss the degeneracies with other parameters, and point out that probes of the matter power spectrum on smaller scales, accurate independent σ_8 measurements, together with better independent measurement of H_0 would help in breaking the degeneracies. As for the neutrino mass, we show that CMB and matter power spectrum at the present time are unable to constrain the neutrino mass below 0.4 eV, unless a strong assumption on the bias parameter is assumed. Adding constraints on σ_8 significantly reduces the uncertainty on the neutrino mass.

Keywords: gravitation – cosmology: theory – large-scale structure of Universe – cosmic microwave background – dark matter: neutrino

1. Constraints on the neutrino background

We describe here the constraints that recent CMB and LSS results have implied on the neutrino background N_{eff}, that is the amount of relativistic energy density at recombination/equivalence.
We consider only CDM adiabatic perturbations, and the following set of parameters, with usual definitions: Ω_{dm}, Ω_Λ, Ω_b, Ω_k, H_0, N_ν, n_s, A_s, z_{re}. We use the WMAP and CBI data together with the 2dF matter power spectrum with no bias assumption, and we discuss the effect of adding other priors.

*Funding provided by NASA grant NAG5-11489.

M. Plionis (ed.), Multiwavelength Cosmology, 93-96.

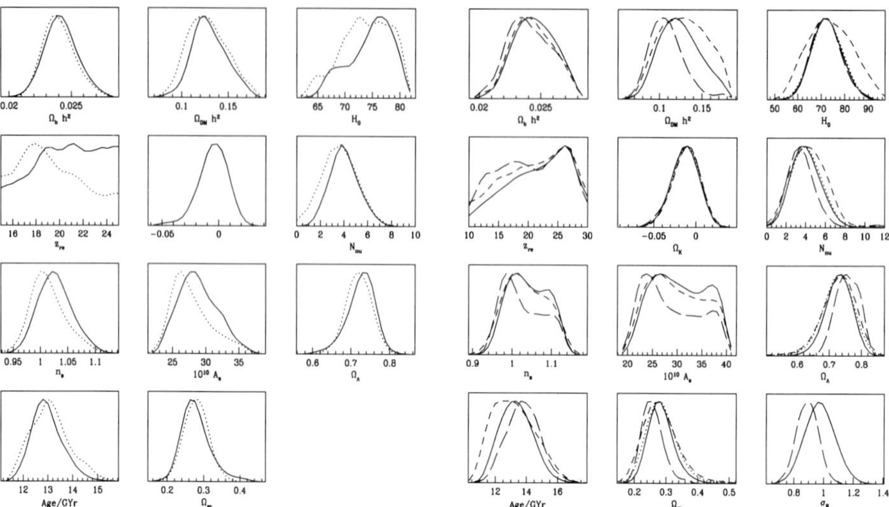

Figure 1. The marginalized likelihoods for the parameters under consideration. We have assumed here a top–hat prior on H_0 corresponding to the 1σ interval allowed by the HST key project results. The solid line is for a general curvature, the dotted corresponds to the flat Universe case. The general curvature tends to push the constraints on N_{eff} toward higher values. The same upper limits on N_{eff} are probably due to the upper limit imposed on H_0 ($<$ 80). Notice that z_{re} in the range considered here is not constrained by the data.

Figure 2. The marginalized likelihoods in the case of a general cosmology. The short–dashed line only consider CMB+2dF data, the dotted includes the H_0 prior from the HST project, and the solid also includes SN data. N_{eff} is restricted to be ≤ 6.6 at 95 per cent C.L., and Ω_k tends to be negative but is still consistent with flat. Note the high n_s and z_{re} values. The long–dashed line is obtained adding an hypothetical prior on σ_8 with the typical scaling from clusters and weak lensing.

First we compare the results for a flat model ($\Omega_k = 0$) with those for a general Universe when H_0 to vary only in the range $64 \leq H_0 \leq 80$ corresponding to the 1σ allowed range of the HST result. In fig.1 we present the likelihoods for each parameter, after marginalization over all the others, for flat and generally curved cases. We find a best fit $N_{eff} = 4.08$ ($N_{eff} = 3.70$) for the general curvature (flat) case, with $2.23 \leq N_{eff} \leq 6.13$ ($1.82 \leq N_{eff} \leq 5.74$) at the 95% C.L. The assumption of a flat Universe biases the result toward low N_{eff} and shrinks the inferred error bars. Notice that the general curvature analysis is important at the present time because of the great improvement that the WMAP results have implied on the CMB power spectrum. Constraints from CMB and LSS on flat Universes previous to the WMAP were much weaker: $N_{eff} \leq 15$ at 95% C.L., rendering the general curvature analysis very little constraining. In fig.1 the flat and curve case present the similar upper limit for

N_{eff} mainly because we imposed a tight top-hat prior on H_0 (< 80). The likelihood indeed seems to prefer high values for H_0, which may allow for extra radiative component.

For the general curvature case, we now proceed to discuss the effect of various priors on the determination of N_{eff}. In fig. 2 we present the results for the general curvature case, where we allow for a wider H_0 (and z_{re}) range. The best fit for N_{eff} in this case is $N_{eff} = 4.31$, with a 95 per cent C.L. range : $1.6 \leq N_{eff} \leq 7.1$ from CMB and 2dF only, and $N_{eff} = 4.08$ with $1.9 \leq N_{eff} \leq 6.62$ when the H_0 prior from the HST project is included.

We note that the inclusion of extra relativistic energy causes a higher n_s values than in the standard case, a higher z_{re} and a slightly closed Universe ($\Omega_k = -0.013 \pm 0.015$). We argue that, since a high N_{eff} boosts the first peak through the early ISW effect, a higher n_s combined with an early reionization of the Universe can still ensure a good fit to the CMB data. Matter power spectrum data on smaller scales than the ones probed by 2dF (e.g. Ly-α forest data) may be used in breaking this degeneracy, because the matter power spectrum is only sensitive to n_s and not to z_{re}. Unfortunately these data are not fully reliable at the present time. In fig. 2 we note that the derived value of σ_8 is quite high (0.97 ± 0.1) yet because of the favoured high n_s values. Actual quotations of the σ_8 value range between 0.75 and 1 (Pierpaoli et al. 2003), so that a broad prior on σ_8 would already decrease the allowed n_s and improve the errors on N_{eff}. Consistent 5 per cent accuracy measurements of σ_8 from both weak lensing and clusters would greatly improve the constraint on N_{eff}. As an example, in fig. 2 we plot the curves obtained with an hypothetical prior on σ_8 with the typical scaling derived from cluster and weak lensing: $\sigma_8\ (\Omega_m/0.3)^{0.6} = 0.85 \pm 0.05$. Such prior would imply a lower Ω_m and $N_{eff} = 3.8 \pm 1.6$. The inclusion of the Supernovae in the analysis doesn't particularly improve the results on N_{eff}, although it slightly improves the Ω_m and Ω_Λ determination.

2. CMB and LSS constraints on m_ν

The scales currently probed by the CMB experiments at $z_{rec} \simeq 1000$ marginally overlap with those probed by the matter matter power spectrum at much lower redshifts. The growth of perturbations between $z_{rec} \simeq 1000$ and now is greatly affected by the presence of massive neutrinos. Considering a small neutrino abundance, the ratio of growth rate between the pure CDM case and the massive neutrino one is:

$$\frac{[\delta_0/\delta_{rec}]_{f_\nu}}{[\delta_0/\delta_{rec}]_{f_\nu=0}} = z_{rec}^{0.25(-1+5\sqrt{1+24/25 f_\nu})-1}; \tag{1}$$

($f_\nu \equiv \Omega_\nu/\Omega_{dm}$) which is the solid curve in fig. 3. The points are a piece of a Markov chain performed on CMB data plus 2dF power spectrum with a flat bias prior. Their density represents the likelihood of that $\sigma_8 - f_\nu$ values, while the gray–scale reflects the $\Omega_{dm}h^2$ value associated with that point. First we notice that the chain density clearly reflects the theoretical curve, indicating that the normalization of the power spectrum plays a crucial role in constraining the neutrino abundance and, in turn, the neutrino mass. For three degenerate neutrinos, $f_\nu \simeq 0.1$ corresponds to a mass $m_\nu \simeq 0.4$ eV ($\sum m_i = 93.8\,\Omega_\nu h^2$ eV, for $\Omega_{dm}h^2 = 0.13$ this is $\sum m_i = 12.2 f_\nu$ eV) . We therefore conclude that CMB plus 2dF data are unable to constrain the neutrino mass below such value, unless a strong assumption on the bias parameter is made (as was the case for the result quoted by the WMAP team). Alternatively, a precise measurement of σ_8 may significantly reduce the uncertainty on m_ν. In particular, values of $\sigma_8 \geq 0.8$ would imply $f_\nu \leq 0.05$, and therefore $m_\nu \leq 0.2$ eV.

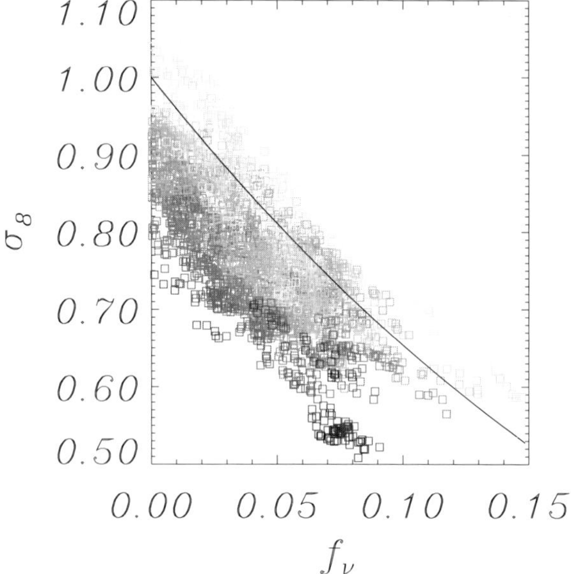

Figure 3. The fraction of neutrino abundance versus the mass variance at 8 h^{-}1Mpc. A precise measurement of σ_8 would greatly improve the neutrino mass limit, currently set by $f_\nu \leq 0.1$ ($m_\nu \leq 0.4$ eV).

References

Pierpaoli E., 2003, MNRAS 342, L63

Pierpaoli E., Borgani S., Scott D., White M., 2003, MNRAS 342, 163

ARCHEOPS: AN INSTRUMENT FOR PRESENT AND FUTURE COSMOLOGY

Matthieu Tristram
LPSC, Grenoble, France
tristram@lpsc.in2p3.fr

on behalf of the Archeops Collaboration
includes scientists from Caltech (USA), Cardiff Univ., CESR (Toulouse), CSNSM (Orsay), CRTBT (Grenoble), IAS (Orsay), IAP (Paris), IROE (Firenze, Italy), ISN (Grenoble), JPL (USA), LAL (Orsay), LAOG (Grenoble), Landau Institute (Russia), La Sapienza Univ. (Roma, Italy), LAOMP (Toulouse), Maynooth Univ. (Ireland), Minnesota Univ. (USA), PCC (Paris), SPP (Saclay)

Abstract ARCHEOPS is a balloon-borne instrument dedicated to measure the cosmic microwave background (CMB) temperature anisotropies. It has, in the millimetre domain (from 143 to 545 GHz), a high angular resolution (about 10 arcminutes) in order to constrain high ℓ multipoles, as well as a large sky coverage fraction (30%) in order to minimize the cosmic variance. It has linked, before WMAP, COBE large angular scales to the first acoustic peak region. From its results, inflation motivated cosmologies are reinforced with a flat Universe ($\Omega_{\rm tot} = 1$ within 3 %). The dark energy density and the baryonic density are in very good agreement with other independent estimations based on supernovae measurements and big bang nucleosynthesis. Important results on galactic dust emission polarization and their implications for PLANCK-HFI are also addressed.

1. Introduction

ARCHEOPS is a CMB bolometer-based instrument using PLANCK-HFI technology that fills a niche where previous experiments were unable to provide strong constraints. Namely, ARCHEOPS seeks to join the gap in ℓ between the large angular scales as measured by COBE/DMR and degree-scale experiments, typically for ℓ between 10 and 200. For that purpose, a large sky coverage is needed. The solution was to adopt a spinning payload mostly above the atmosphere, scanning the sky in circles with an elevation of around 41 degrees. The gondola, at a float altitude above 32 km, spins across the sky at a rate of 2 rpm which, combined with the Earth rotation, produces well sampled sky map at 143, 217, 353 and 545 GHz.

M. Plionis (ed.), Multiwavelength Cosmology, 97-100.

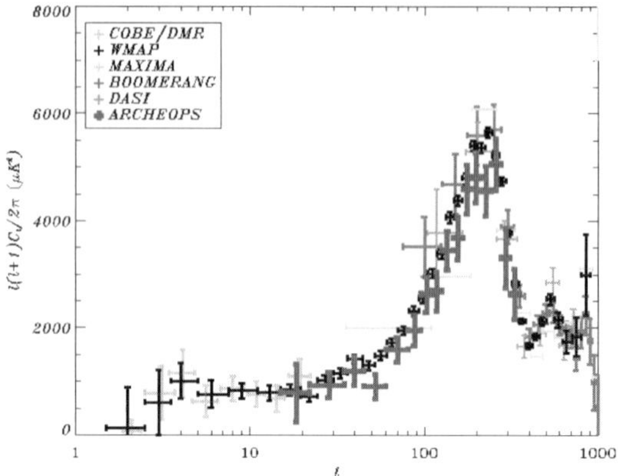

Figure 1. ARCHEOPS CMB power spectrum (Benoît et al. 2003a) in 16 bins along with other recent experiments COBE (Tegmark et al. 1996), WMAP (Bennett et al. 2003), MAXIMA (Lee et al. 2001), BOOMERANG (Netterfield et al. 2002) and DASI (Halverson et al. 2002).

2. Description of the instrument

The instrument was designed by adapting concepts put forward for PLANCK-HFI and using balloon-borne constraints (Benoît et al. 2002) : namely, an open ^{3}He-^{4}He dilution cryostat cooling spiderweb-type bolometers at 100 mK, cold individual optics with horns at different temperature stages (0.1, 1.6, 10 K) and an off-axis Gregorian telescope. The CMB signal is measured by the 143 and 217 GHz detectors while interstellar dust emission and atmospheric emission are monitored with the 353 (polarized) and 545 GHz detectors. The whole instrument is baffled so as to avoid stray radiation from the Earth and the balloon. We report on the first results obtained from the last flight (12.5 night hours) that was performed from Kiruna (Sweden) to Russia in February 2002.

3. Results

After being calibrated with the CMB dipole – in agreement with the FIRAS Galaxy or Jupiter emission – eight detectors (yielding effective beams of typically 12 arcminute FWHM) at 143 and 217 GHz are found to have a sensitivity better than 200 $\mu K_{CMB}.s^{1/2}$. A large part of the data reduction was devoted to removing systematic effects coming from temperature variations on the various thermal stages and atmospheric effects.

Benoît et al. 2003a and Benoît et al. 2003b show the results of a first analysis of the data, which are summarized below. Only the best bolometer of each CMB channel (143 and 217 GHz) was used. The data are cleaned and calibrated, and the pointing is reconstructed from the stellar sensor data. The sky power spectrum above a galactic latitude of 30 °(free of foreground contamination) is deduced using a MASTER-like approach (Hivon et al. 2002). The observed spectrum is shown in Fig. 1 and compared to a selection of other recent experiments. Much attention was paid to the possible systematic effects that could affect the results. At low ℓ, dust contamination and at large ℓ, bolometer time constant and beam uncertainties are all found to be negligible with respect to statistical errors. The sample variance at low ℓ and the photon noise at high ℓ are found to be a large fraction of the final ARCHEOPS error bars in Fig. 1.

4. Cosmological constraints

ARCHEOPS provides a precise determination of the first acoustic peak in terms of position at the multipole $\ell_{\rm peak} = 220 \pm 6$, height and width. Using a large grid of cosmological models with 7 parameters, one can compute their likelihood with respect to the datasets. An analysis of Archeops data in combination with other CMB datasets constrains the baryon content of the Universe to a value $\Omega_b h^2 = 0.022^{+0.003}_{-0.004}$ which is compatible with Big-Bang nucleosynthesis (O'Meara et al. 2001) and with a similar accuracy (Fig. 2). Using the recent HST determination of the Hubble constant (Freedman et al. 2001) leads to tight constraints on the total density, *e.g.* $\Omega_{\rm tot} = 1.00^{+0.03}_{-0.02}$, *i.e.* the Universe would be flat. An excellent absolute calibration consistency is found between COBE, ARCHEOPS and other CMB experiments (Fig. 1). All these measurements are fully compatible with inflation-motivated cosmological models. The constraints shown on Fig. 2 (right), leading to a value of $\Omega_\Lambda = 0.73^{+0.09}_{-0.07}$ for the dark energy content, are independent from and in good agreement with supernovae measurements (Perlmutter et al. 1999) if a flat Universe is assumed.

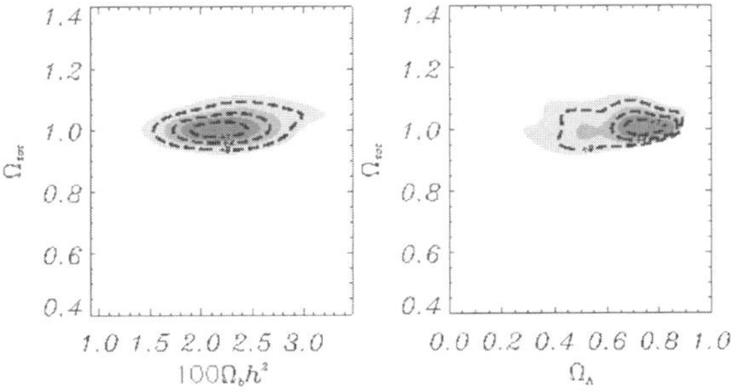

Figure 2. Likelihood contours for 3 of the cosmological parameters: total density versus baryonic density and cosmological constant. Greyscale corresponds to 2-D limits and dashed line to 1-D contours equivalent to 1, 2, and 3 σ thresholds. A Prior on the Hubble constant $H_0 = 72 \pm 8\,{\rm km/s/Mpc}$ (68% CL, Freedman et al. 2001) has been added.

5. Polarized Foregrounds

The polarized channel at 353GHz gives important results on the polarization of the emission of the dust in the galactic plane. Concerning the large scales, we find a diffuse emission polarized at 4-5% with an orientation mainly perpendicular to the galactic plane. We found also several clouds of a few square degrees polarized at more than 10% in the Gemini and the Cepheus regions. It is interesting to note that the brightest region, Cygnus, is not polarized. These results suggest a powerful grain alignment mechanism throughout interstellar medium. All the interpretations are developed in Benoît et al. 2003c. Interstellar dust polarization emission will be a major foreground for the detection of the polarized CMB for PLANCK-HFI.

6. Conclusions

The measured power spectrum (Benoît et al. 2003a) matches the COBE data and provides for the first time a direct link between the Sachs-Wolfe plateau and the first acoustic peak. The measured spectrum is in good agreement with that predicted by inflation models producing scale-free adiabatic perturbations and a flat Universe. Finally note that these results were obtained with only half a day of data.

Use of all available bolometers and of a larger sky fraction should yield an even more accurate and broader CMB power spectrum in the near future. The large experience gained on this balloon-borne experiment is providing a large feedback to the PLANCK-HFI data processing community.

References

Bennett, C. L., et al. , accepted by ApJ, 2003, astro-ph/0302207

Benoît, A., Ade, P., Amblard, A., et al. , Astropart. Phys., 2002, 17, 101-124, astro-ph/0106152

Benoît, A., Ade, P., Amblard, A., et al. , A, 2003a, **399**, 3, L19

Benoît, A., Ade, P., Amblard, A., et al. , A, 2003b, **399**, 3, L25

Benoît, A., Ade, P., Amblard, A., et al. , submitted to A, 2003c, astro-ph/0306222

Freedman, W. L., Madore, B. F., Gibson, B. K., et al. , ApJ, **553**, 47, 2001

Halverson, N. W., Leitch, E. M., Pryke, C. et al. , MNRAS, **568**, 38, 2002

Hivon, E., Gorski, K. M., Netterfield, C. B., et al. , ApJ, **567**, 2, 2002

Lee, A. T., Ade, P., Balbi, A. et al. , ApJ, **561**, L1-L6, 2001

Netterfield, C. B. et al. , ApJ, **571**, 604, 2002

O'Meara, J. M.; Tytler, D., Kirkman, D., et al. , ApJ, **552**, 718, 2001

Perlmutter, S., Aldering, G., Goldhaber, G., et al. 1999, ApJ, 517, 565

Tegmark, M., et al. , ApJ, **464**, L35, 1996

RADIO/SUB-MM WAVELENGTHS

OBSCURED STAR FORMATION IN THE HIGH-Z SUBMILLIMETRE UNIVERSE

David H. Hughes *
Instituto Nacional de Astrofisica, Optica y Electronica
Luis Enrique Erro 1, Santa Maria Tonantzintla, Puebla, Pue., Mexico
dhughes@inaoep.mx

Abstract The availability of large-format filled-array FIR-(sub)millimetre wavelength cameras during the next decade promises to provide unprecedented imaging fidelity. The simultaneous increase in the telescope collecting area of new facilities will also provide the observational data with greater sensitivity and resolution. Together, these instrumental advances will contribute important data to some of the most challenging questions in observational cosmology. In this brief review I highlight a few of the major results from the first 6 years of (sub)millimetre extragalactic surveys, in particular those that have offered constraints on the evolutionary history of high-redshift AGN and optically-obscured dusty starburst galaxies. I also describe some of the difficulties and ambiguities that arise in the interpretation of the existing submillimetre data and their follow-up multi-wavelength observations. Finally I outline why, with the combined capabilities of the future FIR to millimetre wavelength experiments, we can shortly expect to answer the outstanding questions regarding the nature, redshift distribution, luminosity function and large-scale distribution of the (sub)millimetre population of galaxies identified in blank-field surveys.

1. Introduction

Rest-frame far-infrared (FIR) to millimetre wavelength observations have the ability to detect violent star formation in heavily-obscured galaxies which can be "missed" in rest-frame optical-UV searches. Due to a strong negative k-correction, submillimetre and millimetre (hereafter *sub-mm*) wavelength observations, in particular, are able to trace the evolution of star formation in dusty galaxies throughout a large volume of the high-redshift Universe (in principle with as much ease at $z \sim 8$ as at $z \sim 1$), and back to extremely early epochs. Given the large accessible volume of sub-mm surveys it is possible to

*Funding provided by CONACYT grant 39953-F, INAOE and the conference local organising committee is gratefully acknowledged

M. Plionis (ed.), Multiwavelength Cosmology, 103-112.

test whether sub-mm galaxies represent the rapid formation of massive (elliptical) systems in a single violent collapse of the material in the highest-density peaks of the underlying large-scale matter distribution, or whether they are built over a longer period from the continuous merging of lower-mass systems with more modest star formation rates. Eventually, interferometric imaging surveys with ALMA (and also the SMA for the brightest targets) will determine whether massive sub-mm galaxies form as a single object or whether, under higher-resolution observations, they break-up into multiple sources. In the meantime, the source-counts, redshift distribution and clustering information obtained from the current generation of sub-mm surveys can still provide useful, albeit crude, initial tests of competing galaxy formation models (e.g. van Kampen 2003). Two comprehensive descriptions of the properties of the starburst galaxies and AGN identified in sub-mm wavelength surveys by Smail et al. (2002) and Blain et al. (2002) provide useful introductions to this paper.

2. Surveying the extragalactic sky at the sub-mm

The first generation of cosmological surveys at sub-mm wavelengths have been conducted with the SCUBA (Holland et al. 1999) and MAMBO cameras, which both use modest-sized bolometer arrays ($\sim$ 100 pixels), on the 15-m JCMT and 30-m IRAM telescopes respectively. These sub-mm surveys, which covered areas ranging from 0.002 – 0.2 deg^2 (Smail et al. 1997; Hughes et al. 1998; Barger et al. 1999a; Eales et al. 1999, 2000; Lilly et al. 1999; Scott et al. 2002; Fox et al 2002; Cowie et al. 2002; Smail et al. 2002; Borys et al. 2003; Webb et al. 2003), have contributed significantly to the first efforts to understand the history of obscured star formation at high-z. The sub-mm data, however, remain limited in their ability to accurately measure the evolutionary model and large-scale distribution of the sub-mm galaxy population. Fig.1 illustrates the two fundamental reasons why this is the case: first, the extragalactic sub-mm source-counts cover a dynamic-range of only 15–20; second, the errors on the binned data reflect the fact that < 100 galaxies have been detected in all sub-mm surveys at a significance $> 4\sigma$. Thus, with these restricted statistics it is difficult to measure the flux-density at which the faint-end source-counts converge, and therefore determine the contribution of this sub-mm population to the extragalactic background emission. The combined effect of clustering and small survey-areas also has a strong impact on these faint source-count measurements (Hughes & Gaztañaga 2000, Gaztañaga & Hughes 2001). Furthermore, the poor source-count statistics leave enough uncertainty in the steepness of the bright-end counts that it is difficult to ascertain whether this reflects a cut-off in the evolving luminosity function and redshift distribution of the sub-mm population.

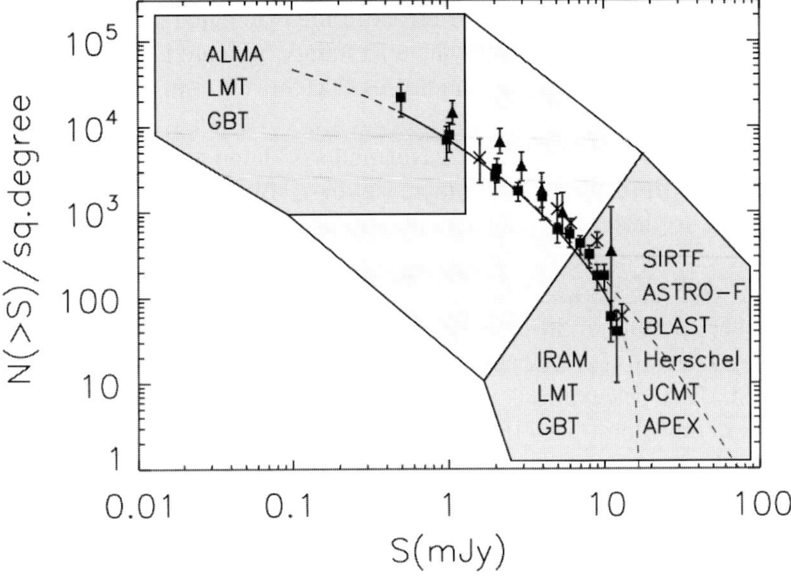

Figure 1. Extragalactic 850μm source-counts measured from the SCUBA surveys summarised in §2. The solid-line represents one of many possible strongly-evolving models that fit the 850μm data. The measured source-counts cover a narrow range of flux densities ($\sim$ 0.5 – 12 mJy) and therefore leave two unexplored regions (shown as dark-grey shaded polygons) populated by the numerous, faint galaxies below the existing observational confusion limits, and the brightest, but rarer galaxies that can only be detected in the widest-area surveys. The combination of both larger single-aperture and interferometric telescopes (e.g. LMT, GBT, ALMA), as well as larger-format cameras on smaller ground-based (JCMT, APEX, IRAM), balloon-borne (BLAST) and satellite telescopes (SIRTF, ASTRO-F, Herschel) are required to provide the necessary increase in resolution, sensitivity and mapping speed to constrain the revolutionary model for the sub-mm population. The shaded-area of the source-count parameter-space will be populated with accurate measurements at 170μm - 3 mm from these new experiments (and existing telescopes with upgraded instrumentation).

The practical reasons for the above limitations have been described elsewhere (Hughes 2000; Hughes & Gaztañaga 2000) and can be summarised as follows: restricted wavelength coverage (enforced by the few ground-based FIR–mm atmospheric windows); low spatial resolution (resulting in both a high extragalactic confusion limit and poor positional accuracy); restricted field-of-view with the current sub-mm and mm bolometer arrays (typically 5 arcmin^2); and low system sensitivity (a combination of instrument noise, size and surface accuracy of telescope aperture, sky transmission and sky noise) which restrict even the widest and shallowest sub-mm surveys to areas < 0.2 deg^2. Hence, in the effort to obtain these blank-field surveys, the existing sub-mm observations are necessarily only sensitive to the most luminous and massive star-forming galaxies ($L_{\rm FIR} \sim 3 > 10^{12} L_{\odot}$, or SFR $> 300 M_{\odot}{\rm yr}^{-1}$). Given the shape of the sub-mm k-correction, the errors in the calculated luminosities and star forming rates (considering only the uncertainty in z) are factors of a few, since the preliminary z information indicates that the sub-mm population is dominated by galaxies at $z > 1$ (§3).

Future observations at sub-mm wavelengths with the next generation of larger telescopes will simultaneously increase the resolution, and provide sufficient sensitivity to detect the fainter population of galaxies that will be accessible with the corresponding reduced confusion-limit. For example, observations at 1.1 mm with the Large Millimeter Telescope (LMT - www.lmtgtm.org), with a primary aperture of 50-m, are expected to reduce the extragalactic confusion limit by a factor of > 20, compared to the measured 850μm limit determined from SCUBA observations with the JCMT. We can expect, therefore, that in the near future mm-wavelength surveys will search for (and find) optically-obscured galaxies with far more modest FIR luminosities ($L_{\rm FIR} \sim 10^{11} L_{\odot}$) and SFRs ($\sim 10 - 50 M_{\odot}{\rm yr}^{-1}$). We must wait for observations with ALMA, however, before fainter and more quiescent galaxies (similar to the Milky Way), can be detected at the highest redshifts (if they exist).

The construction of larger telescopes alone, however, is not sufficient if the goal is to detect (at the confusion-limit) the faintest and most numerous sub-mm galaxies, as well as the brightest and rarest sub-mm sources in the high-z Universe. For example, a search for the most extreme star-forming galaxies (SFR $\gg 5000 M_{\odot}{\rm yr}^{-1}$), associated perhaps with the rapid formation of a massive elliptical in less than a few Gyrs, would require a sub-mm survey covering > 100 sq. degrees before a statistically-significant result could be achieved. Such a survey is well beyond the capability of current instrumentation.

Although the increased sensitivity (using larger telescopes) provides a faster mapping-speed for the current bolometer cameras, the greatest gains will be achieved through the development of the large-format (> 5000 pixel) cameras that can fill the focal-plane of the future large ground-based telescopes (e.g. 100-m GBT, 50-m LMT). The recognition of this fact is prompting the intense activity to build the next-generation of sub-mm and mm cameras, many of which will exploit (i) the speed and sensitivity of superconducting transition-edged sensors (TES), (ii) their multiplexing capability, when combined with a read-out array of SQUIDs, and (iii) new micro-machining techniques that enable these large, fully-sampled monolithic wafers to be fabricated.

3. Multi-wavelength follow-up of sub-mm surveys

To fully understand the nature and physical properties of the sub-mm galaxy population, there is a need for multi-frequency follow-up data (at X-ray, optical, IR, FIR and radio wavelengths) providing valuable information about the galaxy morphology, age, kinematics, dynamics, stellar and gas content, and possible presence of a buried (or heavily-obscured) AGN in the host galaxy.

3.1 Optical and IR imaging

The searches for the optical and IR counterparts of sub-mm galaxies have enjoyed different levels of success which reflects the diversity of their properties (Ivison et al. 2000), and led Smail et al. (2002) to propose a broad classification scheme (class 0, I, II). In some cases (e.g. Ledlow et al. 2002; Smail et al. 2003a) bright class II (typically red) host galaxies have been found at both optical and IR wavelengths ($I < 25$, $K \sim 18.5 - 21$), whilst other searches have only yielded positive results when deeper optical and/or IR follow-up observations have been made, detecting class I (Lutz et al. 2001) or intermediate class I-II counterparts (Frayer et al. 2003). An extreme example is the counterpart to HDF850.1 (Hughes et al. 1998), the brightest sub-mm source in the northern Hubble Deep Field, which avoided detection for many years despite the wealth of multi-wavelength data. Dunlop et al. (2002) finally identified and associated an extremely faint red and optically invisible (class 0) host galaxy ($I > 28.7, K = 23.5$) with HDF850.1. Not only is this counterpart inaccessible to optical and IR spectroscopic observations, but follow-up imaging to these depths can only be conducted over limited areas ($< 1\,\text{deg}^2$). A K-band study by Frayer et al. (2003) finds a median magnitude $K = 22 \pm 1$ for the sub-mm population identified in the SCUBA Cluster Lens Survey. Only 3/15 sub-mm sources do not have an IR counterpart ($K > 23$). Frayer et al. argue therefore, that, in the absence of radio detection, IR colour selection ($J - K > 3$) can also help identify the counterparts to sub-mm galaxies.

3.2 Radio and millimetre interferometric imaging

The high-resolution and sub-arcsec positional accuracy of radio and mm-λ interferometric observations have certainly assisted the majority of the identifications of the optical and/or IR counterparts (e.g. Frayer et al. 2000). For example, the radio follow-up observations of the sub-mm surveys in the ELAIS N2 and Lockman Hole fields (Scott et al. 2002; Fox et al. 2002) detected 60 and 70%, respectively, of the sub-mm sources with 1.4 GHz fluxes (5σ) $> 45\mu$Jy and $> 25\mu$Jy (Ivison et al. 2002). At these depths, given the low surface-density of radio-sources, the detection of a 5σ radio-peak within a few arscecs of a sub-mm source makes a convincing case for a genuine association, and can lead to an attempt to gain an optical spectroscopic redshift (e.g. Chapman et al. 2003). SMM04431+0210 provides a counter-example of a sub-mm source without a radio detection ($3\sigma < 70\mu$Jy at 1.4 GHz), whilst in this instance the strong IR counterpart ($K = 19.4$) provided the key to measuring the rest-frame optical redshift (Frayer et al. 2003). If, however, the value of the radio detection is not only to provide an accurate position, but also to help constrain the photometric redshift (§4), then the accuracy and calibration of the extracted radio fluxes must also be considered. Using the same example, the fraction of sub-mm sources with significant ($S/N > 3$) 1.4 GHz radio-fluxes in the ELAIS N2 and Lockman Hole surveys reduces to 40–50%.

Since the above 1.4 GHz surveys already represent some of the deepest radio observations it will not be easy (until the completion of the e-VLA and e-Merlin) for future radio observations to keep pace with the rapid increase in depth and area ($\sim$ 10 deg^2) of the next generation of confusion-limited sub-mm surveys (§4). Aretxaga et al. (2003) demonstrated that future radio surveys must reach a 5σ sensitivity of 5–8μJy at 1.4 GHz if $\sim$80% of the bright sub-mm population ($>$ 6 mJy at 850μm, or $>$ 2 mJy at 1.1 mm) are expected to have a radio counterpart with a valid measurement of the flux density.

3.3 X-ray, optical, IR and CO spectroscopy

There has been considerable effort to obtain (rest-frame) spectroscopic z's for the optically-bright ($I < 26$) sub-mm galaxies (e.g. Chapman et al. 2003). At the time of this conference (June 2003) only 14 blank-field sub-mm galaxies had spectroscopic redshifts. More recently, Smail et al. (2003b) - see their Fig.3 - presented the spectroscopic-z distribution for 41 sub-mm galaxies which have a median redshift of $z \sim 2.3$, and 50% of the sources are found in the range $1.9 < z < 2.6$. From these optical spectra it has also been possible to measure the fraction of AGN (10–20%) hosted in a sub-mm galaxy (Barger et al. 1999b; Smail et al. 2000). Recent results from the spectral analysis of the 2 Ms Chandra Deep Field North (Alexander et al. 2003), however, suggest that a higher fraction ($\sim$ 50%) of sub-mm galaxies host heavily obscured AGN (observed only in hard X-rays). Considering the depth of the 2 Ms Chandra survey that was required to make the sub-mm–X-ray connection, and the fact that the X-ray AGN spectra generally show Compton-thin absorption, lends further support to the argument that star formation processes (and not AGN) dominate the high bolometric luminosities.

In some cases for which a rest-frame optical spectroscopic z's exists (e.g. Chapman et al. 2003; Frayer et al. 2003), CO receivers, operating at $\sim$ 90 GHz (with bandwidths of $\Delta\nu \sim 1 - 0.5$ GHz, which provides a range $\Delta z \sim 0.02 - 0.06$ at $z \sim 2 - 4$), can be tuned to detect the molecular gas component in these high-z sources. As anticipated, the CO observations indicate that the sub-mm galaxies are massive gas-rich systems with sufficient fuel ($M_{H_2} \sim 10^{10} - 10^{11} M_\odot$) to drive the high-levels of star-formation suggested by their rest-frame FIR luminosities (e.g. Frayer et al. 1999; Neri et al. 2003; Genzel et al. 2003). These CO detections do not come cheaply - even after the

lengthy initial sub-mm, radio, optical and IR imaging and spectroscopic observations, the confirmation of each optical redshift currently takes 15–40 hours. Partly as a result of the required sensitivity and limited receiver bandwidth, only 5 published CO detections, and hence only 5 confirmed optical redshifts (in the range $z \sim 2.4 - 3.3$), exist for sub-mm galaxies.
There is a natural bias towards the optical spectroscopic detection of lower-z (unlensed) sources. The path of (i), initial sub-mm detection, to (ii), radio identification, to (iii), optical and IR imaging and association with the faint radio-source, and finally (iv), to optical spectroscopy and then to CO spectroscopy, breaks down at redshifts > 3 when the level of radio emission drops below even the deepest detection thresholds. This selection bias, which explains the redshift cut-off (upper-quartile $z \sim 2.6$) observed in the sub-mm galaxy population (Chapman et al. 2003; Smail et al. 2003b), prevents the identification of the highest-z sub-mm sources and hence the objects that would present the most extreme test of galaxy formation models. The incompleteness in the spectroscopic redshift distribution ($\geq 30\%$, consisting of galaxies at $z > 3$) is highly dependent on the prediction from the imprecisely-known evolutionary model for the sub-mm population. This same uncertainty in the evolutionary model is one of the reasons the upper-bound of the redshift distribution, determined from photometric colour information (§4), is still poorly constrained.

4. Measuring the evolutionary history of sub-mm galaxies

The initial SCUBA and MAMBO surveys are too small to measure the clustering properties of the sub-mm population on scales $> 9h^{-1}$ Mpc. The necessity for larger-area surveys, and more broad-band spectral information (particularly at 90–500μm) is obvious. Future sub-mm surveys from balloon-borne (BLAST), airborne (SOFIA), satellite (SIRTF, ASTRO-F, Herschel) and ground-based facilities (JCMT, CSO, IRAM, GBT, LMT, APEX) will map significant areas of the FIR – mm sky to their respective confusion-limits, producing larger samples ($\sim 10^3 - 10^6$) of luminous starburst galaxies. Higher resolution experiments (SMA, CARMA, IRAM PdB and ALMA) will provide the more detailed studies over restricted fields. Thus, whilst there have been advances in the number of spectroscopic redshifts (assuming the associations to be correct), there remains a growing need for an alternative efficient and independent measurement of the redshift distribution for large numbers of sub-mm galaxies; one that does not depend on the security of the faint optical or IR identification, or the necessity for a prior radio-detection.

4.1 Photometric redshifts at sub-mm wavelengths

This "*redshift deadlock*" hinders the development of models to explain the evolutionary history of sub-mm galaxies since, without redshifts, one cannot derive robust measurements of the rest-frame FIR luminosities and SFRs, or describe the 3-dimensional clustering distribution for this optically-obscured population. There has been considerable effort in assessing whether broad continuum features in the spectral energy distributions (SEDs) of sub-mm galaxies at rest-frame mid-IR to radio wavelengths can provide photometric-redshifts with sufficient accuracy (Hughes et al. 1998, 2002; Carilli & Yun 1999; Barger et al. 1999b; Smail et al. 2000; Yun & Carilli 2000; Rengarajan & Takeuchi 2001; Aretxaga et al. 2003; Wiklind 2003).

Monte-Carlo simulations (which take into account observational and calibration errors) showed that photometric redshifts can be measured from future 250, 350 and 500μm BLAST (www.chile1.physics.upenn.edu/blastpublic) and Herschel/SPIRE surveys with an accuracy of $\delta z \leq 0.5$ (Hughes et al. 2002). The addition of longer-λ sub-mm and radio data for individual galaxies can reduce this redshift uncertainty (and is one of the reasons to combine 850μm SCUBA data and BLAST data in the SCUBA Half Degree Extragalactic Survey (SHADES - www.roe.ac.uk/ifa/shades). Aretxaga et al. (2003) applied a similar analysis to the catalogues of sub-mm galaxies in completed SCUBA surveys, and derived photometric z's for individual sources using their existing ground-based radio–sub-mm data. The results of this latter work, and other studies (Smail et al. 2002; Ivison et al. 2002), was a redshift probability distribution for the sub-mm galaxy population which places the majority of the sources at $2 < z < 4$, with a median redshift $z \sim 3$ - a result that is to similar to the spectroscopic measurements (§3). Wiklind (2003) concluded that even greater accuracy was possible ($dz \sim \pm 0.3$), based on just the 450/850μm flux ratio, although the range of FIR and radio properties in the adopted SED templates was narrower - which raises an important point.

There is an intrinsic level of degeneracy between the temperature of the dust emission (or shape of the FIR SED) and the photometric z, in the sense that hot high-z sources will have the same observed colours as cold low-z sources. This has led Blain et al. (2003) to argue that, unless a strong luminosity-temperature relationship exists for the sub-mm population and the galaxies used to calibrate the technique, then photometric redshifts will have insufficient accuracy to be useful. These issues will be undoubtedly be revisited until there exist a sufficient number of robust (optical or CO) spectroscopic redshifts, and sensitive, accurately calibrated rest-frame FIR-sub-mm data for the same targets. Only then can we really understand the biases in the method and determine whether the derived photometric redshifts are *useful*, and if they can be used to constrain the statistical properties of the sub-mm population.

4.2 Measuring spectroscopic-z's at millimetre λ's

The primary observational goal of BLAST (Balloon-borne Large Aperture Submm Telescope) is to conduct large extragalactic surveys ($\sim 1-10\,\mathrm{deg}^2$) at 250, 350 and 500$\mu$m (from 2004 onwards) and provide accurate source-counts, photometric-redshifts and star formation rates for luminous sub-mm galaxies ($L_{\rm FIR} > 3 \times 10^{12} L_{\odot}$). The mapping speed (Hughes et al. 2002) is sufficient to detect $\sim$ 30 sources/hr at the extragalactic confusion limit of 20–30 mJy (see fig.23 in Blain et al. (2002) for comparison with other experiments). BLAST expects to detect $>$ 5000 high-z sources in a 10-15 day flight, significantly more than the total number of sub-mm galaxies ($<$ 300) detected todate. The motivation for the extragalactic BLAST surveys (which is similar to that of SHADES) is to address 3 fundamental questions: (i) what is the cosmic history of massive dust-enshrouded star formation activity; (ii) are high-z sub-mm sources the progenitors of massive present-day elliptical galaxies; (iii) what fraction of sub-mm sources harbour an heavily-obscured AGN.

The advantage of having constrained photometric-redshifts from BLAST, or at least robust lower-limits, for large-numbers of individual sub-mm galaxies (some without optical, IR or radio counterparts), is that we can now determine the likelihood that a redshifted rotational CO transition-line falls into the frequency range of a particular mm-wavelength spectral-line receiver. For example, in the case of the LMT, an ultra-wide-bandwidth receiver is under construction for operation in the 90 GHz window. With an instantaneous bandwidth $\Delta\nu \sim$ 35 GHz, this receiver is ideally suited to a search for redshifted CO-lines. At $z > 2.3$, adjacent CO lines will be separated by $\sim$ $115\mathrm{GHz}/(1+z) \leq 35\,\mathrm{GHz}$, and thus, given enough sensitivity, the detection of at least one CO line is guaranteed. The presence of significant masses of molecular gas in a few high-z sub-mm galaxies has already been demonstrated (§3), but only after an accurate optical redshift had been previously obtained. The obvious advantage of the ultra-wide-band receivers is that an accurate redshift is not needed in advance of the CO observations. Although the detection of one CO-line in the 90 GHz window is insufficient to determine a redshift, it will still allow other narrower-band receivers on the LMT or 100-m GBT to tune and then search for additional CO-transitions at higher or lower frequencies. If any of the sub-mm sources lie at $z > 6$ then two CO-lines in principle can be detected in the 90 GHz window, providing an immediate measure of the *true* z. The recent detection of CO(3–2) in a quasar at $z = 6.42$ (Walter et al. 2003) proves the existence of large masses ($2 \times 10^{10} M_{\odot}$) of molecular gas in the early Universe which is available to fuel an AGN, and/or the violent (and possibly dust-obscured) bursts of massive star formation during the first stages of galaxy evolution. To conclude, the diversity of the future multi-wavelength data will answer some of the outstanding cosmological questions.

References

Alexander, D.M. et al. 2003, AJ, 125, 583

Aretxaga, I. et al. 2003, MNRAS, 342, 759

Barger, A.J. et al. 1999a, ApJ, 518, L5

Barger, A.J., et al. 1999b, AJ, 117, 2656

Blain, A.W et al. 2002, Physics Reports, 369, 111

Blain, A.W., Barnard, V., Chapman, S.C. 2003,MNRAS, 338, 733

Borys, C. et al. 2003, MNRAS, 344, 385

Carilli, C. & Yun, M.S. 2000, ApJ, 530, 618

Chapman, S. et al. 2003, Nature, 422, 695

Cowie, L.L. et al. 2002, AJ, 123, 2197

Dunlop, J.S. et al. 2003, astro-ph/0205480.

Eales, S. et al. 1999, ApJ, 515, 518

Eales, S. et al. 2000, AJ, 120, 2244

Fox, M. et al. 2002, MNRAS, 331, 839

Frayer, D. et al. 1999, ApJ, 514, L13

Frayer, D. et al. 2003, astro-ph/0304043

Gaztañaga, E. & Hughes, D.H. 2001, in Deep Millimeter Surveys, ed. J.D. Lowenthal, D.H.Hughes, (World Scientific), 131, astro-ph/0103224

Genzel, R. et al. 2003, ApJ, 584, 633

Glenn, J. et al. 2003, in S.P.I.E., 4855, Millimeter and Submillimeter Detectors for Astronomy, ed. T.Phillips, J. Zmuidzinas (SPIE), 30

Holland, W. et al. 1999, MNRAS, 303, 659

Hughes, D.H. et al. 1998, Nature, 394, 241

Hughes, D.H. 2000, in ASP Conf. Ser., 200, Clustering at High Redshift, ed. F.Favata, A.Kaas, A.Wilson (San Fransisco, ASP), 81, astro-ph/0003414

Hughes, D.H. & Gaztañaga, E. 2000, in 33rd ESLAB symposium, ESA SP 445, Star formation on Large-scales to Small-scales, ed. F.Favata, A.A.Kaas, A.Wilson (ESA), 29, astro-ph/0004002

Hughes, D.H. et al. 2002, MNRAS, 335, 871

Ivison, R.J. et al. 2000, MNRAS, 315, 209

Ivison, R.J. et al. 2002, MNRAS, 337, 1

Ledlow, M.J. et al. 2002, ApJ, 577, L79

Lilly, S. et al. 1999, ApJ, 518, 641

Lutz, D. et al. 2001, A&A, 378, 70

Neri, R. et al. 2003, ApJ, 597, 113

Rengarajan, T.N. & Takeuchi, T. 2001, PASJ, 53, 433

Scott, S. et al. 2002, MNRAS, 331, 817

Smail, I. et al. 2002, MNRAS, 331, 495

Smail, I. et al. 2003a, MNRAS, 342, 1185

Smail, I. et al. 2003b, astrop-ph/0311285

van Kampen, E. 2003, astro-ph/0310297

Walter, F. et al. 2003, Nature, 424, 406

Webb, T.M. et al. 2003, ApJ, 587, 41

Wiklind, T. 2003, ApJ, 588, 736

Yun, M.S. & Carilli, C. 2000, ApJ, 568, 88

DEEP NEAR-INFRARED IMAGING OF SUB-MILLIMETER SELECTED GALAXIES

Naveen A. Reddy,[1] David T. Frayer,[2] Lee Armus,[2] Andrew W. Blain[1], Nick Scoville,[1] & Ian Smail[3]

[1] *Astronomy Department, California Institute of Technology 105-24, Pasadena, CA 91125, USA*

[2] *SIRTF Science Center, California Institute of Technology 220-06, Pasadena, CA 91125, USA*

[3] *Department of Physics, University of Durham, South Road, Durham, DH1 3LE, UK*

Abstract We report on a deep near-infrared survey of submillimeter galaxies (SMGs) using the Near Infrared Camera (NIRC) on the Keck I telescope. Of the 15 background submillimeter sources in the SCUBA Cluster Lens Survey, 12 have candidate $K-$band counterparts to $K \sim 23$ with a lensing-corrected median of $\langle K \rangle = 22 \pm 1$ mag. For SMGs with accurate radio and/or millimeter positions, the median NIR color is $\langle J - K \rangle = 2.6$ mag. Bright SMGs ($K < 19$ mag) show normal colors around $J - K \sim 2$ mag and fainter ones are redder. We propose that the $J - K$ color may indicate likely SMG counterparts that are too faint to be found at optical and/or radio wavelengths. The surface density of $J - K > 3$ mag sources indicates they are more numerous than $S_{850\mu m} > 2$ mJy SMGs and could either represent the faint end of the SMG population or a distinct class of high redshift galaxies. This ambiguity will be clarified by SIRTF observations of the rest-frame near-IR emission from this faint red galaxy population.

Keywords: galaxies: active — evolution — formation — starburst

1. Introduction

Observations with SCUBA have uncovered a large population of high redshift submillimeter (submm) galaxies (SMGs), but identifying their counterparts at shorter wavelengths is difficult due to confusion given the large beam size at 850μm (e.g., Smail, Ivison, & Blain 1997; Hughes et al. 1998; Barger et al. 1999; Smail et al. 2002). The sub-arcsec accuracy provided by radio observations allows for the identification of counterparts, but only for the bright SMG population (Chapman et al. 2002).

Instrumental sensitivity at near-IR (NIR) wavelengths is significantly better than at optical for a red galaxy SED and, as such, we have investigated the vi-

M. Plionis (ed.), Multiwavelength Cosmology, 113-116.

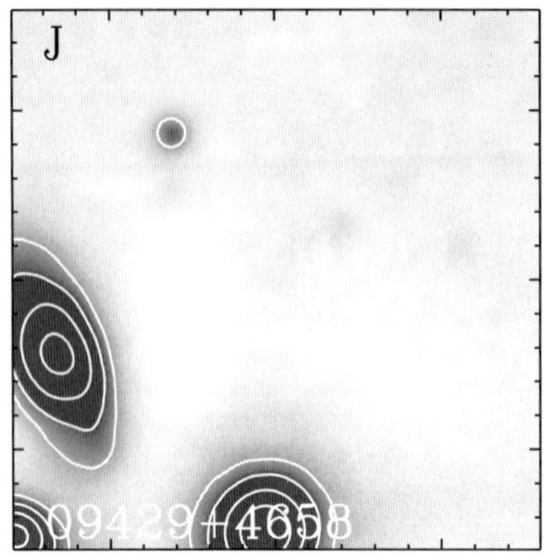

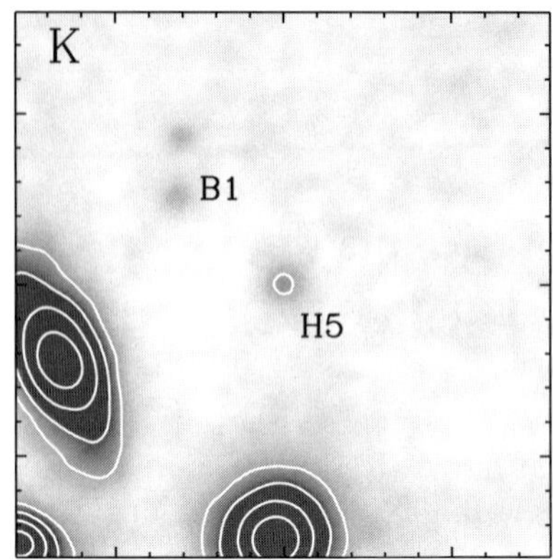

Figure 1. NIRC $16'' \times 16''$ $J-$ and $K-$band images of one of the 15 targeted fields of the SCUBA Cluster Lens Survey (Smail et al. 2000), where contours start at $+5\sigma$ and increase by factors of 2. The source H5 is the reddest NIR galaxy known to date, with $J - K > 4.9$.

ability of using the $J - K$ color to identify SMG counterparts. Using the Near Infrared Camera (NIRC) on the Keck I telescope, we have imaged fields from the SCUBA Cluster Lens Survey (Smail et al. 1997, 2002) to a depth comparable to the deepest existing $K-$band observations ($K \sim 23.5$; Totani et al. 2001a, b; Franx et al. 2003), ~ 100 times deeper than previous $K-$band observations in submm fields. We observed in both $J-$ and $K-$bands for sources with candidate $K-$band counterparts, and observed deeper at $K-$band in fields with no clear counterparts to search for fainter sources (e.g., Frayer et al. 2000). Of the 15 targeted fields, 12 have 3σ $K-$band candidate counterparts; the rest are blank to $K \sim 23.5$ mag. Seven SMGs have confirmed counterparts based on interferometric observations (e.g., Frayer et al. 2000; Smail et al. 2000; Ivison et al. 2000). We have identified another 5 possible counterparts based on red $J - K$ colors. Figure 1 shows one of the possible counterparts identified from its red $J - K$ color. The data are described in more detail in Frayer et al. (2003).
$H_0 = 70$ km s^{-1} Mpc^{-1}, $\Omega_{\rm M} = 0.3$, and $\Omega_\Lambda = 0.7$ are assumed throughout.

2. NIR Properties of Submillimeter Galaxies

Figure 2 shows the photometry results from the survey. Galaxies in the foreground clusters ($z < 0.4$) have typical colors of $J - K \lesssim 1.6$ mag, while background galaxies at $z > 1$ have colors of $J - K > 1.5$ mag (Totani et al. 2001b). Bright SMGs ($K < 19.5$ mag) are bluer by > 1 mag than fainter SMGs which show very red colors of $J - K > 3$ mag. The reddest source in the sample with $J - K > 4.9$ mag (3σ) is the reddest NIR galaxy known to date (see Figure 1). These red galaxies are not distributed uniformly along the

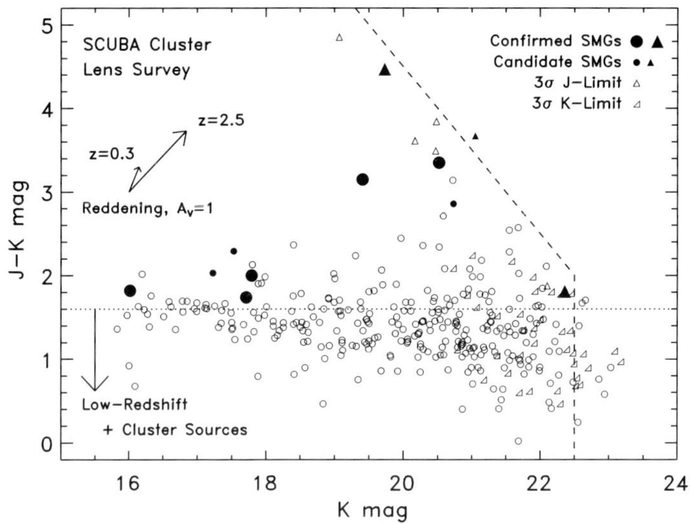

Figure 2. $J-K$ color versus $K-$band magnitude for 339 detected sources in 12 SMG fields. Confirmed and candidate SMG counterparts are denoted by large and small solid symbols, respectively, while arrows indicate limits. The dashed line represents the survey limits.

reddening vector, so extinction may not fully account for the observed distribution. If we assume these galaxies lie at $z \sim 2.5$ (Chapman et al. 2003), their colors could be explained by 2 to 3 magnitudes of visual extinction.

Taking into account the several undetected SMGs, we find a likely median magnitude of $\langle K \rangle = 22 \pm 1$ mag. At $z \sim 2.5$ this implies an unobscured absolute magnitude of $M_{\rm R} = -21.5$ mag, comparable to that found for local IRAS 1 Jy ULIRGs ($M_{\rm R} = -21.9$ or $2L^*$; Kim, Veilleux, & Sanders 2002). The median NIR color of the 6 confirmed SMG counterparts is $J - K =$ 2.6 mag, or rest frame $U - R \sim 1.2$ mag, again comparable to that found for local ULIRGs. Gaussian fitting to the $K-$band images indicates SMG sizes from < 0.5 to 2.1 arcsec, similar to the sizes observed for the central regions of ULIRGs (Surace et al. 1998; Scoville et al. 2000) and bulges of disk galaxies. A number of SMGs show multiple components indicative of merger activity. These results suggest that SMGs are (1) similar to ULIRGs in their optical properties and morphologies, (2) may be a formative phase for the bulges of normal galaxies or the cores of massive ellipticals (Lilly et al. 1999), and (3) are consistent with $> M^*$ galaxies.

3. Faint Red $J - K$ Galaxies

Correcting for lensing, we find a $J - K > 3$ mag source density of 5 arcmin^{-2} to $K \sim 22.5$, 5 times higher than that found in blank-field surveys to similar depths (Totani et al. 2001a, 2001b; Franx et al. 2003) and biased upward since we are targeting fields containing SMGs. Excluding all such red sources from within 10$''$ of the $850\mu\mathrm{m}$ position, the source density decreases to 2 arcmin^{-2}. Strong SCUBA detections ($> 5\sigma$) have $850\mu\mathrm{m}$ positions known to within $\pm 6''$, implying a chance coincidence rate with a $J - K > 3$ mag source of 1%. Consequently, $J - K > 3$ mag sources are likely counterparts if they lie within $\sim 5''$ of the $850\mu\mathrm{m}$ position.

We confirm the result from blank-field surveys that extremely red galaxies become numerous at fainter magnitudes, and may dominate the star formation activity at high redshift if they are moderately luminous ($L \sim 10^{12}$ L$_\odot$). Alternatively, these sources could be very faint sources at $z > 2.5$ or even luminous LBG-type galaxies at $z > 10$ (Dickinson et al. 2000). SIRTF IRAC and MIPS observations should clarify the status of these extremely red galaxies.

Acknowledgments

We thank the Keck Observatory staff who made these observations possible, and D. Thompson for his NIRCtools software. DTF and LA are supported by the Jet Propulsion Laboratory, California Institute of Technology, under contract with NASA. AWB and NAR acknowledge support from the NSF grant AST-0205937 and a NSF Graduate Research Fellowship, respectively.

References

Barger, A. J., et al. 1999, AJ, 117, 2656
Chapman, S. C., et al. 2002, MNRAS, 330, 92
Chapman, S. C., et al. 2003, Nature, 422, 695
Dickinson, M., et al. 2000, ApJ, 531, 624
Franx, M., et al. 2003, ApJ, 587, L79
Frayer, D. T., et al. 2000, AJ, 120, 1668
Frayer, D. T., et al. 2003, AJ, submitted
Hughes, D., et al. 1998, Nature, 394, 241
Ivison, R. J., et al. 2000, MNRAS, 315, 209
Kim, D.-C., Veilleux, S., & Sanders, D. B. 2002, ApJ, 143, 277
Lilly, S. J., et al. 1999, ApJ, 518, 641
Scoville, N. Z., et al. 2000, AJ, 119, 991
Smail, I., Ivison, R. J., & Blain, A. W. 1997, ApJ, 490, L5
Smail, I., et al. 2000, ApJ, 528, 612
Smail, I., et al. 2002, MNRAS, 331, 495
Surace, J. A., et al. 1998, ApJ, 492, 116
Totani, T., et al. 2001a, ApJ, 559, 592
Totani, T., et al. 2001b, ApJ, 558, L87

SIMULATING THE HIGH-REDSHIFT UNIVERSE IN THE SUB-MM

Eelco van Kampen
Institute for Astrophysics, University of Innsbruck
Technikerstr. 25, A-6020 Innsbruck
Austria
Eelco.v.Kampen@uibk.ac.at

Abstract I present various simulations of an on-going large sub-mm survey, *SHADES*, showing how constraints can be put on galaxy formation models and cosmology from this survey.

Keywords: galaxy formation, large-scale structure, cosmology

1. Introduction

An important problem with most current galaxy formation models is how to establish whether a set of model parameters that produces a good match to observations is unique, as there are likely to be degeneracies amongst the various free parameters of the model. As most useful observational data used to constrain the model parameters are obtained from our local universe, this 'uniqueness problem' can be resolved by comparing model predictions and observations at high redshift, which in many respects is independent from a comparison at low redshift. Specifically, the *SCUBA* half-degree extra-galactic survey (*SHADES* for short; see `http://www.roe.ac.uk/ifa/shades` for details) will provide highly valuable observational data for this purpose.

2. A new wide-area sub-mm survey: *SHADES*

SHADES is a major new extragalactic survey with *SCUBA*, the "Submillimetre Common-User Bolometer Array" (Holland et al. 1999), covering 0.5 sq. degrees to a 4σ detection limit of S_{850} = 8 mJy. The data from *SCUBA* will be supplemented with data from *BLAST*, a "Balloon-borne Large-Aperture Submillimeter Telescope" (Devlin 2001), which will undertake a series of nested extragalactic surveys at 250, 350 and 500 μm.

M. Plionis (ed.), Multiwavelength Cosmology, 117-120.

This wide-area survey will yield a substantial ($\simeq$ 200) sample of bright unconfused sub-mm sources with meaningful redshift estimates ($\delta z \simeq \pm 0.5$). The survey is designed to answer three fundamental questions. What is the cosmic history of massive dust-enshrouded star-formation activity ? Are SCUBA sources the progenitors of present-day massive ellipticals ? What fraction of SCUBA sources harbour a dust-obscured AGN ?
The crude but near-complete redshift information provided by *BLAST* is sufficient to answer the second question provided the survey covers sufficient area and contains enough sources to measure the clustering of bright sub-mm sources on scales up to $\simeq$10 Mpc.

3. Modeling the high-redshift sub-mm population

The assumption made here is that the bright sub-mm sources seen by *SCUBA* are dust-enshrouded starburst galaxies, as most show no direct evidence for AGN activity (Almaini et al. 2003). Simulating *SHADES* is thus best done using a phenomenological galaxy formation model, from which lightcones are constructed for all galaxies with an 850 μm flux over 8 mJy.
The merging history of galaxy haloes is taken directly from N-body simulations, which use special techniques to prevent galaxy-scale haloes undergoing 'overmerging' owing to inadequate numerical resolution (van Kampen 1995). When haloes merge, a criterion based on dynamical friction is used to decide how many galaxies exist in the newly merged halo. The most massive of those galaxies becomes the single central galaxy to which gas can cool, while the others become its satellites.
When a halo first forms, it is assumed to have an isothermal-sphere density profile. A fraction Ω_b/Ω of this is in the form of gas at the virial temperature, which can cool to form stars within a single galaxy at the centre of the halo. Energy output from supernovae reheats some of the cooled gas back to the hot phase. Each halo maintains an internal account of the amounts of gas being transferred between the two phases, consumed by the formation of stars, and lost to the environment.
The model includes two modes of star formation: quiescent star formation in disks, and starbursts during merger events. The evolution of the metals is followed, because the cooling of the hot gas depends on metal content, and a stellar population of high metallicity will be much redder than a low metallicity one of the same age. It is taken as established that the population of brown dwarfs makes a negligible contribution to the total stellar mass density, and the model does not allow an adjustable M/L ratio for the stellar population.
The 850 μm flux is assumed to be directly proportional to the star formation rate, with an 8 mJy flux corresponding to 1000 solar masses per year. The relation is taken to be fuzzy, so that a random amount of flux (amounting to about

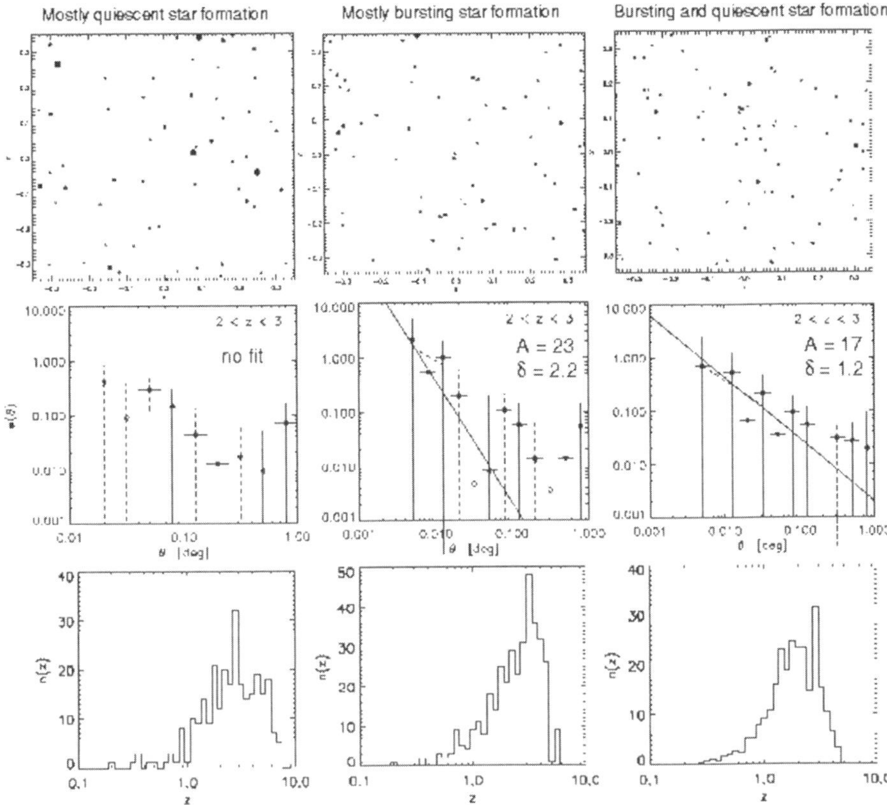

Figure 1. The three columns of panels represent results from three different galaxy formation models, which mainly differ in how stars form (as indicated at the top). The top row shows a single realization of a mock *SHADES*, with the size of the dots corresponding to the flux of each source. The middle row shows the angular correlation functions for these maps, whereas the bottom row displays the redshift distribution of the sources. Differences in these predictions can be clearly observed.

25 per cent) is added/subtracted to reflect differences in dust temperatures and grain sizes which are not yet modeled.

4. Predictions for three different models

Using the phenomenological galaxy formation model described above, three different mock sub-mm surveys resembling *SHADES* were produced. These models only differ in their star formation laws, in that one model has most stars forming quiescently in disks, one has most stars forming in merger-induced starbursts, and the remaining model has a mix of both. The resulting maps and predictions for the angular clustering and redshift distribution are shown in Fig. 1.

While the number counts are similar (not shown), the clustering properties, and to a lesser extent the redshift distributions, are clearly different for these realizations. It seems that the clustering strength depends on which star formation mode dominates, with the burst model yielding relatively strong clustering, and the quiescent model showing very little clustering. It has to be noted that there is a fairly large spread in predicted clustering strengths for different realizations of the *same model*, mainly due to cosmic variance. Still, the trend remains visible after considering 50 realizations for each model. The way forward is to make optimal use of combining the difference in the clustering *and* the redshift distribution. This work is currently in progress (van Kampen et al. 2003).

5. Conclusions and discussion

The on-going large sub-mm survey *SHADES* has the potential to put significant constraints on galaxy formation models, and help resolve the uniqueness problem of such models due to the uncertainties in the assumptions, approximations, and choice of parameters. A potential problem is that of cosmic variance: even though *SHADES* is a the largest extragalactic survey ever undertaken in the sub-mm waveband, the total sky coverage is still much smaller than typically achieved in the optical wavebands. However, two factors work in our favour: the availability of (crude) redshift estimates for each of the sources, and the expectation that clustering of bright sub-mm galaxies is relatively strong (e.g. Percival et al. 2003, Scott et al. 2003, Webb et al. 2003, van Kampen et al. 2003).

Acknowledgments

This work was partly supported by the Austrian Science Foundation FWF under grant P15868, and a PPARC rolling grant.

References

Almaini, O., et al., 2003, MNRAS, 338, 303
Devlin, M., 2001, astro-ph/001232
Holland, W., et al., 1999, MNRAS, 303, 659
van Kampen, E., 1995, MNRAS, 273, 295
van Kampen, E., et al., 2003, in preparation
Percival, W.J., Scott D., Peacock J.A., Dunlop J.S., 2003, MNRAS, 338, 31
Scott, S.E., et al., 2003, in preparation
Webb, T.M., et al., 2003, ApJ, 587, 41

A BAYESIAN PHOTOMETRIC REDSHIFT TECHNIQUE FOR MM-SELECTED GALAXIES

Edward L. Chapin, David H. Hughes, Itziar Aretxaga *
I.N.A.O.E., Aptdo. Postal 51 y 216, Puebla, Mexico
echapin@inaoep.mx

Abstract We present a Bayesian technique, which uses a luminosity function prior, to calculate photometric redshifts for optically-obscured starburst galaxies. An example is compared with previous Monte Carlo simulations.

Multi-wavelength measurements of the SEDs of local galaxies have led to the development of radio-(sub)mm-FIR colour indices in order to derive photometric redshifts for the population of dust-enshrouded starburst galaxies identified in the blank-field SCUBA and MAMBO surveys (Hughes et al. 1998, 2002; Carilli & Yun 2000; Yun & Carilli 2002; Wiklind 2003; Aretxaga et al. 2003). The bland spectral features in the radio-FIR regime lead to a degeneracy between the colours and redshift (e.g. Blain et al. 2003). The inclusion of luminosity information breaks some of that degeneracy, and provides robust redshift estimates with uncertainties of $\Delta z \simeq \pm 0.5$ determined from Monte Carlo simulations. We present a continuation of our work (see also Hughes - these proceedings) which, instead of using simulated catalogues, provides an analytical expression for the redshift probability distribution, given observed data, and prior information - the luminosity function of mm galaxies. These calculations produce a higher resolution redshift probability distribution than the Monte Carlo technique in a fraction of the computation time.

Given a set of flux density measurements $S = \{s_i\}$, and some prior information I the redshift probability density function $p(z|S,I)$ is factored using a general form of Bayes' theorem in three variables:

$$p(z|S,I) = p(S|z,I)p(z|I)/p(S|I) \tag{1}$$

*Partial funding for this work was provided by NSERC grant PGSB-253705-2002, CONACyT grants 39953-F and 39548-F, INAOE, and the conference local organizing committee.

M. Plionis (ed.), Multiwavelength Cosmology, 121-124.

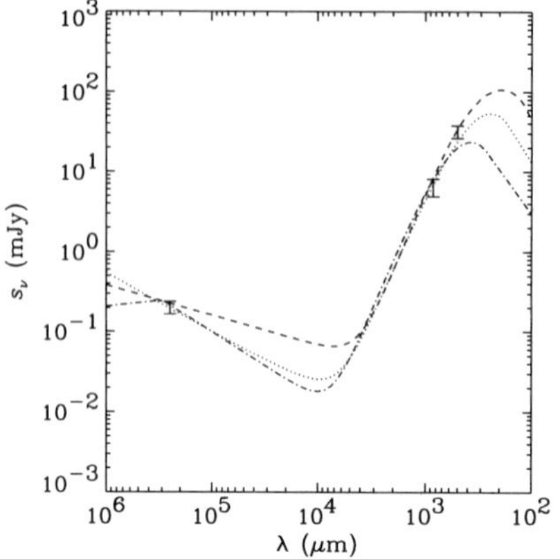

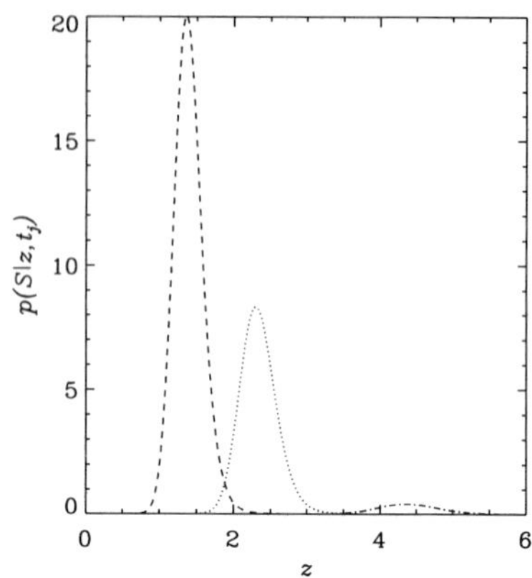

Figure 1. *left panel:* Simulated data at 1.2GHz, 850μm and 500μm with 10% errors, for a galaxy with the same colours as M82 lying at z=2.5, scaled to an observed 850μm flux density of 8 mJy. Observed SED models (*dashed*: Arp220, *dotted*: M82, and *dot-dashed*: Mrk231) are over-plotted using the maximum likelihood models. *right panel:* Likelihood functions integrated over a for the three models. For this realization of the noise, the data is more consistent with the incorrect model (an Arp220-like SED at $z \simeq 1.5$) than the true model (an M82-like SED at $z \simeq 2.5$).

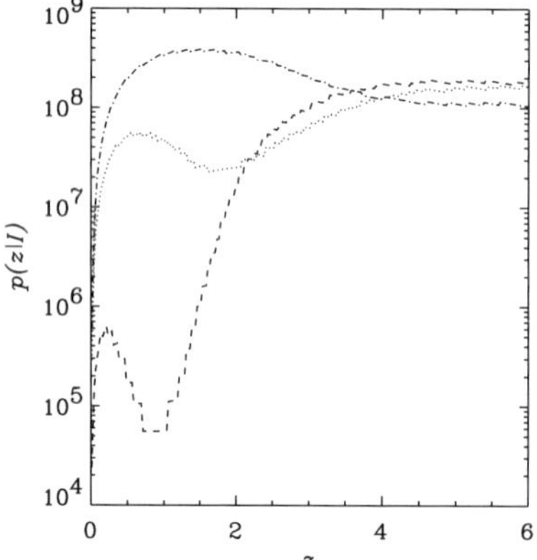

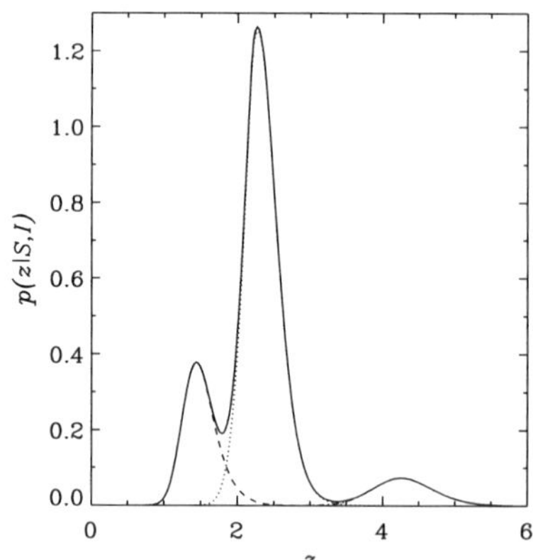

Figure 2. *left panel:* The redshift prior evaluated at the maximum likelihood values of a for each model, as a function of z, illustrating the relative weighting to be applied to the likelihood functions. *right panel:* The Bayesian redshift probability distribution calculated from the product of the prior with the sum of the three likelihood functions, and then integrated over a (solid line). The weighted likelihood functions are plotted beneath the total distribution. Weighting the Arp220 model at lower redshift by the much weaker prior probability with respect to the M82 model correctly assigns the maximum probability density to the correct redshift of $z \simeq 2.5$.

The first term in the numerator is the *likelihood function*. In the absence of prior information, the likelihood is the probability of measuring S given an object at redshift z. Maximizing this function through model fitting yields the *maximum likelihood* solution. This solution does not however identify the best model from the set of all models with which the data are consistent. If some other information is known beforehand, the second *prior* term in the numerator can include an appropriate weight to the likelihood function to find the most probable solution. Ignoring the denominator in Eq. 1 which serves only as a normalization constant, a model for the observed SED is required to calculate the likelihood function, and a prior is constructed from the probability of observing an object with the luminosity of that model.

The observed SED for mm galaxies is modeled by redshifting a rest-frame SED template $t_j(\lambda)$ and multiplying it by a scale factor a, $\tilde{s}_i = at_j(\lambda_i/(1+z))$. Since $\tilde{s}_i$ is determined completely from the model parameters, a standard likelihood function is constructed assuming independent Gaussian errors in the data, independently of the prior, of the form $p(S|z, t_j, a)$. Using simulated data at three FIR-radio wavelengths, Figure 1 illustrates this procedure for three different SED templates.

The prior is constructed from the 60μm luminosity function (Saunders et al. 1990) which undergoes pure luminosity evolution as $(1+z)^{3.2}$ for $0 < z < 2.2$, and constant evolution for $2.2 < z < 10$. Taking a as the 60μm flux density of the model galaxy in the rest-frame:

$$p(z, a|I) = (N(L(a, z), z)/N_T)(dL(a, z)/da) \qquad (2)$$

where $N(L, z)$ is the number density of objects as a function of 60μm luminosity and redshift (the product of the evolving luminosity function and the differential volume element), N_T is the total number of objects in the Universe, and $L(a, z)$ is the 60μm luminosity corresponding to the flux density a at redshift z.

The Bayesian redshift probability is a function of t_j and a. Assuming that the colours of *any* observed galaxy are represented by a unique template t_j from a discrete SED library, the Bayesian redshift distribution may be written as the sum over all the marginal probabilities that the observed galaxy is simultaneously at redshift z and belongs to each template class t_j, and then integrate over a (see also Benitez 2000):

$$p(z|S, I) = (1/p(S|I)) \int \sum_j p(S|z, t_j, a)p(z, a|I)da \qquad (3)$$

In Figure 2 the *best-fit* model derived from the maximum likelihood analysis has been rejected in favor of the *most probable* model according to Bayesian inference, to obtain the redshift of the simulated galaxy. Although the lower

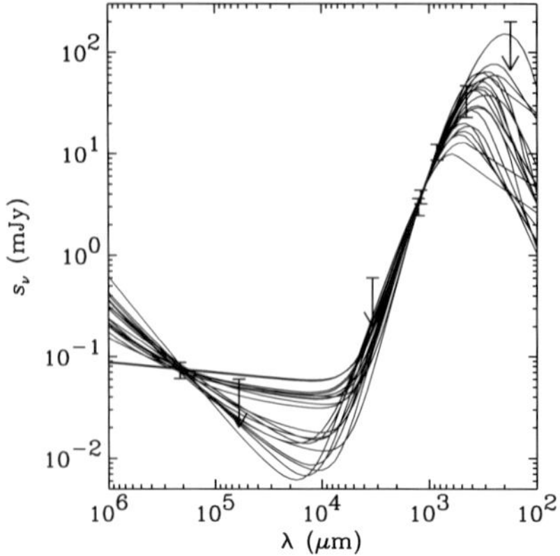

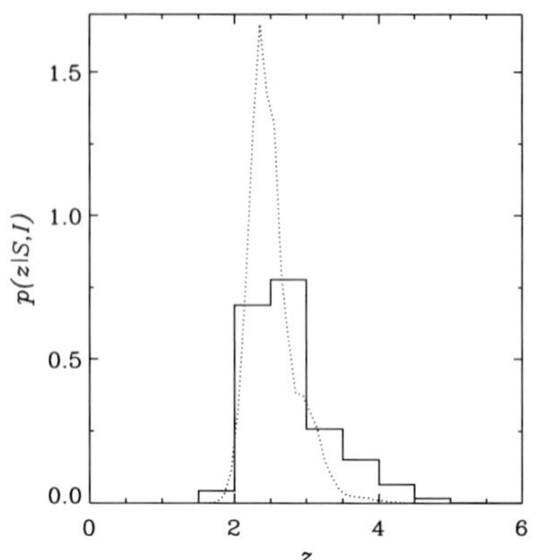

Figure 3. *left panel:* Observed data for the submm source Lockman Hole 850.1 and the most likely SED models using a library of 20 templates (eliminating models that exceed 3-σ upper limits). *right panel:* Comparison between a Bayesian photometric redshift distribution (dotted line) and a Monte Carlo simulation, using the same SED templates and luminosity function prior (Aretxaga et al. 2003). Both methods give qualitatively similar distributions centered about a peak probability of $z \simeq 2.5$.

redshift model is intrinsically dimmer for the same observed flux density (and would therefore seem favorable), the chance of observing it is much smaller than that of the brighter model at higher redshift, given the strong evolution of the luminosity function, and the greater differential volume sampled at that distance.

A photometric redshift has been calculated for the blank-field submm source Lockman Hole 850.1 using a Monte Carlo simulation (Aretxaga et al. 2003). A direct comparison with a Bayesian prediction is provided in Figure 3. The clear advantage of the Bayesian probability distribution is that it may be evaluated with arbitrary redshift precision, and furthermore it can be calculated in a few seconds, whereas the Monte Carlo simulation required several minutes. Clearly this order-of-magnitude improvement in execution time, and resolution, make it the technique of choice for analyzing future surveys with thousands of (sub)mm sources.

References

Aretxaga, I. et al., 2003, MNRAS, 342, 759; Benitez, N., 2000, ApJ, 536, 571; Blain, A.W. et al., 2003, MNRAS, 338, 733; Carilli, C.L. & Yun, M.S., 2000, ApJ, 530, 618; Hughes, D.H. et al., 1998, Nature, 394, 241; Hughes, D.H. et al., 2002, MNRAS, 335, 871; Saunders. W. et al., 1990, MNRAS, 242, 318; Wiklind, T., 2003, ApJ, 588, 736; Yun, M.S. & Carilli, C.L., 2002, ApJ, 568, 88.

EXTREMELY RED GALAXIES IN THE PHOENIX DEEP SURVEY

A. M. Hopkins[1], J. Afonso, A. Georgakakis, M. Sullivan, B. Mobasher, L. E. Cram
University of Pittsburgh, Department of Physics and Astronomy
3941 O'Hara St, Pittsburgh, PA 15206, USA
[1] *Hubble Fellow*

Abstract The Phoenix Deep Survey (PDS) is a multiwavelength survey based on deep 1.4 GHz radio observations used to identify a large sample of star forming galaxies to $z = 1$. Here we present an exploration of the evolutionary constraints on the star-forming population imposed by the 1.4 GHz source counts, followed by an analysis of the average properties of extremely red galaxies in the PDS, by using the "stacking" technique.

Keywords: galaxies: evolution — galaxies: starburst — radio continuum: galaxies

1. Introduction

The study of galaxy evolution in recent years has included a strong focus on the star formation properties of galaxies. Many of these studies are based primarily on selection at ultraviolet (UV) and optical wavelengths, and are known to be strongly affected by obscuration due to dust. It has been shown that selection at these wavelengths results in samples of star forming systems that omit a significant fraction of heavily obscured galaxies 11. There have in addition been suggestions that the most strongly star forming systems suffer the most obscuration 1; 8; 3; 12; 7. Using radio selection to construct a star forming galaxy sample allows the detection of such systems, and the average obscuration in this case is significantly higher than in optically selected samples 1; 8.

To identify a homogeneously selected catalogue of star forming galaxies, unbiased by obscuration effects, and spanning a broad redshift range ($0 < z < 1$), the Phoenix Deep Survey (PDS) uses a deep 1.4 GHz mosaic image, based on observations with the Australia Telescope Compact Array. This has been used to construct one of the largest existing deep 1.4 GHz source catalogues 6 (see http://www.atnf.csiro.au/people/ahopkins/phoenix/) from which the star forming galaxy sample will be drawn. The PDS has already been highly successful

M. Plionis (ed.), Multiwavelength Cosmology, 125-128.

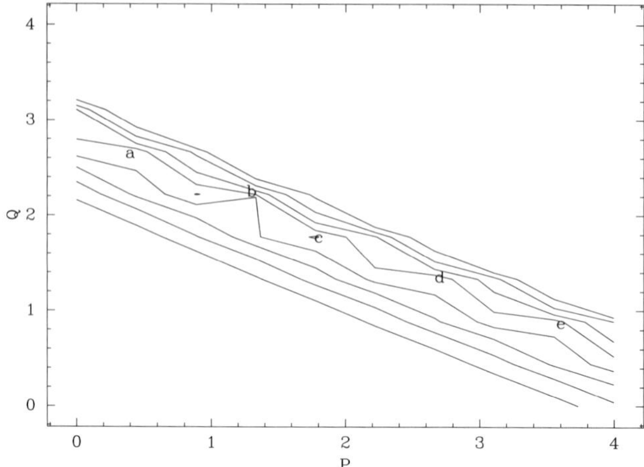

Figure 1. Likelihood contours in the (P, Q) plane, indicating the degeneracy between rates of luminosity and density evolution for star-forming galaxies.

in providing a basis for several investigations of the nature of star forming galaxies and their evolution (see references in 6). Throughout the present investigation we assume a ($\Omega_M = 0.3, \Omega_\Lambda = 0.7, H_0 = 70$) cosmology.

2. 1.4 GHz source counts and models

The PDS 1.4 GHz source counts have already been explored in detail 6. Here we explore source count models based on local luminosity functions (LFs) and their assumed forms of evolution. We assume an evolving LF for AGNs 5 (converted to our assumed cosmology), since we are interested here in exploring the properties of the star-forming (SF) galaxies. We then use a measured local 1.4 GHz LF for SF galaxies 10, and compare the observed source counts with the model predictions subject to a range of both luminosity evolution, $L(z) \propto (1 + z)^Q$, and density evolution, $\phi(z) \propto (1 + z)^P$, for the SF population. The χ^2 estimator for each combination of (P, Q), gives the statistical likelihood, and contours showing the region of maximum likelihood in the (P, Q) plane are shown in Figure 1. There is a clear degeneracy, with maximum likelihood occuring for any (P, Q) satisfying $Q \approx 2.7 - 0.6\,P$ (note that only positive (P, Q) have been considered in this analysis). This sort of degeneracy has also been shown in investigations of optically selected galaxies 9. To illustrate the effect of the degeneracy, Figure 2 shows the predicted source counts compared with the observations for five (P, Q) pairs, indicated by the positions a-e in Figure 1. The redshift distributions predicted by these models, however, are significantly different, and photometric redshifts are now being utilised in order to differentiate between the different evolutionary models.

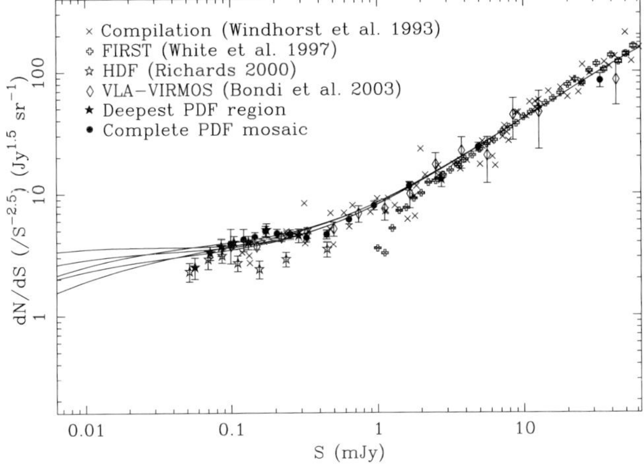

Figure 2. Observed 1.4 GHz source counts compared with model source count predictions (solid lines). The models, while all predicting similar source count distributions, span a broad range of combinations for the luminosity and density evolution parameters.

3. Extremely red galaxies

A sample of over 400 extremely red galaxies (ERGs) has been compiled from PDS observations, using the colour criterion $R - K > 5$. Of these, about 20 are detected at 1.4 GHz, while the majority (over 90%) remain undetected. In an effort to understand more about the properties of these systems, the "stacking analysis" technique often used at X-ray wavelengths has been adopted. We extract sub-images from the radio mosaic at the location of the non-detected ERGs, and construct the weighted average of the sub-images (weighted by 1/rms^2, to maximise the resulting signal-to-noise, since the radio mosaic has a varying noise level over the image). Sub-images where low S/N emission ($> 1.5\,\sigma$) is present at the location of the non-detected source are excluded from the stacking, in order to avoid biasing the stacking signal result by the presence of a small number of low S/N sources. A further 22 (out of 399 candidate non-detected ERGs) were excluded in this manner, leaving 377 to contribute to the stacking signal. The stacked image, shown in Figure 3, has an rms noise of $1.1\,\mu$Jy, and a $\approx 6\,\sigma$ detection at $6.5\,\mu$Jy. While the actual redshift distribution of these objects is unknown, a significant fraction of ERGs appear to be dusty starbursts at $z \gtrsim 1$ 11. Assuming that the average redshift of these sources is $z \approx 1$, the inferred average ERG 1.4 GHz luminosity is 3.5×10^{22} W Hz^{-1}. This corresponds to an average star formation rate (assuming the ERGs are all star-forming systems) of $\approx 20\ M_{\odot}$yr^{-1} (adopting the recent calibration of Bell 2003 2; this value doubles if the calibration of Condon 1992 4 is used). Clearly these results rely on a large number of assumptions, but they serve the useful

Figure 3. Stacked image, incorporating 377 ERO candidates, from the PDS radio mosaic. The image rms is about 1.1 μJy, and the stacked signal is detected at $\approx 6\,\sigma$ with a flux density of 6.5 μJy.

purpose of providing a preliminary estimate for the average properties of these systems in order to support proposals for further, more detailed investigation.
AMH acknowledges support provided by the National Aeronautics and Space Administration (NASA) through Hubble Fellowship grant HST-HF-01140.01-A awarded by the Space Telescope Science Institute (STScI). JA gratefully acknowledges the support from the Science and Technology Foundation (FCT, Portugal) through the fellowship BPD-5535-2001 and the research grant POCTI-FNU-43805-2001.

References

Afonso, J., Hopkins, A., Mobasher, B., & Almeida, C. 2003, ApJ, (in press; astro-ph/0307175)

Bell, E. F. 2003, ApJ, 586, 794

Buat, V., Boselli, A., Gavazzi, G., & Bonfanti, C. 2002, A&A, 383, 801

Condon, J. J. 1992, ARA&A, 30, 575

Dunlop, J. S. & Peacock, J. A. 1990, MNRAS, 247, 19

Hopkins, A. M., Afonso, J., Chan, B., Cram, L. E., Georgakakis, A., & Mobasher, B. 2003, AJ, 125, 465

Hopkins, A. M., Connolly, A. J., Haarsma, D. B., & Cram, L. E. 2001, AJ, 122, 288

Hopkins, A. M. et al. 2003, ApJ, (submitted; astro-ph/0306621)

Lin, H., et al., 1999, ApJ, 518, 533

Sadler, E. M. et al. 2002, MNRAS, 329, 227

Smail, I., Owen, F. N., Morrison, G. E., Keel, W. C., Ivison, R. J., & Ledlow, M. J. 2002, ApJ, 581, 844

Sullivan, M., Mobasher, B., Chan, B., Cram, L., Ellis, R., Treyer, M., & Hopkins, A. 2001, ApJ, 558, 72

A PHYSICAL MODEL FOR THE JOINT EVOLUTION OF QSOS AND SPHEROIDS

G.L. Granato,[1] L. Silva,[2] G. De Zotti,[1] A. Bressan[1] and L. Danese[3]
[1]*INAF, Padova, Italy,* [2]*INAF, Trieste, Italy,* [3]*SISSA, Trieste, Italy*

1. The model

We present a detailed, physically grounded, model for the early co-evolution of spheroidal galaxies and of active nuclei at their centers. The model is based on very simple recipes, that can be easily implemented. The main components and the transfer processes accounted for are depicted in Fig. 1, while we defer to Granato et al. (2004) for a full description. In summary, we start from the diffuse gas within the dark matter halo falling down into the star forming regions at a rate ruled by the dynamic and the cooling times. Part of this gas condenses into stars, at a rate again controlled by the local dynamic and cooling times. But the gas also feels the feedback from supernovae and from active nuclei, heating it and possibly expelling it from the potential well. Also the radiation drag on the cold gas decreases its angular momentum, causing an inflow into a reservoir around the central black hole. Viscous drag then causes the gas to flow from the reservoir into the black hole, increasing its mass and powering the nuclear activity. Among the most novel point of our work with respect to other semi-analytic approaches, we point out the treatment of the evolution of QSO activity in the forming spheroids, and of the resulting effects on galaxy evolution.

2. Results

In the shallower potential wells (corresponding to lower halo masses and, for given mass, to lower virialization redshifts), the supernova heating is increasingly effective in slowing down the star formation and in driving gas outflows, resulting in an increase of star/dark-matter ratio with increasing halo mass. As a consequence, the star formation is faster within the most massive halos, and the more so if they virialize at substantial redshifts. Thus, in keeping with the proposition by Granato et al. (2001), physical processes acting on baryons effectively reverse the order of formation of galaxies compared to that of dark-matter halos.

M. Plionis (ed.), Multiwavelength Cosmology, 129-132.

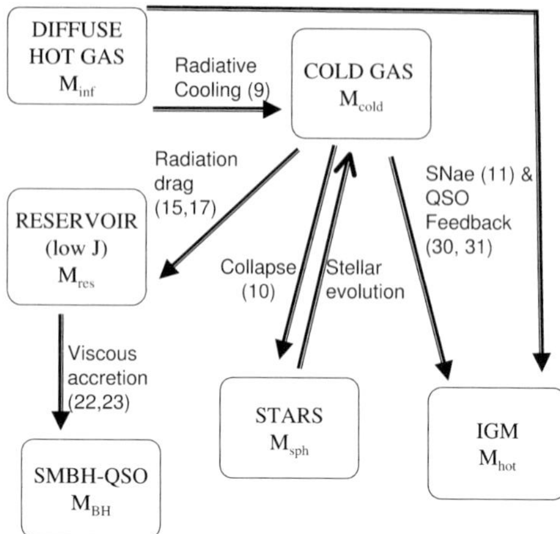

Figure 1. Scheme of the baryonic components included in the model (boxes), and of the corresponding mass transfer processes (arrows).

A higher star-formation rate also implies a higher radiation drag, resulting in a faster loss of angular momentum of the gas (Umemura 2001) and, consequently, in a faster inflow towards the central black-hole. In turn, the kinetic energy carried by outflows driven by active nuclei through line acceleration is proportional to $L_{\rm QSO}^{3/2}$, and this mechanism can inject in the interstellar medium a sufficient amount of energy to unbind it. The time required to sweep out the interstellar medium, thus halting both the star formation and the black-hole growth, is again shorter for larger halos. For the most massive galaxies ($M_{\rm vir} \gtrsim 10^{12}\,M_{\odot}$) virializing at $3 \lesssim z_{\rm vir} \lesssim 6$, this time is $< 1\,$Gyr, so that the bulk of the star-formation may be completed before type Ia supernovae can significantly increase the Fe abundance of the interstellar medium; this process can then account for the α-enhancement seen in the largest galaxies.
The interplay between star formation and nuclear activity determines the relationship between the black-hole mass and the mass, or velocity dispersion, of the host galaxy, as well as the black-hole mass function. As illustrated by Figs. 2 and 3, the model predictions are in excellent agreement with the observational data. A specific prediction of the model is a substantial steepening of the $M_{\rm BH}$–σ relation for $\sigma \lesssim 150\,{\rm km\,s^{-1}}$: the mass of the BH associated to less massive halos is lower than expected from an extrapolation from higher masses, because of the combined effect of SN heating, which is increasingly effective with decreasing galaxy mass in hindering the gas inflow towards the central BH, and of the decreased radiation drag.
The ratio between QSO and SNae feedback is an increasing function of the mass. At low mass the cumulative effect of QSO is almost negligible with re-

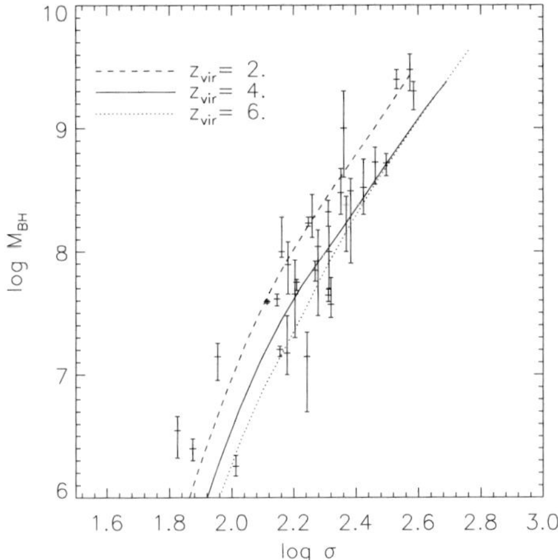

Figure 2. Predicted relationship between black-hole mass and line-of-sight velocity dispersion of the host galaxy for different virialization redshifts.

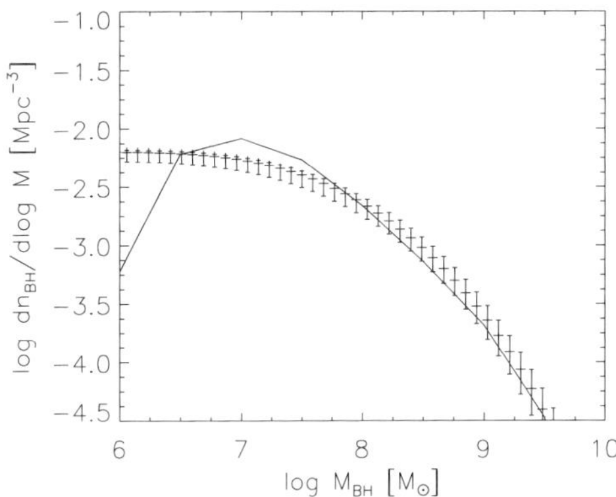

Figure 3. Predicted local black-hole mass function compared with the recent estimate by Shankar et al. (in preparation). The total mass density in BHs they derive is $\rho_{BH} \simeq 4 \times 10^5 M_\odot \mathrm{Mpc}^{-3}$, 25% less than our model. The decline of the model at low $M_{\rm BH}$ is due to having considered only objects with $M_{\rm vir} \geq 2.5 \times 10^{11} M\odot$.

spect to that of SNae, but it becomes dominant at intermediate mass (typically by a factor of a few) and high mass (by a factor $\gtrsim 10$). Note that the QSO effect usually increases exponentially with time, while that of SNae increases more slowly. Thus the instantaneous QSO effect becomes dominant, if ever, only a few e-folding times before the maximum of QSO activity.

Coupling the model with GRASIL (Silva et al. 1998), the code computing in a self-consistent way the chemical and spectrophotometric evolution of galaxies over a very wide wavelength interval, we have obtained predictions for the sub-mm counts and the corresponding redshift distributions as well as for the redshift distributions of sources detected by deep K-band surveys, which proved to be extremely challenging for all the current semi-analytic models. The results, shown by Fig. 4, are again very encouraging.
The cosmic history of BH accretion, is in keeping with the common notion that quasar activity peaks at redshift 2-3. Also, the formation rate of spheroids derived in this paper is consistent with that we computed in Granato et al. 2001 from a deconvolution of the high-z luminosity function of QSO.

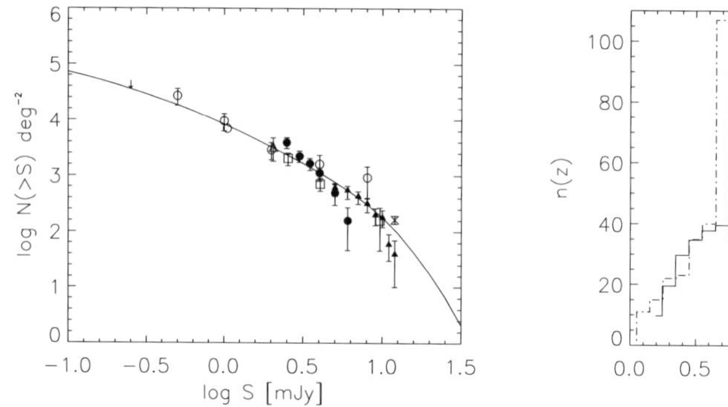

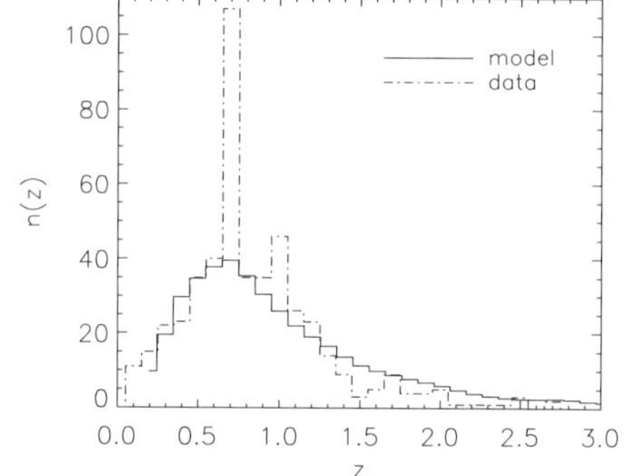

Figure 4. Left: predicted 850 μm extragalactic counts compared with observed SCUBA counts. Right: predicted redshift distribution of galaxies brighter than $K = 20$ compared with the results of the K20 survey

References

Granato, G.L., Silva, L., Monaco, P., Panuzzo, P., Salucci, P., De Zotti, G., & Danese, L. 2001, MNRAS, 324, 757

Granato, G.L., De Zotti, G., Silva,L., Bressan, A. & Danese, L. 2004, ApJ, in press

Silva, L., Granato, G.L., Bressan, A., & Danese, L. 1998, ApJ, 509, 103

Umemura, M. 2001, ApJ, 560, L29

THE LOCAL SUB-MM LUMINOSITY FUNCTIONS AND PREDICTIONS FROM ASTRO-F/SIRTF TO HERSCHEL

Stephen Serjeant[1], Diana Harrison[2]
[1]*School of Physical Sciences, University of Kent, Canterbury, Kent, CT2 7NR*
[2]*Dept. of Physics, Cavendish Laboratory, Madingley Road, Cambridge CB3 0HE, UK*

Abstract We present new determinations of the local sub-mm luminosity functions. We find the local sub-mm luminosity density converging to $7.3 \pm 0.2 \times 10^{19}\ h_{65}^{-1}$ W Hz^{-1} Mpc^{-3} at 850μm solving the "sub-mm Olbers' Paradox." Using the sub-mm colour temperature relations from the SCUBA Local Universe Galaxy Survey, and the discovery of excess 450μm excess emission in these galaxies, we interpolate and extrapolate the IRAS detections to make predictions of the SEDs of all 15411 PSC-z galaxies from $50 - 3000\mu$m. Despite the long extrapolations we find excellent agreement with (a) the 90μm luminosity function of Serjeant et al. (2001), (b) the 850μm luminosity function of Dunne et al. (2000), (c) the mm-wave photometry of Andreani & Franceschini (1996); (d) the asymptotic differential and integral source count predictions at $50 - 3000\mu$m by Rowan-Robinson (2001). Remarkably, the local luminosity density and the extragalactic background light together strongly constrain the cosmic star formation history for a wide class of evolutionary assumptions. We find that the extragalactic background light, the 850μm 8mJy source counts, and the Ω_* constraints all independently point to a decline in the comoving star formation rate at $z > 1$.

1. Introduction

The evolution of the sub-mm galaxy population can be strongly constrained by the integrated extragalactic background light, the local multiwavelength luminosity functions, and the source counts. The local 850μm luminosity function was derived in the SCUBA Local Universe Galaxy Survey (SLUGS, Dunne et al. 2000) from their SCUBA photometry of the IRAS Bright Galaxy Survey. A curious aspect of their luminosity function was that the faint end slope was not sufficiently shallow for the local luminosity density to converge, which the authors referred to as the "sub-mm Olbers' paradox". This is a pity from the point of view of modeling the high redshift population, since the integrated extragalactic background light is a key constraint. In order to find the expected flattening of the luminosity function slope at lower luminosities, the SLUGS sur-

M. Plionis (ed.), Multiwavelength Cosmology, 133-136.

vey is currently being extended with SCUBA photometry of optically-selected galaxies. Meanwhile, several authors have attempted to use the colour temperature – luminosity relation found in SLUGS to transform the $60\mu m$ luminosity function to other wavelengths, and hence constrain the high-redshift evolution (e.g. Lagache, Dole & Puget 2003, Chapman et al. 2003). The discovery of an additional excess component at $450\mu m$ (Dunne & Eales 2001) relative to their initial colour temperature – luminosity relation has not so far been included in such models.

2. Models of PSC-z galaxies

In this paper we take an alternative approach to determining the multiwavelength local luminosity functions. We model the spectral energy distributions (SEDs) of all 15411 PSC-z galaxies (Saunders et al. 2000), constrained by all available far-infrared and sub-mm colour-colour relations from SLUGS and elsewhere. This guarantees the correct local population mix at every wavelength and minimises the assumptions about the trends of SED shape with luminosity, and is sufficient to determine the $450\mu m$ and $850\mu m$ luminosities to better than a factor of 2 in individual galaxies. This predicted photometry is more than sufficient to determine local luminosity functions. For interpolations between these bands (e.g. predicted photometry from SIRTF, ASTRO-F, or Herschel) we express the SED as the sum of two $\beta = 2$ grey bodies, which the measured and/or predicted bands are just sufficient to determine uniquely. More details can be found in Serjeant & Harrision (2003).

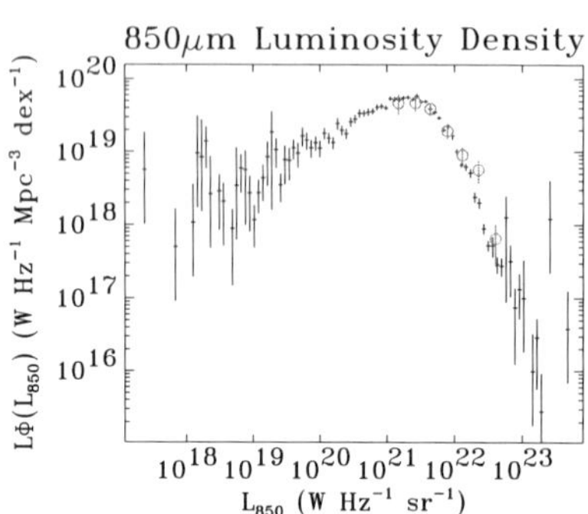

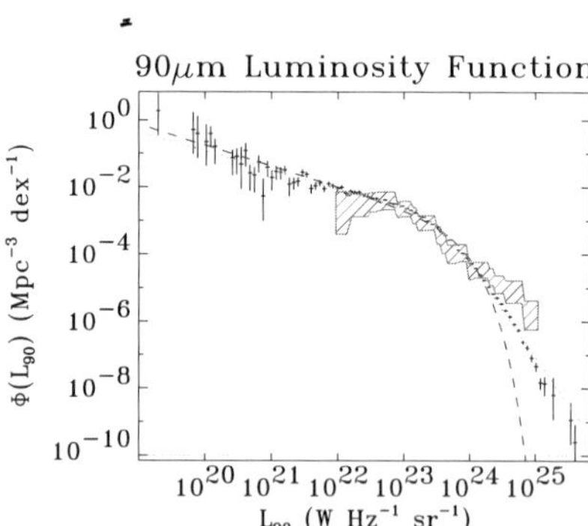

Figure 1. Left: projected local $850\mu m$ luminosity density from PSC-Z (error bars) assuming pure luminosity evolution of $(1+z)^3$, compared with the directly measurements from Dunne et al. (2000). Right: projected $90\mu m$ luminosity function, compared to the direct determination of Serjeant et al. 2001 (shaded area).

3. Results

There are few local galaxies with abundant multi-wavelength data to test our models, but our models are consistent with the observed 1.25mm photometry of Andreani & Franceschini (1996), the 175μm ISO Serendipity Survey (e.g. Stickel et al. 2000, also demonstrating that the cool dust excess reported by Stickel et al. 2000 is identical to that reported by Dunne & Eales 2001), and the SLUGS galaxies with multi-wavelength data. The projected bright-end source counts are also in excellent agreement with source count models (Rowan-Robinson 2001).

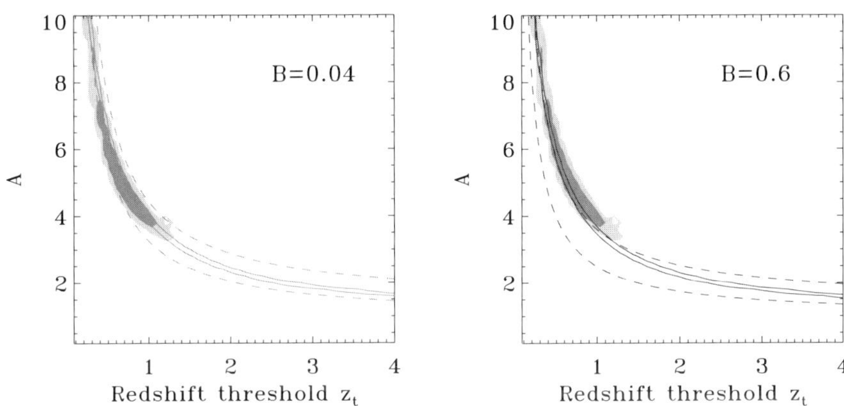

Figure 2. Constraints on the parameterisation of the $z \lesssim 2$ cosmic star formation history. The definitions of the parameters A and z_t are given in the text. The shaded regions show the 68% and 95% confidence regions (inner and outer region respectively) for A and z_t, marginalised over B. These constraints are derived from only the extragalactic background light and our determination of the local spectral luminosity density. The dashed lines show the *independent* constraints from $\Omega_* = 0.003 \pm 0.0009h^{-1}$ (e.g. Lanzetta, Yahil, & Fernandez-Soto 1996), and the full line shows the predicted 8mJy 850μm source count constraint of $N(> S) = 320^{+80}_{-100}$ deg^{-2} (Scott et al. 2002) assuming pure luminosity evolution.

4. Discussion

The PSC-z 850μm luminosity function is in excellent agreement with the direct determination of Dunne et al. (2000) and clearly shows the convergence in the luminosity density (fig. 1). Other interpolated luminosity functions can also be generated (fig. 1). The extragalactic background light depends only on these local luminosity densities and the cosmic star formation history, which we parameterise as $\dot{\rho}(z)/\dot{\rho}(z = 0) = (1 + z)^A$ at $z < z_t$, and $(1 + z_t)^A B^{z-z_t}$ at $z > z_t$. Not only do we obtain strong constraints on the parameters A, B, z_t

(figs. 2 and 3) requiring a decline in comoving star formation rate at $z > 1$, but the 850μm survey source counts and Ω_* constraints both point to a similar constraint. To reconcile this with other constraints on the cosmic star formation history requires differential evolution in the sub-mm population and/or a top-heavy IMF at high-z.

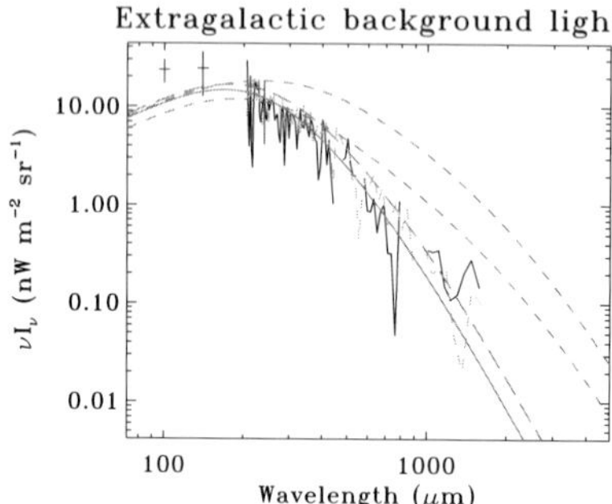

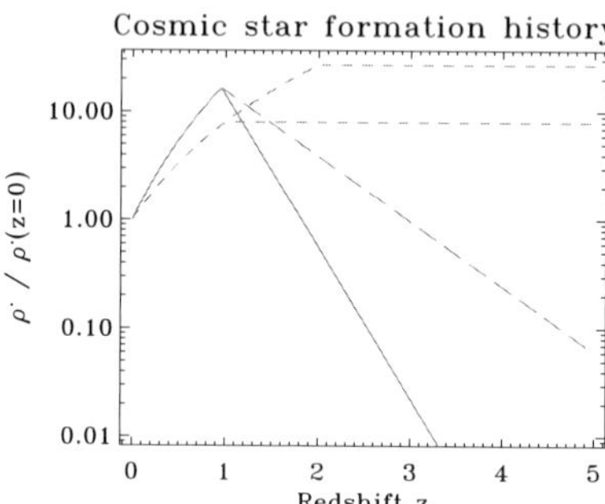

Figure 3. Left figure shows the extragalactic background light (data from Lagache et al. 1999) modelled by the cosmic star formation history described in the text. The data longward of 200μm plotted as broken lines is the FIRAS spectrum: full line is whole sky, and dashed line is Lockman Hole only. Also plotted are the DIRBE data points, not included in our fitting. The smooth curves are models of this spectrum, corresponding to cosmic star formation histories plotted in the right hand figure. The full line is the global maximum likelihood fit, and the long-dashed line has the same parameters except an enhanced high-z star formation rate ($B = 0.25$) which is marginally inconsistent with the extragalactic background. The two short-dashed lines demonstrate selected alternative models: $z_t = 1$ and $z_t = 2$, both with $A = 3$ and $B = 1$. Note that in general the models with the higher star formation rates at $z > 2$ are also the models predicting the larger background at wavelengths $\lambda \sim 1$mm (including in particular a comparison of the two short-dashed curves). In general, models with high volume-averaged star formation rates at $z > 2$ overpredict the sub-mm/mm-wave background.

References

Andreani, P., & Franceschini, A., 1996 MNRAS 283, 85
Chapman, S.C, et al., 2003, preprint astro-ph/0301233
Dunne, L., et al., 2000 MNRAS 315, 115
Dunne., L., & Eales, S.A., 2001 MNRAS 327, 697
Lagache, G., et al., 1999, A&A 344, 322
Lagache, G., Dole, H., Puget, J.-L., 2003, MNRAS 338, 555
Saunders, W., et al., 2000 MNRAS 317, 55
Scott, S.E., et al. 2002 MNRAS 331, 817
Serjeant, S., et al., 2001, MNRAS 322, 262
Serjeant, S., & Harrison, D., 2003, MNRAS submitted
Stickel., M., et al., 2000, A&A 359, 865

MULTICOLOUR PHOTOMETRY OF THE VIRMOS-VLA RADIO SOURCES

P. Ciliegi and H. J. McCracken
on the behalf of the VIRMOS and TERAPIX Consortium
INAF - Osservatorio Astronomico di Bologna - Italy
University of Bologna - Italy
ciliegi@bo.astro.it

Abstract In this paper we present the optical identifications of the 1.4 GHz radio sources detected in the VIRMOS VLA Deep survey. Using optical (U, B, V, R and I) data, we found an optical identification for 718 of the 1054 radio sources. From the analysis of a color-colour diagram, we find that the median $(V-I)_{AB}$ colour of the galaxies which show radio emission is redder than the median $(V-I)_{AB}$ colour of the radio-quiet galaxies. This result suggests a higher redshift for the radio sources in comparison to the radio-quiet galaxies with similar optical magnitude at least in the magnitude slice 20< I_{AB} <22.

Keywords: Radio continuum, radio survey, cosmology

1. Introduction

Deep 1.4-GHz counts show an upturn below a few millijansky (mJy), corresponding to a rapid increase in the number of faint sources. However, despite many dedicated efforts (see for example Benn et al. 1993, Hammer et al. 1995, Gruppioni et al. 1999, Richards et al. 1999, Prandoni et al. 2001) the relative fraction of the populations responsible of the sub-mJy radio counts (AGN, starbursts, late and early type galaxies), are not well established yet.
The VIRMOS VLA Deep survey is one of the best available surveys to investigate the nature of the sub-mJy population. In fact, coupled with a 1 deg^2 deep 1.4 GHz radio data obtained with the Very Large Array down a 5σ flux limit of $\sim$80 μJy (Bondi et al. 2003), a deep optical photometric survey is already available down to B_{AB} $\sim$26.5, V_{AB} $\sim$26.2, R_{AB} $\sim$25.9 , I_{AB} $\sim$25.0 (McCracken et al. 2003) and U_{AB} $\sim$24.75 (Radovich et al. 2003 in preparation) and a spectroscopic survey down to I_{AB} $\sim$24.0 is being performed with the VIMOS spectrograph at the VLT (see Le Fevre et al., these proceedings). In

M. Plionis (ed.), Multiwavelength Cosmology, 137-140.

this paper we present the optical identifications of the VIRMOS radio sources obtained with the photometric data in the U, B, V, R and I bands.

2. Optical properties of the radio sources

2.1 Optical identification

Almost the whole square degree of the VLA-Virmos Deep Field has been observed in the B, V, R and I bands with the CFH12K wide-field mosaic camera during the CFH12K-VIRMOS deep imaging survey, while a U band survey has been carried out with the wide field imaging (WFI) mosaic camera mounted on the ESO MPI 2.2 meter telescope at La Silla, Chile. The total effective area covered by the U band survey is $\sim$0.73 deg^2; 637 radio sources have U photometry.

Fifty-seven of the 1054 radio sources in the complete radio sample are outside the B, V, R, I CCD area, while 24 radio sources are within the masked regions in the optical catalogue, since they are spatially coincident with bright stars or their diffraction spikes.

The total number of radio sources for which optical data are available in the VIRMOS CCD catalogue is therefore 973. Using a Likelihood Ratio (LR) technique (see Ciliegi et al. 2003 for more details on the LR technique) we find 718 radio sources with a likely identification, 339 of which have also a U band counterpart.

2.2 Magnitude distribution

In absence of spectroscopic data, the magnitude and color distributions of the optical counterparts can be used to derive some information on the nature of faint radio sources. In Figure 1 we show the magnitude distribution in the I_{AB} bands of the optical counterparts of the radio sources as a filled histogram. The empty histogram shows the magnitude distribution of the whole data set.

From Figure 1 it is clear that the magnitude distribution of the optical counterparts of the radio sources reaches a maximum at a magnitude well above our optical limiting magnitude, consistently with the fact that a large fraction ($\sim$74%) of the radio sources have an optical identification. A similar result has been obtained by Richards et al. (1999) in the identification of the μJy radio sources in the Hubble Deep Field region. Approximately 76% of the HDF radio sources (84 out of 111) have been identified to I=25 mag, with the bulk of the sample identified with relatively bright ($I \leq$22 mag) galaxies.

2.3 Colour-colour diagrams

In Figure 2 we compare the optical colors ($(V-I)_{AB}$ vs. $(B-V)_{AB}$) of the optical counterparts of the radio sources with those of the whole optical

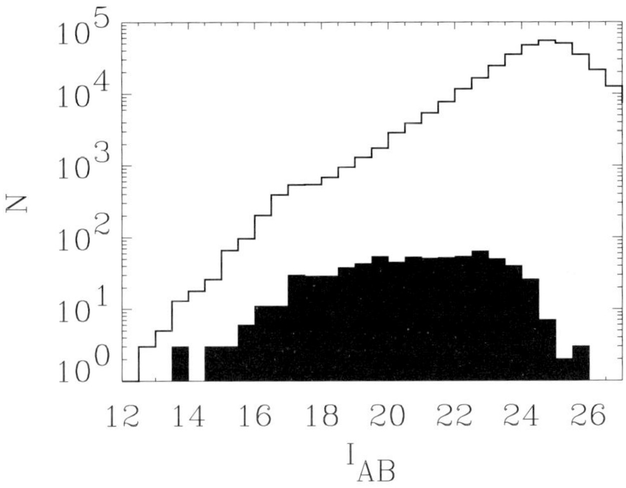

Figure 1. Magnitude distributions in the I_{AB} band for the optical counterparts of the radio sources (filled histogram) and for the whole optical sample (empty histogram).

data set, in two different magnitude slices: 20< I_{AB} <22 (left panel) and 22< I_{AB} <24 (right panel).
As shown in Figure 2, the whole optical data set in the magnitude slice 20< I_{AB} <22 (see McCracken et al 2003 for a detailed discussion) occupies two distinct loci, with a reasonably well defined separation between high (z≥0.4) and low (z≤0.4) redshift galaxies, while the radio sources seem to populate only the locus of the high redshift (z∼0.4-0.5) galaxies.
In the fainter magnitude slice (22< I_{AB} <24; right panel of Figure 2) the colours are bluer than in the brighter magnitude slice both for the radio sources and the whole optical data set. Even if less evident from the figure, also in this magnitude slice there is a statistically significant difference between the colour distributions of the radio sources and the radio quiet galaxies. Its interpretations, however, is less straightforward than in the brighter magnitude slice and will require a more detailed analysis of the full photometric data set (Ciliegi et al. 2004 in preparation). In this paper we will give a full description of the photometric properties of the radio sources in the VIRMOS VLA Deep Survey.

References

Benn C.R., Rowan-Robinson M., McMahon R.G., Broadhurst T.J. & Lawrence A., 1993, MNRAS, 263, 98

Bondi M., Ciliegi P., Zamorani G. et al., 2003, A&A, 403, 857

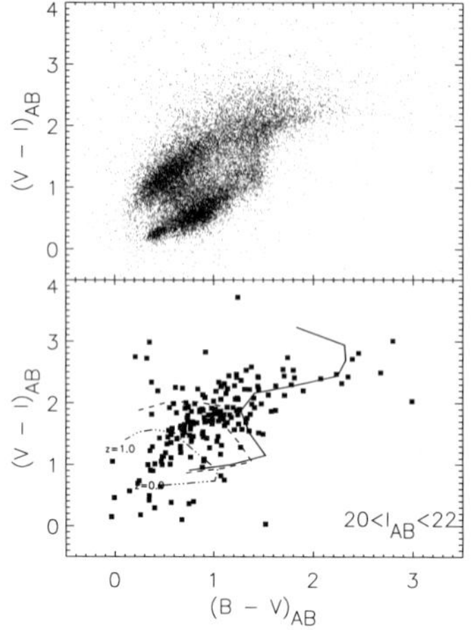

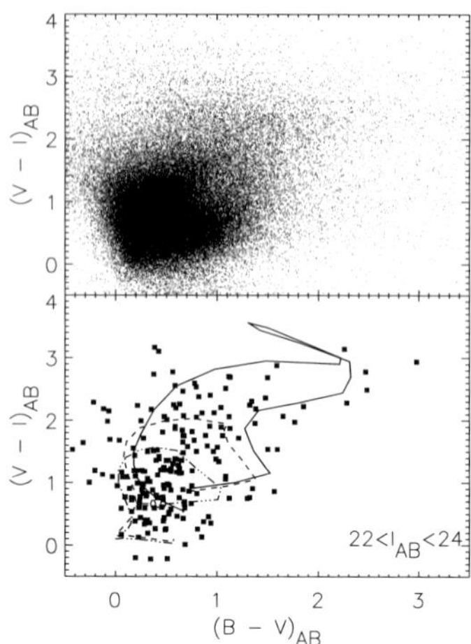

Figure 2. $(V - I)_{AB}$ versus $(B - V)_{AB}$ colours for the whole data set (small dots in the upper panels) and for the optical counterparts of the radio sources (filled squares in the lower panels). The expected colours of early-type, Sa and Sbc galaxies (solid, dashed, dot-dot-dot dashed lines, respectively) are shown as a function of redshift, up to z=1 for 20< I_{AB} <22 and up to z=3 for 22< I_{AB} <24

Ciliegi P., Zamorani G., Hasinger G., Lehmann I., Szokoly G. & Wilson G., 2003, MNRAS, 398, 901

Gruppioni C., Mignoli M. & Zamorani G., 1999, MNRAS, 304, 199

Hammer F., Crampton D., Lilly S. J., Le Fevre O. & Kenet T., 1995, MNRAS, 276, 1085

McCracken H.J., Radovich M. , Bertin E., 2003 in press, astro-ph/0306254

Prandoni I., Gregorini L., Parma P., et al., 2001, A&A, 369, 787

Radovich M., Arnaboldi M., Ripepi et al. 2003, A&A, in press

Richards E. A., Fomalont E.B., Kellermann R.A., et al., 1999, ApJL., 526, 73

PROTO-CLUSTERS ASSOCIATED WITH RADIO GALAXIES FROM $Z = 2$ TO $Z = 4$

Jaron Kurk,[1,2] Bram Venemans,[1] Huub Röttgering,[1] George Miley[1] and Laura Pentericci[3]

[1] *Sterrewacht Leiden, P.O. Box 9513, 2300 RA, Leiden, The Netherlands*

[2] *INAF, Osservatorio Astrofisico di Arcetri Largo Enrico Fermi 5, 50125, Firenze, Italy*

[3] *Max-Planck-Institut für Astronomie, Königstuhl 17, D-69117, Heidelberg, Germany*

kurk@arcetri.astro.it

Abstract

We have carried out narrow band imaging targeted at the Lyα line of emitters associated with radio galaxies at $2 < z < 5$. Subsequent spectroscopy has confirmed the identity of > 120 Lyα emitters and led to the discovery of six proto-clusters up to $z = 4.1$ with velocity dispersions between 300 and 1000 km s^{-1}. The number of emission line galaxies in the observed fields is four to fifteen times higher than in blank fields. These results strongly support the idea that high redshift radio galaxies are the progenitors of central brightest cluster galaxies located in the progenitors of clusters of galaxies.

1. Introduction

Observational studies of cosmological structures can never be complete without the full picture provided by multi wavelength explorations, but in the case of the search for very high redshift clusters or their progenitors the use of multi wavelength observations is particularly fruitful. The detection of galaxy aggregations using conventional optical and X-ray becomes difficult at $z > 1$ due to the presence of many fore- and background objects and/or the surface brightness dimming of the extended emission. Several lines of observation indicate that powerful radio galaxies, as are found up to $z = 5.2$, are located within regions of high galaxy density. The technique of narrow band imaging can detect a group of emission line galaxies at a narrow high redshift range. Applying this technique to fields containing high redshift galaxies, we succeed to find galaxy groups at high redshift. The groups will be used to study cluster progenitors, its galaxies and the massive radio galaxy hosts.

M. Plionis (ed.), Multiwavelength Cosmology, 141-144.

2. Sample and observations

We had initially selected ten high redshift radio galaxies (HzRGs, $z > 2$) from our compendium of about 150 $z > 2$ radio sources, many of which have been found in the Leiden survey of steep spectrum radio sources (Røttgering et al. 1994; De Breuck et al. 2000). We have successfully observed six fields and during an extension of the program, also the field of the most distant known radio galaxy, at $z = 5.19$, bringing the total number of nights allocated to this VLT Large Program to twenty. The observing time was more or less evenly distributed over imaging and the subsequent multi object spectroscopy, both carried out with the FORS2 instrument. The detector of this instrument has a field of view of almost 7×7 arcminutes (about 3×3 Mpc at $z > 2$). For the two highest redshift sources custom narrow band filter were manufactured. For the others we have used available FORS filters. In the following, two observed fields will be described in more detail and a summary of the results of the program will be presented.

3. Early results

I: MRC 1138-262 at z = 2.16: This radio galaxy was the target of our pilot study carried out with FORS1 at the VLT. It is exceptionally suitable as it manifests many of the properties which indicate a rich environment, namely the highest radio rotation measure in a sample of 80 HzRGs and a very distorted radio morphology (Carilli et al. 1997), a very clumpy UV morphology (Pentericci et al. 1997), a large and luminous Lyα halo ($\sim$ 150 kpc, Kurk et al. 2000a), and extended X-ray emission (Carilli et al. 2002). In addition, its redshift of 2.16 is near the lower limit where detection of Lyα emission with ground based facilities is possible. Narrow band imaging of the field of 1138-262 resulted in the detection of 50 candidate Lyα emitters (LAEs, Kurk et al. 2000b). Multi object spectroscopy of these candidates confirms the presence of a single emission line at the expected wavelength for fifteen objects (Pentericci et al. 2000). One of these is an AGN as shown by the broadness of its emission line (FWHM $\sim$ 6000 $\mathrm{km\,s^{-1}}$). The others seem to form two groups with velocity dispersions of $\sim$ 200 and 400 $\mathrm{km\,s^{-1}}$(if regarded as one group the dispersion is $\sim$ 1000 $\mathrm{km\,s^{-1}}$). The volume density of LAEs is about four times higher than that measured for the blank field population of LAEs (Steidel et al. 2000). In addition to the observations necessary to identify the LAEs, we have obtained imaging in a number of optical and infrared bands with FORS and ISAAC at the VLT, one of which is a narrow band filter in the NIR targeted at redshifted Hα emission of galaxies associated with 1138$-$262. Excess narrow band flux testifies the presence of 40 candidate Hα emitters in the field (Kurk et al. 2003). Long slit ISAAC spectroscopy of 9 candidates shows that three objects exhibit emission lines identified as Hα and [NII] while the remaining 6 objects have fainter single lines consistent with the identification with Hα, one of which has a FWHM of $\sim$ 5000 km s^{-1}revealing another AGN at $z = 2.16$. The two AGN discovered with the narrow band technique are part of a larger group of five AGN probably associated with the radio galaxy as shown by Chandra observations of the field (Pentericci et al. 2002). The number density of soft X-ray sources detected by Chandra is in excess of that found in other fields. The detected overdensities of Lyα, Hα and X-ray emitters draw a consistent picture of a powerful radio galaxy in an aggregation of galaxies which may form a cluster of galaxies.

TN J1338-1942 at z = 4.11: With regard to both its continuum and Lyα emission, this galaxy is amongst the most luminous radio galaxies known. Its Lyα line profile and radio morphology are very asymmetric indicating strong interaction with dense gas (De Breuck et al. 1999). Based on the line equivalent width derived from the narrow and broad (R) band images and also on the absence of emission in the broad B band, 28 objects in the field of 1338$-$1942 were selected as candidate LAEs (Venemans et al. 2002). Of the 23 candidates subsequently observed spectroscopically, 20 show a single emission line in the expected wavelength range. The identification of the observed features with lines other than Lyα can be excluded in almost all cases. The velocity distribution of the emitters has a dispersion of 326 km s^{-1}, significantly smaller than expected from a random distribution of emitters selected with the narrow band filter. Compared with the LALA survey (Rhoads et al. 2000), the volume density of LAEs is about 15 times larger than in a blank field. The spatial distribution of emitters is not homogeneous over the observed field. The radio galaxy is not in the center of the observed structure but rather at the northern edge.

4. High redshift proto-clusters and radio galaxies

At the time this talk was delivered, six HzRG fields were observed sufficiently deep to assess the density of companion LAEs. In the fields of HzRGs at z = 2.86, 2.92, 3.13 and 3.14 the number of candidate (and spectroscopically confirmed) emitters are: 52 (37), 70 (30), 78 (31) and 20 (11), which amounts to over-densities of 4 to 15. The velocity dispersion of the ensembles of emitters observed decreases with increasing redshift: from $\sim$ 1000 km s^{-1}at $z = 2$ to $\sim$ 325 km s^{-1}at $z = 4$. It would be premature to draw conclusions on this result, however, as there are only six data points. The size of the structures of LAEs is in all cases larger than the field sampled: $>$ 3 Mpc (comoving) and seems sometimes bound in one direction. We have compared the number of spikes in the redshift distribution of Lyman Break Galaxies (LBGs) at $2.7 < z < 3.4$ (Steidel et al. 1998) with the number of powerful ($P_{2.7\,GHz} > 10^{33}$ erg s^{-1} Hz^{-1} sr^{-1}) radio galaxies in this redshift range (

Dunlop and Peacock, 1990). Assuming that the radio sources are active only once for a period of $\sim 10^7$ year, the numbers are consistent with every LBG redshift spike being associated with a massive galaxy that has been or will become a luminous radio source once. From the preliminary results we derive the following conclusions: narrow band imaging is an efficient technique to find galaxy over-densities in a narrow redshift range, HzRGs are excellent tracers of these over-densities and may be the progenitors of central brightest cluster galaxies located in the progenitors of clusters of galaxies.

References

Carilli, C.L., Røttgering, H.J.A., van Ojik, R., Miley, G.K., and van Breugel, W.J.M. (1997), *ApJ Suppl.*, 109:1.

Carilli, C.L., Harris, D.E., Pentericci, L., Røttgering, H.J.A., Miley, G.K., Kurk, J.D., and van Breugel, W. (2002), *ApJ*, 567:781.

De Breuck, C., van Breugel, W., Minniti, D., Miley, G., Røttgering, H., Stanford, S. A., and Carilli, C. (1999). VLT spectroscopy of the z=4.11 Radio Galaxy TN J1338-1942, *A&A*, 352:L51

De Breuck, C., van Breugel, W., Røttgering, H. J. A., and Miley, G. (2000). *A&ASS*, 143:303.

Dunlop, J. S. and Peacock, J. A. (1990). The Redshift Cut-Off in the Luminosity Function of Radio Galaxies and Quasars, *MNRAS*, 247:19.

Kurk, J. D., Røttgering, H. J. A., Pentericci, L., and Miley, G. K. (2000a). Observations of radio galaxy mrc 1138-262: a giant halo of lya and ha emitting ionized gas. In Henney, Will, Steffen, Wolfgang, Binette, Luc, and Raga, Alejandro, editors, *Emission Lines from Jet Flows, Isla Mujeres, November 13-17, 2000*, volume 13 of *Conference Series of Revista Mexicana de Astronom"a y Astrof"sica*, page 21.

Kurk, J.D., Pentericci, L., Røttgering, H.J.A., and Miley, G.K. (2003), *A&A*, page submitted.

Kurk, J.D., Røttgering, H.J.A., Pentericci, L., Miley, G.K., van Breugel, W., Carilli, C.L., Ford, H., Heckman, T., McCarthy, P., and Moorwood, A. (2000b), *A&A*, 358:L1.

Pentericci, L., Kurk, J. D., Carilli, C. L., Harris, D. E., Miley, G. K., and Røttgering, H. J. A. (2002). A Chandra study of X-ray sources in the field of the z=2.16 radio galaxy MRC 1138-262, *A&A*, 396:109

Pentericci, L., Kurk, J.D., Røttgering, H.J.A., Miley, G.K., van Breugel, W., Carilli, C.L., Ford, H., Heckman, T., McCarthy, P., and Moorwood, A. (2000), *A&A*, 361:L25.

Pentericci, L., Røttgering, H.J.A., Miley, G.K., Carilli, C.L., and McCarthy, P. (1997), *A&A*, 326:580.

Rhoads, J. E., Malhotra, S., Dey, A., Stern, D., Spinrad, H., and Jannuzi, B. T. (2000), *ApJ Lett.*, 545:L85–L88.

Røttgering, H.J.A., Lacy, M., Miley, G.K., Chambers, K.C., and Saunders, R. (1994), *A&ASS*, 108:79.

Steidel, C. C., Adelberger, K. L., Dickinson, M., Giavalisco, M., Pettini, M., and Kellogg, M. (1998), *ApJ*, 492:428.

Steidel, C. C., Adelberger, K. L., Shapley, A. E., Pettini, M., Dickinson, M., and Giavalisco, M. (2000), *ApJ*, 532:170

Venemans, B. P., Kurk, J. D., Miley, G. K., Rottgering, H. J. A., van Breugel, W., Carilli, C. L., Breuck, C. De, Ford, H., Heckman, T., McCarthy, P., and Pentericci, L. (2002), *ApJ Lett.*, 569:L11

HIGH REDSHIFT RADIO GALAXIES AS TRACERS OF GALAXY CLUSTERS. *XMM-NEWTON* OBSERVATIONS

Elena Belsole, Diana M. Worrall, Martin J. Hardcastle
H.H. Wills Physics Laboratory
University of Bristol, Bristol BS8 1TL, U.K.
e.belsole@bristol.ac.uk

Abstract Distant powerful radio galaxies are useful tracers of galaxy groups and clusters, since through their very existence they point to gaseous atmospheres on scales of at least the radio source diameter. We present new results from *XMM-Newton* observations of three luminous radio galaxies, where evidence for both radio-related and cluster mission is seen.

Keywords: Radio galaxies, clusters of galaxies, X-ray, radio

1. Introduction

Distant powerful radio galaxies should be a useful tracer of galaxy groups and clusters, since the confinement of the jet requires a gaseous environment which should radiate in the X-ray.

The selection of cluster samples through their X-ray flux is biased towards the most luminous objects at high redshift, whereas radio selection should give a more representative sample of cluster X-ray properties. This may be useful for testing theories of structure formation. With the new-generation satellites *XMM-Newton* and *Chandra* we have an unprecedented opportunity to make X-ray spectral measurements of radio galaxies at $z > 1$, and separate spatially the different X-ray components.

New observations of FRII radio galaxies and quasars in the 3CRR catalog with $z > 0.6$ (Crawford & Fabian 2003; Donahue et al. 2003) have found that any cluster emission is of lower luminosity than measured with ROSAT (but this does not apply to 3C220.1; Worrall et al. 2001). Much of the extended X-ray emission is associated with the radio lobes and hotspots in some sources (e.g. Hardcastle et al. 2002).

Here we present results for three powerful radio galaxies at redshifts between 0.7 and 1.6 observed with *XMM-Newton*. The sources (3C 184, $z = 0.996$;

M. Plionis (ed.), Multiwavelength Cosmology, 145-148.

3C 292, $z = 0.71$; 3C 322, $z = 1.68$) were selected from the 3CRR catalog (Laing, Riley & Longair 1983) and have linear size > 50 kpc, which should be a signpost of clusters with significant atmospheres. For further details see Belsole et al. (2003).

2. Observations & Imaging

Calibrated event files for all three sources from the EPIC camera were provided by the *XMM-Newton* SOC and were filtered in a standard way. The net exposure time for results presented here are: 58 ks (MOS) 16.4 (pn) for 3C 184, 20 ks for 3C 292, and 9 ks for 3C 322. Adaptively smoothed images in the energy band 0.3 - 7.0 keV are shown in Figure 1.
The image of 3C 184 is strongly dominated by point-source emission. However, some degree of extended emission is detected up to 45$''$(300 kpc). The X-ray peak coincides, within 2$''$, with the core of the radio galaxy.
The X-ray emission of 3C 292 is interestingly elongated in the same direction as the radio emission, but the correspondence is not one-to-one. Since the resolution of the radio map at 1.4 GHz is roughly the same size as the *XMM-Newton* PSF, we can make a direct morphological comparison. The extended X-ray emission is detected up to 1.6$'$ (580 kpc) to the north-west/south-east direction. In the orthogonal direction, the smoothed image indicates X-ray emission surrounding the centre up to 38$''$, marginally consistent with the 90% encircled energy of a point source.
The X-ray morphology shown by the smoothed image of 3C 322 suggests that the emission is extended. The X-ray and radio emission coincide, with an indication that the radio emission is more extended than the X-ray. However this is the highest redshift source and the X-ray statistics are not good enough for a detailed comparison of the radio and X-ray emission.

3. Spectra

Spectra were extracted from the vignetting-corrected data, using a local background. We were unable to separate the different X-ray components of 3C 184 and thus extracted a spectrum in a circle of 40$''$ about the source. The data and folded model are shown in Figure 2. We obtain a best fit adopting a model composed of a power law of spectral index $\Gamma = 1.30^{+0.40}_{-0.37}$ and an absorbed power law of spectral index $\Gamma = 1.8^{+0.20}_{-0.08}$ and intrinsic absorption $N_{\rm H} = 6.9^{+7.4}_{-2.1} \times 10^{23}$ cm^{-2}. The goodness of the fit is χ^2 = 36.29 / 41 d.o.f. and errors are quoted at 1σ for one significant parameter.
In the case of 3C 292, we accumulated spectra in different regions in order to separate the possible components associated with the nucleus, the lobes, and cluster emission. The extraction region corresponding to the radio lobes was defined on the basis of the radio map. The lobe spectrum (Figure 2) is well

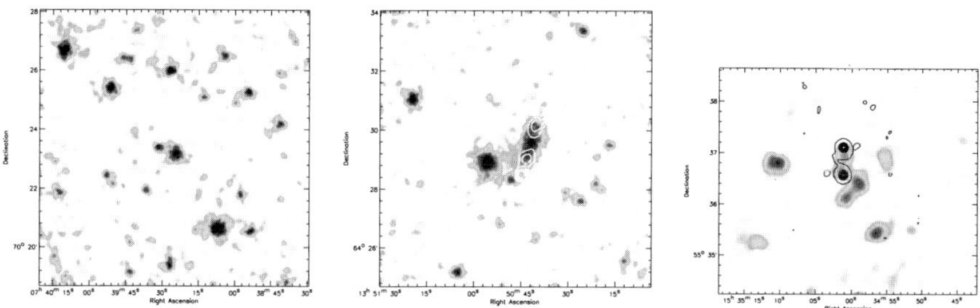

Figure 1. XMM/EPIC adaptive smoothed images of the three sources. Contours correspond to the radio emission. Left: 3C 184; the radio contours are not superimposed since the radio galaxy is small in angular size. Centre: 3C 292: contours correspond to a VLA 1.4 GHz image. Right: 3C 322; contours are from a 4.86 GHz image.

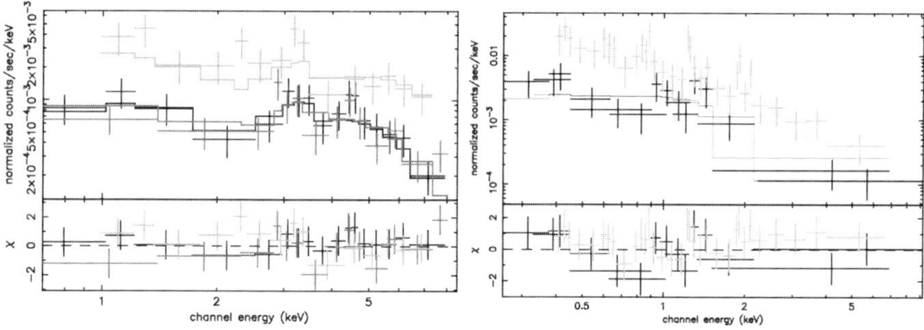

Figure 2. Left: Background-subtracted spectrum and folded model of the central 40$''$ of 3C 184. The pn is in gray, the MOS in black. Right: Spectrum and folded model of the lobes of 3C 292.

fitted with an unabsorbed power law of spectral index $\Gamma = 1.8 \pm 0.1$. Adopting this model the unabsorbed flux at 1 keV is 407 nJy. This is in fairly good agreement with the flux expected from inverse Compton (IC) scattering of the CMB photons if the lobes are radiating at minimum energy. By extracting a global spectrum, which includes the nucleus, the lobes, and more extended emission, we separate spectrally the cluster emission and find it to be well fitted with a thermal model of $kT = 1.9 \pm 0.3$ keV.

A detailed spectral analysis for 3C 322 is difficult because of the low photon statistics ($\sim$80 photons from the whole EPIC). Despite the poor statistics, the spectrum in an elliptical region about the source clearly displays two components, as for the other two sources. The soft component is better described by a thermal model of $kT = 1.5^{+1.8}_{-0.7}$ keV than a power law.

4. Discussion and preliminary conclusions

The XMM observation of 3C 184 indicates that the X-ray spectrum is dominated by the hidden nucleus and a soft component related to radio emission. If the soft component is produced by IC scattering of photons in the lobes, then the magnetic field of the source should be $\sim$2 times lower than the minimum energy condition. Deltorn et al. (1997) observed the field of 3C 184 with HST finding an overdensity of galaxies of which 11 were at redshift $z = 1$. They furthermore detected a gravitational arc. The extraction of a radial profile (not discussed here) gives us an indication of the detection of extended emission up to a radius of 0.8 Mpc. Using the photon statistics of this extended emission and the $L_X - T$ relation for low redshift galaxy clusters (Arnaud & Evrard 1999) we obtain an estimated temperature of $\sim$2 keV for this extended component.

The X-ray emission from the lobes of 3C 292 is likely to be related to the radio emission. In this case, from the measured X-ray flux and the expected flux for IC scattering of the CMB photons, we find that the source is very close to minimum energy. Another soft component contributes to the spectrum on larger scales. Its temperature ($kT \sim 2$ keV) is too hot for its luminosity. Since we cannot separate spatially the two components, confusion with other components, such as the nucleus, is likely. However, at this redshift, other physical conditions such as mergers or evolution of the $L_X - T$ relation could play a role.

The poor photon statistics of 3C 322 do not allow us to constrain the origin of the X-ray emission. The source is detected with each of the three cameras and we observe a correspondence between the X-ray emission and the radio emission from the galaxy. We have an indication, from the spectral analysis, that the soft emission is more likely to be thermal in nature than non-thermal. From our spectral results, we find that if a cluster-like environment is the origin of the soft emission, the cluster temperature and X-ray luminosity are in agreement with the $L_X - T$ relation for nearby luminous clusters..

References

Arnaud, M., and Evrard, A., 1999, MNRAS, 305,631

Belsole, E., et al., 2003, MNRAS, in preparation

Crawford, C.S. and Fabian, A.C., 2003, MNRAS, 339, 1163

Deltorn, J-M., Le Fęvre O., Crampton, D., Dickinson, M., 1997, ApJ, 483, L21

Donahue, M., Daly, R.A., and Horner, D.J., 2003, ApJ, 584, 643

Hardcastle, M.J., Birkinshaw, M., Cameron, R.A., Harris, D.E., Looney, L.W., Worrall, D.M., 2002, ApJ, 581, 948

Laing, R. A., Riley, J. M., Longair, M. S., 1983, MNRAS, 204, 151

Worrall, D.M, Birkinshaw, M., Hardcastle, M.J., and Lawrence, C.R., 2001, 326, 1127

A NEW DEEP SCUBA SURVEY OF GRAVITATIONALLY LENSING CLUSTERS

Kirsten K. Knudsen and Paul P. van der Werf
Leiden Observatory
P.O. Box 9513, 2300 RA Leiden
The Netherlands
kraiberg@strw.leidenuniv.nl

Abstract We have conducted a new deep SCUBA survey, which has targeted 12 lensing galaxy clusters and one blank field. The total area surveyed is $70\,\mathrm{arcmin}^2$ in which we have detected 60 sources. We have detected several sub-mJy sources after correcting for the gravitational lensing by the intervening clusters. Here we present the preliminary results and point out two highlights.

Keywords: Survey — Infrared: Galaxies — Submillimetre

1. Introduction

The extragalactic background light determined for the whole spectrum shows that from the structures in the universe most of the energy is emitted in the IR and in the optical/UV (e.g., Fixsen et al., 1998; Madau & Pozzetti 2000). The latter is the stellar light, while the former is the processed light, optical/UV photons absorbed by dust and re-emitted in the IR. The total energy output at the wavelengths ranges are comparable. Studies resolving the IR background have found that it is powered by a relatively small number of dusty star forming galaxies. These galaxies have large star formation rates, $100 - 3000\,\mathrm{M_{\odot}yr^{-1}}$, and are very luminous in the IR, typically $10^{11-13}\,\mathrm{L_{\odot}}$, thus they are a.k.a. (ultra-)luminous IR galaxies.
As half of the cosmic star formation appears to be hidden behind dust, studies of the high-z universe in the IR are imperative. The redshifted far-IR emission is observable at the submillimetre (submm) wavelengths with ground based telescopes and instrumentation. An important aspect of submm cosmology is the negative k correction, which allows observations of objects out to a redshift of 10. The observed flux density is approximately constant with redshift, which is of great benefit for detection experiments, but as a result it does not give an indication of redshift. Even if the source counts are known, different models

M. Plionis (ed.), Multiwavelength Cosmology, 149-152.

with largely different redshift distributions can be used to reproduce the source counts (Blain et al. 1999). As a consequence, to study the star formation history of these objects the redshift distribution must be determined by measuring the redshift of the individual objects. The negative k-correction is unique to only a few wavelengths. At most other wavelengths the positive k-correction makes it difficult to observe objects at high redshift. This of course affects the follow-up and identification studies the submm sources.

2. A new deep SCUBA survey

We have conducted a survey with SCUBA (Submillimeter Common User Bolometer Array; Holland et al. 1999), mounted at the J.C. Maxwell Telescope, Hawaii, aimed at studying the faint submm population. We have targeted 12 galaxy cluster fields and one blank field. The clusters are strongly gravitationally lensing and have redshifts between 0.03 and 0.88. The cluster fields have been observed to a depth of $1\sigma_{\rm rms} \sim 1 - 2\,{\rm mJy/beam}$ (one field shallower). 55 sources have been detected in those fields. The blank field is the NTT Deep Field, which was observed to $1\sigma_{\rm rms} \sim 1\,{\rm mJy/beam}$. Here 5 sources were detected. The total area surveyed is $70\,{\rm arcmin}^2$. The analysis and source extraction involved Mexican Hat Wavelets (Cayon et al. 2000) and Monte Carlo simulations (Knudsen et al. *in prep.*, Barnard et al. *in prep*).

3. Lensing and counts

The gravitational lensing by the foreground clusters amplifies the background sources. Furthermore, it magnifies the regions behind the clusters. As a result the area surveyed in the source plane is 2-3 times smaller than the area seen in the image plane. The effect of lensing is only of benefit when the counts are steep, which they are in the case of submm sources. The lensing moves the confusion limit to fainter flux density levels, hence allowing us to reliably pick out the faint sources. Most of the previous surveys made in the submm have targeted blank fields, and thereby studied the bright and intermediate population (e.g., Scott et al., 2002; Webb et al., 2003). Only a limited number of surveys have been able to study the faint population (e.g., Smail et al., 1998). In our survey, after correcting for the gravitational lensing, we have been able to observe objects with sub-mJy fluxes and survey an area large enough for us to study the counts of the faint population. The preliminary number counts based on half of our survey are shown in Fig. 1. Tentatively we find a slope $\alpha = 1.5$, when assuming a power law $N(>S) \propto S^{-\alpha}$. This is comparable to other surveys.

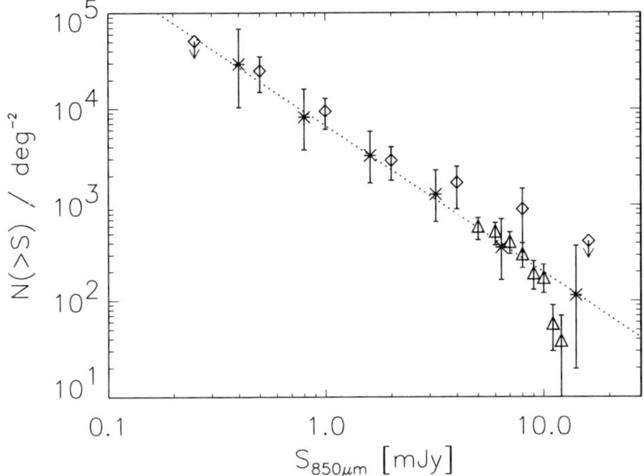

Figure 1. The cumulative $850\,\mu\mathrm{m}$ source counts based on half of our survey. The *asterisks* represents this work. For comparison the counts from Smail et al. (2002) (*diamonds*) and from Scott et al. (2002) (*triangles*) have been included. The *dotted line* shows the best fit to our data points.

4. Follow-up and Identifications

We have done follow-up observations with ISAAC at the VLT, obtaining deep Ks images (limiting magnitude $\sim 21.5\,\mathrm{mag}$ (Vega)). This has been combined with archival, deep optical data from HST and VLT, and when possible also with observations at other wavelengths (radio, mid-IR). A substantial fraction of the plausibly identified submm sources turn out to be very or extremely red objects (the EROs are often defined to have $I - K > 4$). Remarkably, in many cases, counterparts identified as red objects have close neighbours which are also red objects. Examples are shown in Fig. 2.

We finally point out two highlights of the project: a multiple imaged galaxy and a type-1 quasar.

The multiple imaged galaxy is found in the field of A2218 (van der Werf et al. *in prep*). Three submm sources were identified with three images of the same very red galaxy, SMM J16359+6612, detected both in optical and near-IR. The submm source is detected both at $850\,\mu\mathrm{m}$ and $450\,\mu\mathrm{m}$ with the same colour for all three images. The magnification factor in total for all three images is 40. The lensing corrected flux density is $f_{850} = 0.9\,\mathrm{mJy}$. Based on detailed lensing models the galaxy has an estimated redshift of 2.5. Its spectral energy distribution is similar to that of Arp220 and HR10, though about a factor 5 fainter. A quasar was identified in the field of A478. The submillimetre source, SMM J04135+10277, has $f_{850} = 25\,\mu\mathrm{m}$ and $f_{450} = 55\,\mu\mathrm{m}$ – these fluxes should be corrected with a magnification factor of 1.3. This is the brightest

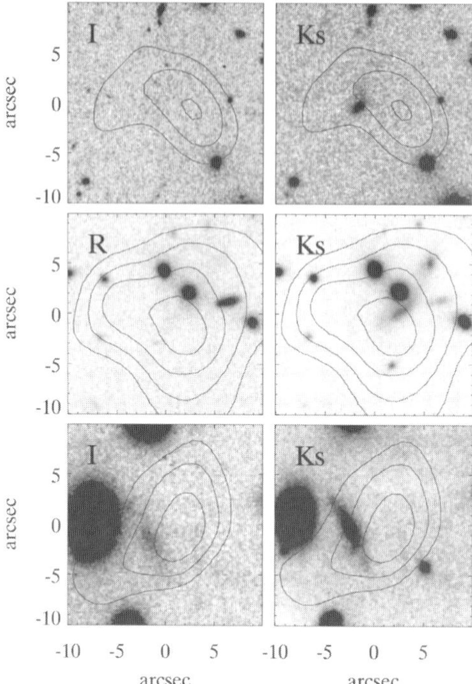

Figure 2 In this mosaic are shown three examples of submm sources with one or more very or extremely red objects nearby. The contours represent the $850\,\mu\text{m}$ signal-to-noise levels (3,4,5,6 and 7).

source in the entire survey and also bright for other similar surveys. The quasar has a redshift of 2.837. CO has been detected at $z = 2.84$ (Hainline et al, in prep). An analysis of the SED suggests that the quasar has a larger submm/far-IR emission than other quasars. The IR luminosity is $L_{IR} = 3 \cdot 10^{13}\,\mathrm{L}_{\odot}$. An analysis of the optical spectrum suggests that the viewing angle from the relativistic beam is large, which can have consequences for the identification of other submm sources (Knudsen et al., 2003).

References

Blain, A.W., Smail, I., Ivison, R.J. & Kneib J.-P., 1999, MNRAS, 302, 632

Cayon, L., et al., 2000, MNRAS, 315, 757

Fixsen D.J., et al., 1998, ApJ, 508, 123

Holland, W.S., et al., 1999, MNRAS, 303, 659

Knudsen, K.K., van der Werf, P.P. & Jaffe, W., 2003, accepted by A&A, astro-ph/0308438

Madau, P. & Pozzetti, L., 2000, MNRAS, 312, L9

Scott, S.E., et al., 2002, MNRAS, 331, 817

Smail I., Ivison, R.J., Blain, A.W. & Kneib, J.-P., 1998, ApJL, 507, 21

Smail I., Ivison, R.J., Blain, A.W. & Kneib, J.-P., 2002, MNRAS, 331, 495

Webb, T.M., et al., 2003, ApJ, 587, 41

INFRA-RED WAVELENGTHS

GALAXY EVOLUTION IN THE IR AND THE PROMISE OF SIRTF (*INVITED*)

Carol J. Lonsdale

Abstract The launch of SIRTF on August 25, 2003, opens an exciting new era for the infrared. Building on the legacy of IRAS, COBE & ISO, SIRTF will image the sky from $3.6-160\mu$m in tiered surveys, from wide ($\sim$80 sq. deg.) reaching z$\sim$1 in depth for L^*_{FIR} galaxies, to small, deep, confusion-limited surveys. SIRTF will measure the accumulation of stellar mass at high redshift, and the evolution of dusty systems (disks, starbursts & AGN) since z$\sim$4, on size scales up to several hundred Mpc. The next decade will also see two major all-sky IR surveys, ASTRO-F and WISE (Wide-field Infrared Survey Explorer), and the launch of Herschel and Planck.

Keywords: Galaxies, infrared; Galaxies, evolution; Instruments, SIRTF

1. The Evolving Dusty Universe

A large fraction of the emission of stars and AGN in galaxies is absorbed and re-emitted by dust in the thermal infrared. The Cosmic Infrared Background (CIB) measured by COBE indicates that 50% or more of the total integrated light in the Universe, exclusive of the CMB, emerges longward of 1μm Puget *et al* 1996; Hauser & Dwek 2001. Much higher IR emission fractions than this can occur in luminous starbursts and AGN, and the UV-optical emission of many IR-luminous galaxies can come from completely different regions than the IR, emphasizing the importance of ISM geometry, extremely high optical depth in young star-forming regions & highly embedded AGN. Thus if much of the star formation history of galaxies occurred in bursts we can only assess it properly in the IR/submm.

ISO undertook several extragalactic surveys from 6.7 to 170μm (see reviews by Genzel & Cesarsky 2000; Franceschini et al. 2001), from the large-shallow ELAIS survey Oliver et al. 2000; Rowan-Robinson et al. 2003 to small-deep surveys reaching $\sim 50\mu$Jy at 15μm and $\sim 15\mu$Jy at 6.7μm Altieri et al. 1999; Oliver et al. 2003; Sato et al. 2003. Deep ISOCAM surveys are dominated by a peak in the number counts near 0.3mJy, which is caused by the 7.7μm PAH feature passing though the filter at redshifts of $\sim 0.8-1$ coupled

M. Plionis (ed.), Multiwavelength Cosmology, 155-163.

with declining counts to fainter fluxes due to a strong k-correction Elbaz et al. 2002. About 70-80% of the mid-IR CIB is resolved by the deepest ISO 15μm surveys, and an M82-like spectral energy distribution (SED) at z$\sim$0.8$-$1 can match the broader CIB in shape quite well, except longward of $\sim$200μm where higher redshift and/or cooler sources are also required. Indeed, luminous IR/submillimeter sources at significantly higher redshifts dominate the submm CIB (D. Hughes, this volume).

ISOPHOT 60μm to 170μm observations have been limited to brighter fluxes, and resolve only 5-10% of the CIB directly Kawara et al. 1998; Dole et al. 2001. Redshifts and luminosities are in the range z<0.3; $L_{IR}\sim10^{9-11}L_{\odot}$, with some relatively cool IR SEDs Serjeant et al. 2001; Kawara et al. 1998; Chapman et al. 2002. Some higher redshift sources are found, up to z=1.2 with $L_{IR}\sim10^{11-12}L_{\odot}$.

Several ISO studies conclude that AGN contribute 10$-$20% of the CIB Fadda et al. 2002; Franceschini et al. 2002; Polletta et al. 2003. The fraction of IR-luminous systems hosting AGN is of much interest in light of the Chandra and XMM discoveries that the hard X-ray background is dominated by obscured AGN which peak at redshift $\sim$0.7-1 Hasinger et al. 2003.

At still longer wavelengths very luminous ULIRGs at redshifts $\sim$2-4 are surprisingly numerous (D. Hughes, this volume). This population is of fundamental importance to understanding galaxy formation and evolution, yet hierarchical structure formation models cannot explain them Somerville et al. 2001 without resort to dramatic assumptions more in keeping with monolithic formation scenarios Granato et al. 2002; Balland et al. 2003. AGN could also contribute directly to the luminosity of these systems, decreasing the conflict with hierarchical models; indeed many submm sources with spectroscopy available do show evidence of embedded AGN Chapman et al. 2003a, although this does not necessarily mean that the AGN is a dominant power source for the IR emission.

Taken together, the IRAS, ISO, COBE and submm surveys available to date show there has been a rapid decline in the infrared energy density of the Universe since z$\sim$1, probably steeper than that seen in the UV-NIR Rowan-Robinson et al. 1997. Locally the IR Universe is dominated by quiescent disk star formation and modest starbursts, and ULIRGs are very rare. Most evolution models for the IR/submm number counts require strong luminosity evolution to a peak at z=1 or higher Chary & Elbaz 2001; Franceschini et al. 2001; Xu et al. 2003; Lagache et al. 2003, with a marked shift towards the importance of luminous ULIRGs by z$\sim$2. However the discovery of some relatively cool sources in the deep ISOPHOT surveys is of considerable note Chapman et al. 2002, since dust masses and temperatures in submm surveys are poorly constrained due to limited SED information, and an overestimation of the dust temperature will lead to an over-estimation of the total IR/submm luminosity, and hence the derived star formation rate, by a factor of 10 or more Kaviani et al. 2003; Rowan-Robinson et al. 2001; Chapman et al. 2003b; Efstathiou et al. 2003. ISOPHOT has found other evidence for cool disk emission: about half the nearby galaxies in the ISOPHOT Serendipity Survey are colder than IRAS, with its longest band at 100μm, could determine Stickel et al. 2003, while local submm surveys show a similar result Dunne & Eales 2001. Interestingly, large disks at z$\sim$2 have recently been reported in deep NICMOS images Labbe et al. 2003. This important issue can be addressed directly by far-IR color and morphological analyses of deep SIRTF samples.

2. Stellar Mass Accumulation

Studies of the accumulated mass in stars have been limited to the K-band, and thus the rest-frame optical at high redshift. The ideal rest observing window is $\sim$1μm, where the SED of low mass stars peaks and where extinction effects are low; SIRTF's IRAC camera was in part optimized to directly observe evolved stellar populations at their SED peak to high redshift Simpson & Eisenhardt 1999.

Optical-NIR determinations of the stellar accumulation history are limited to small fields and suffer the effects of cosmic variance. The global increase in stellar mass since z$\sim$3 agrees reasonably well with predictions of CDM hierarchical merging models Dickinson et al. 2003a; Rudnick et al. 2003, consistent with $\sim$50% of the stellar mass in the Universe having formed by z$\sim$1, most of it since z$\sim$3. Other studies find evidence that some massive systems have formed earlier in a more monolithic fashion, based on numbers of luminous K-selected galaxies and high clustering levels of faint EROs Cimatti 2003; Pozzetti et al. 2003; Daddi et al. 2003. A common speculation is that the high SFR ULIRGs at z=2-4 are the progenitors of these massive clustered spheroids, which have already formed most of their stars well before z=1. This is not to say that the CDM-based hierarchical structure formation paradigm is seriously challenged, but that more sophisticated physics is likely required to accelerate the formation of massive systems at early times Somerville et al. 2003; Cimatti 2003.

The uncertainties in these optical/NIR studies are large, due to the short wavelength and uncertainties in the modeling, including the IMF and its universality, extinction, metallicity, starburst lifetimes, and to selection effects, including missing high M/L and low mass & low surface brightness systems, and field-to-field variance. Improvements that might be expected in mass estimation from the 3$-$8μm SIRTF surveys are illustrated by Berta et al. (2003)

3. The Promise of SIRTF and Future Missions

Cosmogonists would like to fully understand the assembly history of galaxies, how the Hubble sequence came into being, and how galaxies of all types and activities came to be distributed in space the way we now see them, all in relation to the underlying dark matter distribution and the cosmological model. We would also like to understand the role of AGN in galaxy evolution. The key questions for IR observers are the roles of starbursts, cool quiescent (disk) star formation & AGN as a function of redshift and environment. This is especially relevant for the high redshift ULIRG population, where we would like to know the SED shapes and luminosities much more accurately, the IMF and burst timescales for bursting systems, AGN contributions to the luminosity, and connections with other high redshift objects: Lyman Break Galaxies (LBGs), radio galaxies and quasars, and absorption systems. Observational goals to address these questions include fully resolving the CIB; statistically robust luminosity functions and clustering measurements in many volume cells across time & variations in the matter density field; stellar masses for large samples out to $z>4$ from 1-8μm photometry; X-ray to radio SEDs for subsamples of all populations at all redshifts; and deep images of subsamples to determine multi-wavelength morphology.

Table 1. SIRTF Extragalactic Surveys

Survey	*Theoretical sensitivity, 5σ, without confusion*							*Area*
	3.6	*4.5*	*5.8*	*8*	*24*	*70*	*160μm*	
	μJy	*μJy*	*μJy*	*μJy*	*mJy*	*mJy*	*mJy*	*sq. deg.*
SWIRE	7.3	9.7	27.5	32.5	0.45	2.75	17.5	65 (7 fields)
FLS	12	15	43	49	0.46	3.0	24.7	5
GTO-wide	8	11	32	38	0.5	3.0	25	9
GTO-deep	2	4	11	16	0.11	1.0	15	2.5 (6 fields)
GTO-ultra	0.6	1.0	3.5	4.8	0.035	0.7	-	0.3 (0.02 MIPS)
GOODS-deep	0.15	0.25	0.8	1.2	0.02	-	-	0.08 (2 fields)
GOODS-ultra	0.1	0.15	0.45	0.65	-	-	-	0.014
Confusion	3	3	3	3	0.1	4	40	

SIRTF Gallagher 2003 launched on Aug 25, 2003, into an Earth-trailing orbit, carrying a helium-cooled 85-cm telescope and three instruments: the Multiband Imaging Photometer for SIRTF, MIPS, (PI G. Rieke; Young 2003), the Infrared Array Camera, IRAC, (PI G. Fazio; Hora et al. 2003) and the Infrared Spectrometer, IRS, (PI J. Houck; Roellig 1998). SIRTF is expected to have a 3-5 year lifetime. The IRAC imaging array sizes are 256$\times$256, giving FOVs of $5'\times5'$ with pixel size of 1.2$''$. The MIPS FOVs are also $5'\times5'$ at 24

& 70μm, and $0.5' \times 5'$ at 160μ, while the physical array sizes are 128×128, 32×32 & 2×20 respectively, with pixel sizes 2.6″, 9.9″ & 16″. IRS also has 15μm imaging capability, using the 33×45 pixel peak-up array, with an FOV of $1' \times 1.2'$ and pixel size of 1.8″.

The SIRTF extragalactic surveys will directly measure far-IR luminosities of the major galaxy populations to z~0.5-1, when the ~80-120μm peak of the FIR SED of typical starbursts/disk systems passes out of the 70μm & 160μm filters. Mid-IR emission will be traced to greater redshifts, and may be used to a certain extent as an L_{IRbol} estimator Papovich & Bell 2002. SIRTF will excel, in particular, as an AGN "machine", due to its superb sensitivity at 8, 24 & 70μm coupled with the mid-IR rest wavelength peak of AGN dust tori SEDs. At 160μm, SIRTF will chart new ground since its sensitivity at this wavelength is unprecedented.

Details of the major SIRTF surveys are given in Table 1: the First Look Survey (FLS), the Guaranteed Time Observer (GTO) surveys and 2 of the 3 extragalactic Legacy surveys: The SIRTF Wide-area InfraRed Extragalactic survey, SWIRE, (PI C. Lonsdale; Lonsdale et al. 2003) and the Great Observatories Origins Deep Survey, GOODS, (PI M. Dickinson; Dickinson et al. 2003b). Also in Table 1 is an estimate of the likely confusion noise in each band from the integrated effect of faint extragalactic source counts, based on several different count models and different methods for estimating confusion noise Vaisanen et al. 2001; Dole et al. 2003; Shupe et al. 2003. The values quoted are generous: the confusion noise may be more serious, especially taking into account realistic pixel sampling and source extraction methods Shupe et al. 2003. All the surveys are expected to be confusion-limited at 160μm, and possibly also 70μm. The shortest IRAC bands may become confusion-limited for the GTO deep survey, and the remaining bands may become confused before the full depth of the GTO-ultra and GOODS surveys is reached. Thus SIRTF confusion limits will be reached in all bands in one survey or another. The fraction of the CIB that might be resolved by SIRTF is about 20% at 160μm, ~50% at 70μm and perhaps ~70% at 24μm Lagache et al. 2003.

All the SIRTF survey fields have been selected to optimize coverage of previous extragalactic surveys and to minimize noise from cirrus and zodiacal emission. Even the shallowest tier (the GTO wide survey in the Bootes region, the FLS survey and the SWIRE Legacy survey, which together cover ~80 sq. deg. in all IRAC & MIPS imaging bands) reaches cosmological distance owing to the superb sensitivity of SIRTF. The goals of these wide surveys are to map dusty galaxies and evolved stellar populations to z>1 across scales of 100 Mpc or more, and to determine luminosity functions in 100s of independent volume cells, detecting over 10^6galaxies and AGN. Of particular importance is the ability to measure clustering accurately on many size scales, in relation

to the dark matter density field. The suite of 9 different fields provides robust protection
against cosmic variance, from local to cosmological distances. An intermediate tier is provided by coordinated IRAC and MIPS GTO deep surveys in 6 different regions, each $25' \times 60'$ in size except the 2 deg long Groth strip which is extended in order to sample several correlation lengths at z$\sim$1. The deepest tier comes from the GTO ultradeep survey and the GOODS Legacy survey, the ultimate depth of which depend on the confusion noise they confront. GOODS will not observe at 70 or 160μm, and at 24μm is contingent on performance relative to the GTO ultradeep program. The goals of GOODS are to determine stellar masses from the IRAC data to $\sim$1 mag. fainter than L^* at z$\sim$3, and to reach L^* at z$\sim$5. At 24μm, GOODS aims to detect the 7.7μm PAH feature at z$\sim$2 to the same luminosity as the ISO 15μm data achieved in the HDF at z$\sim$1, and to detect "ordinary" Lyman Break Galaxies at z$\sim$2.5 Dickinson et al. 2003b. GOODS observes two separate fields (HDFN & Chandra DFS) to guard against cosmic variance.

Extensive X-ray to radio complementary programs are underway for all SIRTF surveys, necessary for fully constraining the accretion and star formation histories of the Universe, and for optical/NIR identifications, photometric redshifts, morphologies and environments. However current X-ray, NIR and radio sensitivities can match neither the width nor the depth of the shallowest/deepest SIRTF surveys in reasonable integration times, and some faint luminous IR sources will also remain unidentified in the optical. Thus much multiwavelength follow-up of SIRTF samples will await future large observatories, from JWST to ALMA (Atacama Large Milimeter Array) & SKA (Sq. Kilometer Array).

The third extragalactic Legacy program, SINGS (SIRTF Infrared Nearby Galaxy Survey; PI R. Kennicutt; Kennicutt et al. 2003), will not directly observe the distant Universe, but will provide vital local "anchor points" for it by characterizing star formation in 75 local galaxies, tracing the processing of energy from young stars through the dusty ISM. Numerous smaller GTO programs will also contribute detailed studies of other known galaxies and AGN, from the local Universe to the most distant known objects; the best resource for exploring SIRTF GTO program is the Reserved Objects Catalog (http://sirtf.caltech.edu/SSC/roc/).

Beyond SIRTF we can look forward to ASTRO-F & WISE, all-sky mapping missions in the near- to far-IR. ASTRO-F is a Japanese-led mission, observing from 2-175μm and due to launch in Spring 2005 (S. Serjeant, this volume). The Wide-field Infrared Survey Explorer, WISE (E. Wright, PI), is currently in extended review at NASA for launch about 2008. It will survey the entire sky at 3.5, 4.7, 12 & 23μm with 3-6 orders of magnitude better sensitivity than previous allsky surveys. Carrying a cold 50-cm telescope in sun-synchronous

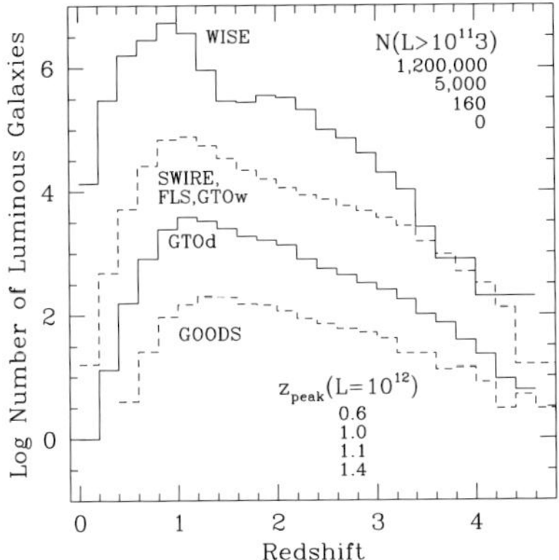

Figure 1. Estimated number of galaxies with $L_{IR} > 10^{11.75}$ in the total area covered by the SIRTF surveys, and the future Explorer WISE, taking into account estimated confusion limits. The number with $L>10^{13}$, and the peak redshift for $10^{12}L_\odot$ systems, are listed in the same sequence as the histograms. Based on the simulation of Xu et al. (2003).

orbit it will deliver over 10^6 images over the entire sky and catalogs of $\sim 5\times10^8$ objects. WISE will map the local Universe to z$\sim$0.6, and has enough volume to detect ULIRGs 16 times more luminous than SWIRE to z$\sim$4 (Fig. 1). Herschel and Planck launch together towards the end of the decade. Herschel, with a 3.5-m mirror Pilbratt 2001, will provide imaging and spectroscopy, reaching submJy photometric sensitivities for deep surveys at 250, 350 & 500μm with SPIRE, and at 70, 110 & 170μm with PACS. Given the larger mirror size compared to SIRTF, Herschel will be able to provide similar sensitivities and resolutions as SIRTF at complementary longer wavelengths. PACS may resolve $\sim$80% of the CIB at 100μm, determining with great accuracy the nature of the dominant population Lagache et al. 2003. SPIRE will not be able to resolve a large fraction of the longer wavelength background, but will provide sensitive measurements in the rest-frame far-IR to z$\sim$3 or higher. The brighter Planck all sky survey at 350μm-20GHz (http://sci.eso.int/planck) will be very sensitive to rare and luminous submm/mm sources, especially cool dusty objects, extending the IRAS, ASTRO-F & WISE all-sky views of the Universe to much longer wavelengths.

4. Summary

ISO resolved a large fraction of the mid-IR CIB, showing it to be dominated by LIRGs at z$\sim$0.8, while at longer wavelengths, luminous, rare, distant systems are important. Star formation is the dominant energy source, with important contributions from heavily dust-obscured AGN. Dust temperatures are not well constrained at high redshift, raising the possible importance of quiescent star formation in cool disks, *vs.* starbursts, at early times. NIR studies of evolved systems also show evidence for both slow hierarchical assembly since z$\sim$3, and more rapid formation of massive systems at early times. Each of the SIRTF surveys will trace both evolved stellar populations (with IRAC) and dusty systems (with MIPS) over matched volumes, providing a supreme opportunity to connect the evolution of these populations in time and space. The shallowest SIRTF surveys will have superb volume and area coverage, sensitive to structures on 100s of Mpc and ULIRGs to z$\sim$4, while the deepest studies will trace L^* systems to z$\sim$2.5, and reach the confusion limits in all SIRTF imaging bands: 3.6 to 160μm. SIRTF may resolve about 70, 50 and 20% of the CIB at 24, 70 & 160μm respectively.

Acknowledgments

I thank the SWIRE team, M. Dickinson, H. Dole, D. Farrah, G. Fazio, K. Ganga, R. Kennicutt, H. Smith, M. Pahre & M. del Carmen Polletta for slides and discussion. This work is supported by NASA through the SIRTF Legacy program under contract 1407 with the Jet Propulsion Laboratory.

References

Altieri, B. et al. 1999, A&A, 343, 65
Balland, C., Devriendt, J. & Silk, J. 2003, MNRAS, 343, 107
Berta, S. et al. 2003, A&A, submitted
Chapman, S. C., Blain, A. W., Ivison, R. J. & Smail, I. R. 2003a, Nature, 422, 695
Chapman, S. C., Helou, G., Lewis, G. F., Dale, D. A. 2003b, ApJ, 588, 186
Chapman, S. C. et al. 2002, ApJ, 573, 66
Chary, R. & Elbaz D. 2001, ApJ, 556, 562
Cimatti, A. 2003, Ap&SS, 285, 231
Daddi, E. et al. 2003, ApJ, 588, 50
Dickinson, M., Papovich, C., Ferguson, H. C. & Budavári, T. 2003a, ApJ, 587, 25
Dickinson, M. et al. 2003b, in *The Mass of Galaxies at Low and High Redshift*, ESO, p324.
Dole, H., Lagache, G. & Puget, J.-L. 2003, ApJ, 585, 617
Dole, H., Gispert, R., Lagache, G., Puget, J.-L., Bouchet, F.R. et al. 2001, A&A, 372, 364
Dunne, L. & Eales, S. 2001, MNRAS, 327, 697
Efstathiou A. & Rowan-Robinson, M. 2003, MNRAS, 343, 293
Elbaz, D. et al. 2002, A&A, 384, 848
Elbaz, D. et al., 1999, A&A, 351, L37
Fadda, D. et al. 2002, A&A, 383, 838
Franceschini, A. et al. 2003, A&A, 403, 501
Franceschini, A. et al. 2002, ApJ, 568, 470
Franceschini, A. et al. 2001, A&A, 378, 1
Gallagher, D. 2003, SPIE, 4850, 17
Genzel, R. & Cesarsky, C. 2000, ARAA, 38, 761
Granato, G. L. et al. 2002, Ap&SS, 281, 497

Hasinger, G. 2003, AIP Conf. Proc., 666, 227
Hauser, M. G. & Dwek E. 2001, ARAA, 39, 249
Hora, J. et al. 2003, SPIE, 4850, 83
Kaviani A., Haehnelt, M. G. & Kauffmann, G. 2003, MNRAS, 340, 739
Kawara, K. et al. 1998, A&A, 336, L9
Kennicutt, R. et al. 2003, PASP, 115, 928
Labbe I., et al. 2003, ApJ, 591, 95
Lagache, G., Dole, H., Puget, J.-L. 2003, MNRAS, 338, 555
Lonsdale, C. J., et al. 2003, PASP, 115, 897
Rowan-Robinson, M. et al. 2003, MNRAS, submitted
Rowan-Robinson, M. 2001, ApJ, 549, 745
Rowan-Robinson, M. et al. 1997, MNRAS, 289, 490
Oliver, S. et al. 2003, MNRAS, 332, 536
Oliver, S. et al. 2000, MNRAS, 316, 749
Papovich, C. & Bell, E. 2002, ApJ, 579, L1
Pilbratt, G. 2001, ESA SP0160, 13; http://sci.eso.int/herschel
Polletta M., Lonsdale, C.J., Xu, C. K. & Wilkes, B. J. 2003, ApJ, submitted
Pozzetti, L. et al. 2003, A&A, 402, 837
Puget, J. L., et al. 1996, A&A, 308, L5.
Roellig, T. et al. 1998, SPIE, 3354, 1192
Rudnick, G. et al. 2003, ApJ, in press
Sato, Y. et al. 2003, A&A, 405, 833
Serjeant, S. et al. 2001, MNRAS, 322, 262
Serjeant, S. et al. 2000, MNRAS, 316, 768
Shupe, D. et al. 2003, in preparation
Simpson C. & Eisenhardt, P. 1999, PASP, 111, 691
Somerville, R. et al. 2003, ApJ, submitted
Somerville R. S. et al. 2001, MNRAS, 320, 289
Stickel, M., Lemke, D., Klaas, U. & Egner, S. 2003, IAUS, 216, 145
Vaisanen, P., Tollestrup, E. & Fazio, G. 2001, MNRAS, 325, 1241
Xu, C. K. et al. 2003, ApJ, 587, 90
Young, E. et al. 2003, SPIE, 4850, 98

SWIRE: THE SIRTF WIDE-AREA INFRARED EXTRAGALACTIC SURVEY

Harding E. Smith & Carol J. Lonsdale

Abstract The largest of the SIRTF Legacy programs, SWIRE will survey 65 sq. deg. in seven high latitude fields selected to be the best wide low-extinction windows into the extragalactic sky. SWIRE will detect millions of spheroids, disks and starburst galaxies to $z > 3$ and will map L* and brighter systems on scales up to 150 Mpc at z∼0.5-1. It will also detect $\sim 10^4$ low extinction AGN and large numbers of obscured AGN. An extensive program of complementary observations is underway. The data are non-proprietary and will be made available beginning in Spring 2004.

Keywords: Galaxies, infrared; Galaxies, evolution; Instruments, SIRTF

1. Introduction

The SIRTF Wide-area InfraRed Extragalactic Survey (*SWIRE*, Dr. Carol Lonsdale, P.I.) is the largest of the six SIRTF Legacy Surveys (900 hours), surveying approximately 65 square degrees in all 7 SIRTF imaging bands. A current description of the SWIRE Survey is given by Lonsdale *et al.* (2003) and on the SWIRE WebPages: *http://www.ipac.caltech.edu/SWIRE*. Table 1 lists the Survey sensitivities.

The Survey will cover seven high-latitude fields, selected to be the most transparent, lowest-background wide-area (> 5sq. deg.) fields in the sky. The fields, covering between 5 and 15 sq. deg. include previously well-known IR extragalactic survey fields (e.g. Lockman Hole and 3 ELAIS ISO Survey Fields) and x-ray fields (Chandra Deep South and XMM Large Scale Survey) are shown in Table 2.

The SWIRE science goal is to enable fundamental studies of galaxy evolution in the infrared for $0.5 < z < 3$:

⊙ evolution of star-forming and passively evolving galaxies in the context of structure formation and environment.

⊙ spatial distribution and clustering of evolved galaxies, Starbursts, & AGN.

⊙ the evolutionary relationship between galaxies and AGN and the contribution of AGN accretion energy to the cosmic backgrounds.

M. Plionis (ed.), Multiwavelength Cosmology, 165-168.

Table 1. SWIRE Sensitivity Limits (Pre-launch est. 5σ; w/o confusion)

IRAC			MIPS		
λ	Sensitivity	Resolution	λ	Sensitivity	Resolution
3.6μm	7.3μJy	$0.9''$	24μm	0.45mJy	$5.5''$
4.5μm	9.7μJy	$1.2''$	70μm	2.75mJy	$16''$
5.8μm	27.5μJy	$1.5''$	160μm	17.5mJy	$36''$
8.0μm	32.5μJy	$1.8''$			

Table 2. SWIRE Survey Fields

Field	Center (J2000) RA	Center (J2000) Dec	Area (sq deg)	Background (100μm; MJy/Sr)	Prob. Obs. Date
ELAIS S1	$00^h38^m\ 30^s$	$-44^\circ\ 00'$	14.6	0.42	2004 Jun/Jul
XMM-LSS	$02^h21^m\ 20^s$	$-04^\circ\ 30'$	9.2	1.3	2004 Jul/Aug
Chandra-S	$03^h32^m\ 00^s$	$-28^\circ\ 16'$	7.8	0.46	2004 Aug/Sep
Lockman	$10^h45^m\ 00^s$	$+58^\circ\ 00'$	14.4	0.38	2004 Apr/May
Lonsdale	$14^h41^m\ 00^s$	$+59^\circ\ 25'$	6.9	0.47	2005 Feb
ELAIS N1	$16^h11^m\ 00^s$	$+55^\circ\ 00'$	9.2	0.44	2004 Jan
ELAIS N2	$16^h36^m\ 48^s$	$+41^\circ\ 02'$	4.7	0.42	2004 Jun/Jul

Galaxy evolution models which match the IRAS/ISO galaxy counts at all wavelengths from 7–100μm as well as the CIRB (Xu *et al* 2001, 2003) predict that SWIRE will detect of the order of 2 million galaxies — spheroids and evolved stellar systems with IRAC, and active star-forming systems with MIPS. SWIRE will also detect about 25,000 classical AGN, and an unknown number, perhaps several times as many, dust-enshrouded AGN.

Recent estimates of the "Universal Star-formation History (*SFH*)" suggest that the bulk of cosmic evolution occurs between redshifts, $0.5 < z < 3$, the redshift interval for which SWIRE is optimized. The median SWIRE redshift for starbursts is predicted to be, $\langle z\rangle \sim 1$, where many estimates find a peak in the *SFH*; luminous infrared galaxies will be detected by *SWIRE* out to $z \sim 3$. Previous estimates of the *SFH* have varying, frequently large, and uncertain corrections for extinction. *SWIRE* will measure the star-formation rates and modes as a function of redshift and environment over this critical epoch.

A key element in the *SWIRE* Survey design is to enable galaxy evolution studies in the context of large-scale structure/environment. One of the *SWIRE* Survey fields covers the deep survey areas of the XMM-LSS Survey (Pierre *et al.* 2003) so that the infrared galaxy census may be directly tied to the presence of rich x-ray clusters to $z > 1$. *SWIRE* will sample several hundred, 100 Mpc scale co-moving volume cells enabling a variety of large-scale structure measures from correlation functions, power spectra, and counts-in-cells to direct comparison with model calculations. SWIRE's measures of the star-formation as a function of environment will be important input for CDM simulations which have been exceedingly successful in simulating the development of structure in the early Universe, but perhaps less so in simulating galaxy evolution within that structure owing to the complexity of the physics of star formation (*e.g.* Kay *et al.* 2002).

The similarities of many AGN SEDs in the mid-far infrared suggests that *SWIRE* will be less biased with respect to AGN types and ages than many other surveys, enabling a more complete census of AGN out to $z > 1$. Although the detection rates should be unbiased, the similarity between the SEDs of obscured AGN to those of Starbursts, and the extreme optical depths will make *identifying* the obscured AGN population very challenging. Low-frequency radio surveys will, of course, identify radio-loud AGN, but these make up only 10–15% of the AGN population. For this reason the XMM-LSS Survey, along with current and planned deeper surveys in hard x-rays will be vital to identifying *SWIRE* AGN.

2. Supporting Observations

An aggressive program of ground-based optical, NIR and radio observations is planned in support of the SWIRE Survey and we are actively pursuing other programs with HST, Chandra, XMM and GALEX. As already described Chandra and XMM Surveys will be important for discovering the obscured AGN population. SWIRE has entered into cooperation with the GALEX team so that the SWIRE fields will be included in the GALEX Deep Survey.

The SWIRE Optical-NIR goal is to obtain moderate-depth optical multi-band ($g' \sim 25.7$, $r' \sim 25$, $i' \sim 24$; Vega magnitudes, 5σ for a $2''$ galaxy) data for the entire Survey area. At these limits we expect to detect approximately $2/3$ of SWIRE sources detected by both MIPS and IRAC. The ELAIS N1, N2 fields have been imaged to somewhat shallower limits ($r' \sim 24$) as part of the INT Wide Field/ELAIS Surveys and have now been released (Rowan-Robinson *et al* 2003). Efforts continue to push deeper in EN1/2 in the optical and into the NIR as part of UK SWIRE-related Programs. An extensive observing program of ELAIS S1 is being undertaken at ESO. Optical imaging of the Lockman, Lonsdale and CDFS fields are being undertaken at KPNO and CTIO with the Mosaic cameras and at Palomar with the $200''$ Large Format Camera.

Two major SWIRE radio surveys are planned. The median 20cm flux density predicted for SWIRE Starburst galaxies is $\sim 50\mu$Jy — too faint to survey the entire area to this depth. We have therefore carried out a deep pencil-beam VLA Survey (F. Owen, PI) and are planning an extended shallow VLA survey (J. Condon, PI).

$\odot$ SWIRE Lockman Deep VLA Survey — 3μJy rms @ 20cm; $\alpha = 10^h\,46^m$ $\delta = +59^\circ\,01'$; $30'$ VLA primary beam is the deepest VLA field at 20cm.

⊙ Cosmic Windows VLA Survey — 50μJy rms @ 20cm in the combined fields of SWIRE, Galex and XMM-LSS which are accessible to the VLA. This Survey is being proposed for the next VLA large survey program.

Finally, in addition to the XMM-LSS coverage of our XMM field, plus individual existing Chandra/XMM fields in CDFS, Lockman, EN1 and EN2, SWIRE is carrying out a large Chandra survey of our Lockman Deep Radio-Optical Field: Chandra Lockman Survey — 630 ks (9×70ks ACIS pointings) is being devoted in Cycle 5 to deep ($\sigma \sim 10^{-15} erg\, cm^{-2} s^{-1}$).

Figure 1. Simulation of SWIRE images in 7 SIRTF bands (Xu et al. 2003) model; 10′ FOV).

3. Summary

The *SWIRE* Legacy Survey is a community Survey; the large dataset which is being accumulated reflects the synergies between the Legacy program and other major public surveys. With a couple of million galaxies and several tens of thousands of AGN, many with redshift estimates and SEDs from x-ray to radio, SWIRE's SIRTF database will be released to the community through the SIRTF Science Center (SSC) Archive with the ancillary data released through IPAC's Infrared Science Archive (IRSA). We hope that *SWIRE* will provide a rich datamine and will provide answers to many of the questions posed here. If you have projects that you would like to do with *SWIRE* data, visit the *SWIRE* WebPages and/or contact one of the team members.

References

Lonsdale, C. J., et al. 2003, PASP, 115, 897.

Kay, S., Pearce, F., Frenk, C., & Jenkins, A. 2002, MNRAS, 330, 113.

Pierre, M. *et al.* 2003, A& A, submitted (astro-ph/0305191)

Rowan-Robinson M., *et al.* 2003, MNRAS, submitted (astro-ph/0308283).

Xu, C., Lonsdale, C., Shupe, D., O'Linger, J., & Masci, F. 2001, ApJ, 562, 179.

Xu, C. K. *et al.* 2003, ApJ, 587, 90.

SUBARU/XMM-NEWTON DEEP SURVEY (SXDS)

K. Sekiguchi[1], M. Akiyama[1], H. Furusawa[1], C. Simpson[2], T. Takata[1], Y. Ueda[3], M.W. Watson[4], and The SXDS Team

[1]*Subaru Telescope, NAOJ, 650 North A'ohoku Place, Hilo, HI 96720, USA*
[2]*Department of Physics, University of Durham, Durham DH1 3LE, UK*
[3]*Institute of Space and Astronautical Science, Kanagawa 229-8510, Japan*
[4]*Department of Physics and Astronomy, University of Leicester LE1 7RH, UK*

Abstract The Subaru/XMM-Newton Deep Survey (SXDS) is a new deep Optical/X-ray survey which aim to take full advantage of the capabilities of the Subaru Telescope and the XMM-Newton. The SXDS is part of an ambitious project to obtain extensive multi-wavelength data across a $\sim 1.3\,\mathrm{deg}^2$ region of sky. Other wavelength data (infrared, submillimeter, and radio) are being pursued through various facilities. Combined with suitably deep images at other wavelengths the SXDS would provide an accurate census of the contents of the Universe without suffering from the biasing effects of large-scale structures. This paper describes the location, size and depth of this survey.

Keywords: surveys, large-scale structure, optical imaging, X-rays:galaxies, active galaxies

1. Introduction

We have initiated a multiwavelength wide area and deep survey to provide an accurate census of objects in the Universe hence advance our understanding of the formation and evolution of structure. The existing surveys are either wide area but relatively bright limiting magnitude or deep limiting magnitude but small area. Detailed simulations (Gaztanaga and Hughes 2001) indicates that we must cover a large enough area (> 1 deg^2) to avoid the effects by the cosmic variance. At the same time we must observe sufficiently deep so that we can accurately determine the global properties of different class of objects.

Our initial survey consists of optical imaging (B, R, i', z')with Suprime-Cam on Subaru Telescope (Miyazaki et al. 2002), and X-ray imaging with the EPIC camera (Strüder et al. 2001) on the ESA XMM-Newton satellite. The VLA 1.4GHz, the JCMT/SCUBA 850μm (as a part of SCUBA HAlf-Degree Extraglactic Survey (SHADES), which will survey two 0.25 deg^2 regions of sky to depths of 5 mJy r.m.s.), the optical spectroscopy using both the Faint Ob-

M. Plionis (ed.), Multiwavelength Cosmology, 169-172.

ject Camera And Spectrograph (FOCAS; Kashikawa et al. 2002) on Subaru Telescope and the 2-degree field spectrograph (2dF; Lewis et al. 2002) on the Anglo-Australian Telescope and some near infrared observations using the Simultaneous InfraRed Imager for Unbiased Surveys (SIRIUS; Nagayama et al. 2001) are being pursued. The SXDS field will be imaged in the U and V bands to limiting magnitudes of ~ 27 by the VLT/VIMOS as an ESO Public Imaging Survey. The UK Infrared Deep Sky Survey (UKIDSS) Ultra-Deep Survey (UDS; Adamson 2001) will use the UKIRT Wide-Field Camera (WFCAM; Casali et al. 2001) to survey an area 0.77 deg^2 in the center of the SXDS field to depths of $J = 25$, $H = 24$, $K = 23$. The *SIRTF* Wide-area InfraRed Extragalactic Survey (SWIRE; Lonsdale 2001; Franceschini & Lonsdale 2003) will include a $3 \times 3\,\mathrm{deg}^2$ region containing the entire SXDS field, with both IRAC (Fazio et al. 1998) and MIPS (Heim et al. 1998). Observations are also planned with the Balloon-borne Large Aperture Submillimeter Telescope (BLAST; Scott et al. 2001), which will operate at 350, 450, and 750μm. The combination of these observations will produce an unprecedented study of a large enough volume of the Universe to not be subject to cosmic variance.

2. The survey location

Our approach makes use of the wide range of currently available observational facilities to observe the survey area over the widest possible range of wavelength. We therefore chosen the survey field would be accessible and suitable for observations at all wavelengths. To permit the deepest possible follow-up observations at other wavelengths, the field must have a low Galactic neutral hydrogen column density and low Galactic reddening, and should also be relatively free of bright stars, X-ray sources, and radio sources and should have low cirrus emission. The field we chose is centered at J0218−05; its properties are described in detail in Table 1.

Table 1. Properties of the SXDS field.

Equatorial coordinates (J2000.0)	$02^h18^m00^s$	$-05°00'00''$
Galactic coordinates	$169°45'32''$	$-59°45'07''$
Ecliptic coordinates	$30°28'12''$	$-17°43'22''$
Galactic reddening	$E(B-V) < 0.04$	
Neutral hydrogen column	$N_{\mathrm{H\,I}} < 2.23 \times 10^{20}\,\mathrm{cm}^{-2}$	
Cirrus	$I_{100\,\mu\mathrm{m}} < 1.3\,\mathrm{MJy\,sr}^{-1}$	

3. The survey design

The X-ray observations were conducted as part of the XMM Survey Science Centre (SSC; Watson et al. 2001) guaranteed time. The X-ray survey consists of mosaic of 7 partially overlapping pointings of the EPIC-pn camera. The nominal exposure times of 100 ksec for the central field, and 50 ksec each for the six flanking fields. The image of the central field approaches the confusion limit for the XMM-Newton in the soft (0.5-2 ke V) band, while together these data form the largest contiguous area over which deep X-ray observations have been performed.

Most of the optical imaging observations were taken as Subaru Telescope's Observatory Key Project. Five Suprime-Cam fields, as indicated in Figure 1, were observed to cover a $\sim 1.3\,\mathrm{deg}^2$ region of sky. Our observations reach the limiting magnitudes at $B = 28.1$, $R = 27.5$, $i' = 27.4$ & $z' = 26.8$ (AB magnitude, 3σ detection with 2" aperture).

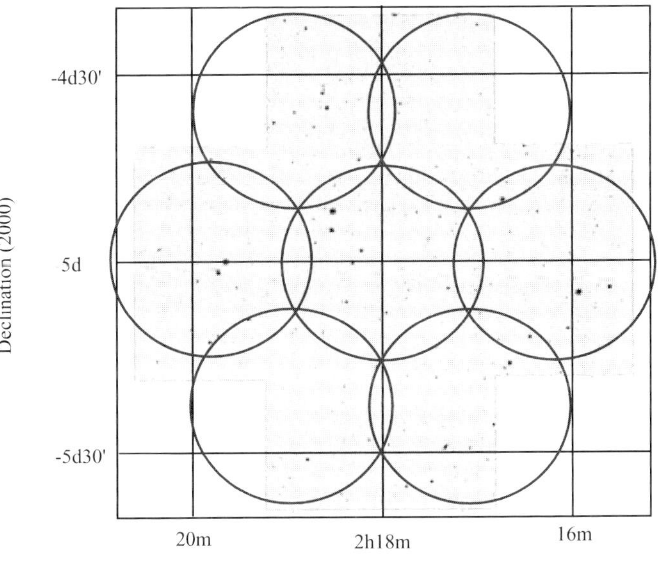

Figure 1. The SXDS field. Five pointings of the Suprime-Cam images were combined to cover a $\sim 1.3\,\mathrm{deg}^2$ field centered at R.A. = 02h18m, decl. = -05d. Circles represent 7 pointings of the XMM-Newton observation.

4. The survey output

The output of the SXDS will be an extensive deep multicolour database. It will contain $\sim 10^6$ extra-galactic objects, as well as a few 10^4 galactic objects. With $\sim 10^6$ galaxies, photometric redshifts will be especially important to quantify the evolution of galaxies, detect distant clusters, and select samples for detailed spectroscopic study. Combined with the deep near-infrared UKIDSS/UDS data, accurate photometric redshift determination will be possible for the vast majority of galaxies (Connolly et al. 1997, Massarotti et al. 2001, Rowan-Robinson 2003). A detailed description of the X-ray (Ueda et al. 2003), the optical imaging (Furusawa et al. 2003), the near-infrared imaging (Takata et al. 2003), and the VLA (Rawlings et al. 2003) data will be presented in future papers. These data will become available to the world astronomy community via the SXDS Web Site in the first half of 2004.

References

Adamson, A. J. 2001, in ASP 232: The New Era of Wide Field Astronomy, eds R. G. Clowes, A. J. Adamson, & G. E. Bromage (San Francisco: ASP), 364

Casali, M., et al. 2001, in ASP 232: The New Era of Wide Field Astronomy, eds R. G. Clowes, A. J. Adamson, & G. E. Bromage (San Francisco: ASP), 357

Fazio, G. G., et al. 1998, in Proc. SPIE 3354: Infrared Astronomical Instrumentation, ed. A. Fowler (Bellingham: SPIE), 1024

Franceschini, A., & Lonsdale, C. 2003, in The Mass of Galaxies at Low and High Redshift, eds R. Bender & A. Renzini (Berlin: Springer–Verlag), astro-ph/0202463

Furusawa, H. et al. 2003, in preparation

Gaztanaga, E., & Hughes, D. 2001, in Deep millimeter surveys: implications for galaxy formation and evolution, eds. J. D. Lowenthal & D. Hughes (Singapore: World Scientific Publishing), 131

Heim, G. B., et al. 1998, in Proc. SPIE 3354: Infrared Astronomical Instrumentation, ed. A. Fowler (Bellingham: SPIE), 985

Kashikawa, N., et al. 2002, PASJ, 54, 819

Lewis, I. J., et al. 2002, MNRAS, 333, 279

Lonsdale, C. J. 2001, BAAS, 198, 25.02

Massarotti, M., Iovino, A., Buzzoni, A., & Valls-Gabaud, D. 2001, A&A, 380, 425

Miyazaki, S., et al. 2002, PASJ, 54, 833

Nagayama, T., et al. 2001, in ASP 232: The New Era of Wide Field Astronomy, eds R. G. Clowes, A. J. Adamson, & G. E. Bromage (San Francisco: ASP), 389

Rawlings, S., Simpson, C., Ivison, R. J., & Sekiguchi, K. 2003, in preparation

Scott, D., et al. 2001, in ESA-SP 460: The Promise of the Herschel Space Observatory, eds G. L. Pilbratt et al. (Noordwijk: ESA), 305

Steidel, C. C., & Adelberger, K. L. 1999, in ASP 193: The Hy-Redshift Universe, eds A. J. Bunker & W. J. M. van Breugel (San Francisco: ASP), 462

Strüder, L., et al. 2001, A&A, 365, L18

Takata, T. et al. 2003, in preparation

Ueda, Y., et al. 2003, in preparation

Watson, M. G., et al. 2001, A&A, 365, L51

DUSTY STARBURSTS AND THE GROWTH OF COSMIC STRUCTURE

D. Elbaz, E. Moy
CEA Saclay/DSM/DAPNIA/Service d'Astrophysique
Orme des Merisiers, F-91191 Gif-sur-Yvette Cedex
France
delbaz@cea.fr

Abstract Dusty starbursts were more numerous around $z \sim 1$ than today and appear to be responsible for the majority of cosmic star formation over the Hubble time. We suggest that they represent a common phase within galaxies in general which is triggered by the growth of cosmic structure.

Keywords: Deep surveys - infrared - galaxy formation

1. Introduction

Dusty starbursts producing stars at a rate of about 50 $M_{\odot}$ yr^{-1} were very efficiently detected by ISOCAM onboard the Infrared Space Observatory (ISO) at 15 μm below $z \sim 1.3$ where the broad bump due to aromatic features in the mid-IR (PAHs, polycyclic aromatic features) remains within the ISOCAM filter. These luminous infrared (LIR) galaxies with infrared luminosities greater than $L_{\rm IR}(8 - 1000\,\mu{\rm m}) = 10^{11}$ $L_{\odot}$ were producing more than thirty times their present-day comoving infrared luminosity density at $z \sim 1$ than today, when it is only a few percent of the bolometric luminosity radiated by galaxies in the optical to near-infrared (Elbaz *et al.* 2002, Chary & Elbaz 2001). More distant LIR galaxies were detected in the sub-millimeter with SCUBA (Chapman et al. 2003 and references therein), with IR luminosities of several 10^{12} $L_{\odot}$ due to sensitivity limits, but their large number density confirms the same trend as observed with ISO, i.e. that the bulk of the cosmic history of star formation was dominated by dusty star formation and can only be derived when carefully accounting for this dusty starbursting phase taking place in most if not all galaxies. This is also suggested by the fact that these galaxies contribute to about two thirds of the cosmic infrared background (CIRB, Elbaz et al. 2002), which contains about half or more of the total energy radiated by galaxies through the Hubble time, and dominate the cosmic history of star for-

M. Plionis (ed.), Multiwavelength Cosmology, 173-176.

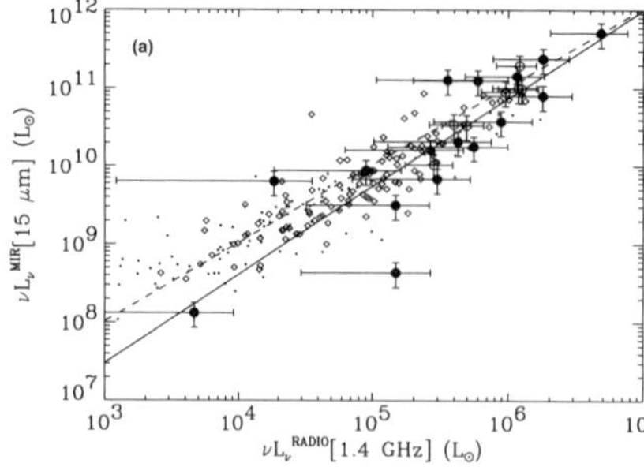

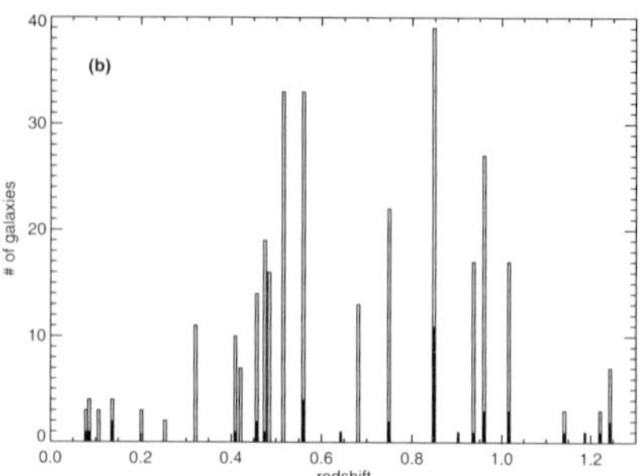

Figure 1. **a)** 15 μm versus radio continuum (1.4 GHz) rest-frame luminosities. Small filled dots: sample of 109 local galaxies from ISOCAM and NVSS. Filled dots with error bars: 17 HDFN galaxies (z ~0.7, radio from VLA or WSRT). Open dots with error bars: 7 CFRS-14 galaxies (z ~0.7, Flores et al. 1999, radio from VLA). Open diamonds: 137 ELAIS galaxies (z ~ 0-0.4). **a)** histogram of the redshift distribution of field (white) and ISOCAM (dark) galaxies which belong to redshift peaks as defined by Cohen et al. (2000).

mation (Chary & Elbaz 2001). Are these galaxies dominated by star formation or nuclear activity ? What is triggering their strong activity ? Is it triggered by external interactions or did it happen naturally within isolated galaxies ?

2. Origin of the luminosity of luminous infrared galaxies at z=1

Among the fields surveyed in the mid-IR with ISOCAM, the Hubble Deep Field North (HDFN) and its Flanking Fields (total area of 27 $arcmin^2$), is the best one to study the contribution of active galactic nuclei (AGN) to the IR luminosity of these LIR galaxies. Thanks to the deepest soft to hard X-ray survey ever performed with Chandra in the HDFN, it is possible to pinpoint AGNs including those affected by dust extinction. Among a total number of 95 sources detected at 15 μm in this field (Aussel et al. 1999, 2003 in prep.; with 47 above a completeness limit of 0.1 mJy), only 5 sources were classified as AGN dominated on the basis of their X-ray properties (Fadda et al. 2002). Hence, unless a large number of AGNs are so dust obscured that they were even missed with the 2 Megaseconds Chandra survey, the vast majority of ISOCAM LIR galaxies are powered by star formation. We are then left with the following question: how can one derive a total IR luminosity on the sole basis of a mid-IR measurement ? We showed in Elbaz et al. (2002) that in the local universe, there exists a very strong correlation between the mid-IR and total IR luminosity of galaxies over three decades in luminosity. In order to test whether these correlations remain valid up to $z \sim 1$, it is possible to use

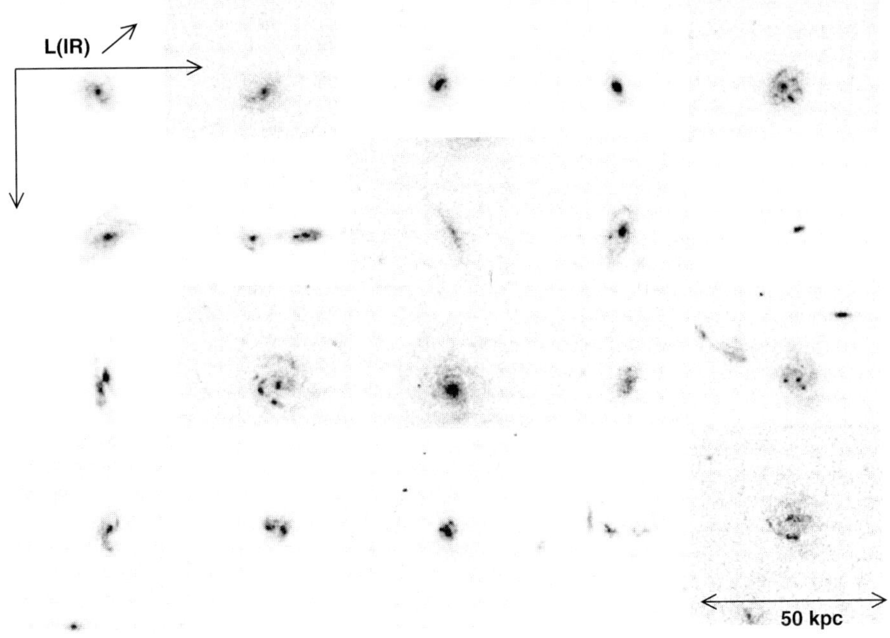

Figure 2. HST-ACS images of LIR galaxies with $11 \leq \log(L_{IR}/L_{\odot}) \leq 12$ (LIRGs) and $z \sim 0.7$. The double-headed arrow indicates the physical size of 50 kpc. The IR luminosity increases from left to right and from top to bottom.

another correlation existing between the radio and total IR luminosity in the local universe and check whether both the mid-IR and radio give a consistent prediction for the total IR luminosity. In the Fig. 1a, we have reproduced the plot from Elbaz et al. (2002) complemented with galaxies detected within the ELAIS survey (Rowan-Robinson et al. 2003). Except at low luminosities were the contribution of cirrus to the IR luminosity becomes non negligible, the 1.4 GHz and 15 μm rest-frame luminosities are correlated up to $z \sim 1$ and therefore predict very consistent total IR luminosities from which star formation rates as well as the contribution of these objects to the CIRB can be computed, leading to the results mentioned in the introduction.

The next question concerns the physical origin of this dusty starburst phase in galaxies (see Elbaz & Cesarsky 2003). In order to address it we have plotted in the Fig 1b the histogram of the redshift distribution of the ISOCAM galaxies with a spectroscopic redshift and above the completeness limit of 0.1 mJy in the HDFN. For comparison, only the redshifts where Cohen et al. (2000) found a redshift peak is plotted for field galaxies. One can clearly see that except for 3 ISOCAM galaxies, all are located within these redshift peaks. We have performed Monte-Carlo simulations to compute the probability that a random sample of optical galaxies of the same size than that of ISOCAM galaxies would fall in as many redshift peaks and we found that it is of the order of one percent, and even less if we restrict ourselves to the biggest redshift peaks (Moy et al., in prep). Is this result in agreement with expectations ? As showed by Franceschini et al. (2001), ISOCAM galaxies are relatively massive, hence more subject to belong to denser regions. Moreover LIR galaxies are mostly triggered by interactions in the local universe and one would naturally expect the same to take place in the more distant universe, hence in connection with cosmic structure formation. Thanks to the deepest existing survey with the ACS camera onboard the HST (GOODS survey, Giavalisco et al. 2003), we were able to study the morphology of ISOCAM galaxies in this field centered on the HDFN. A first result follows our expectations: most of these galaxies present a disturbed morphology and this is truer as a function of increasing IR luminosity (see Fig. 2). A second result was less expected: several LIR galaxies appear to be relatively big disk galaxies, which could be affected by a neighboring dwarf galaxy or a passing-by galaxy but they are not major mergers in the main phase of interacting. This is a very important result that should be carefully considered by authors of galaxy formation simulations. A large fraction of present-day stars may originate from a phase of violent star formation triggered by passing-by galaxies even within disk galaxies which could remain as such in the present-day universe. Another important consequence of this type of physical origin for the dusty starburst phase is that there are not enough major mergers in simulations to explain the large excess of LIR galaxies in the distant universe and the fact that they dominate the cosmic star formation history. Indeed each individual galaxy could have experienced several encounters with other galaxies at the time of structure formation which would induce a series of bursts of star formation, each of them responsible for only a few percent of the final stellar mass of the galaxy.

References

Aussel, H., Cesarsky, C.J., Elbaz, D., Starck, J.L. 1999, A&A 342, 313
Chapman, S.C., Blain, A.W., Ivison, R.J., Smail, I. 2003, Nature 422, 695
Chary, R.R., Elbaz, D. 2001, ApJ 556, 562
Cohen, J.G., Hogg, D.W., Blandford, R., et al. 2000, ApJ 538, 29
Elbaz, D., Cesarsky, C.J., Fadda, D., et al. 1999, A&A 351, L37
Elbaz, D., Cesarsky, C.J., Chanial, P., et al. 2002, A&A, 384, 848
Elbaz, D., Cesarsky, C.J. 2003, Science 300, 270
Fadda, D., Flores, H., Hasinger, G., et al. 2002, A&A 383, 838
Flores, H., Hammer, F., Thuan, T.X., et al. 1999, ApJ 517, 148
Franceschini, A., Aussel, H., Cesarsky, C.J., Elbaz, D., Fadda, D. 2001, A&A 378, 1
Giavalisco, M., et al. 2003, ApJ (in press, astro-ph/0309105)
Rowan-Robinson, M., et al. 2003, MNRAS (submitted, astro-ph/0308283)

FINAL ANALYSIS OF ELAIS 15 μM FIELDS

Mattia Vaccari[1,2], Carlo Lari[3], Luca Angeretti[3], Carlotta Gruppioni[4,5], Francesca Pozzi[6], Oliver Prouton[1], Alberto Franceschini[1] and Gianni Zamorani[5]

[1]*Dipartimento di Astronomia, Universita di Padova, Vicolo dell'Osservatorio 2, I-35122, Padova, Italy*

vaccari@pd.astro.it

[2]*CISAS "G. Colombo", Universita di Padova, Italy*

[3]*Istituto di Radioastronomia, Consiglio Nazionale delle Ricerche, Bologna, Italy*

[4]*Osservatorio Astronomico di Padova, INAF, Padova, Italy*

[5]*Osservatorio Astronomico di Bologna, INAF, Bologna, Italy*

[6]*Dipartimento di Astronomia, Universita di Bologna, Bologna, Italy*

Abstract The Final Analysis of ELAIS 15 μm observations carried out with the ISOCAM instrument onboard the ISO satellite is described.
The production and scientific potential of the resulting catalogue are discussed, including the latest enhancements to the employed data reduction technique.
The catalogue includes about 2000 sources in the 0.5 – 100 mJy flux range over an area of about 10 deg^2 and is thus the largest non-serendipitous extragalactic source catalogue obtained to date from ISO data.

Keywords: infrared: galaxies – galaxies: formation, evolution, active, starburst – cosmology: observations – methods: data analysis – catalogues.

1. Introduction

The IRAS mission (Neugebauer et al. 1984; Soifer et al. 1987) was extremely successful in characterizing for the first time the global properties of the mid and far infrared sky, leading to relevant discoveries such as Luminous, Ultraluminous and Hyperluminous Infrared Galaxies (LIRGs, ULIRG and HLIRGs, respectively), a substantial population of evolving starburts and to the detection of large-scale structure in infrared galaxy distribution.
Although conceived as an observatory-type mission, the Infrared Space Observatory (ISO, Kessler et al. 1996) was in many ways the natural successor to IRAS, bringing a gain of a factor $\sim$ 1000 in sensitivity and $\sim$ 10 in angular resolution. A substantial amount of ISO observing time was therefore devoted

M. Plionis (ed.), Multiwavelength Cosmology, 177-180.

to field surveys aimed at detecting faint infrared galaxies down to cosmological distances.

2. ELAIS and its 15 μm Dataset

The European Large Area ISO Survey (ELAIS, Oliver et al. 2000; Rowan-Robinson et al. 2003) was the most ambitious non-serendipitous survey and the largest Open Time project carried out with ISO, aimed at bridging the flux gap between IRAS all-sky survey and ISO deeper surveys mapping areas of about 10 deg^2 at 15 and 90 μm and smaller areas at 7 and 170 μm with the ISOCAM (Cesarsky et al. 1996) and ISOPHOT (Lemke et al. 1996) cameras. Thanks to an extensive multi-wavelength coverage, the ELAIS fields have now become the best studied sky areas of their size, and natural targets of on-going or planned large-area surveys with the most powerful ground and space-based facilities.
The ELAIS 15 μm main dataset covers a total area of about 10 deg^2 divided into 4 fields, one (S1) in the southern hemisphere and three (N1, N2 and N3) in the northern one. The fields were selected on the basis of their high Ecliptic latitude ($|\beta| > 40°$, to reduce the impact of Zodiacal dust emission), low cirrus emission ($I_{100\mu m} < 1.5$ MJy/sr) and absence of any bright ($S_{12\mu m} > 0.6$ Jy) IRAS 12 μm source.

3. Data Reduction

Reduction of data obtained with ISO instrumentation has always proved very difficult for a number of reasons. As far as ISOCAM observations carried out using its Long Wavelength (LW) detector are concerned, the two most important instrumental phenomena one has to deal with are the qualitatively very different effects produced on the detector's electronics by the frequent and severe cosmic ray impacts, which have long been known and referred to as *glitches*, and its sizable transient behaviours after changes in the incident photon flux, which we will hereafter simply refer to as *transients*. In both cases, the cryogenic operational temperatures of the detector caused it to very slowly respond after these events. Lack of an accurate modeling of these effects can lead to incompleteness, spurious detections and errors in flux determination.
The data reduction described here was carried out using the LARI method (Lari et al. 2001; Lari et al. 2003a; Vaccari et al. 2003), a technique developed to overcome these difficulties and provide a fully-interactive technique for the reduction and analysis of ISOCAM and ISOPHOT data, particularly suited for the detection of faint sources and thus for the full exploitation of their scientific potential. The method describes the sequence of readouts, or time history, of each pixel of ISOCAM LW detector in terms of a mathematical model for the charge release towards the contacts. Such a model is based on the assumption

of the existence, in each pixel, of two charge reservoirs, a short-lived one Q_b (*breve*) and a long-lived one Q_l (*lunga*), evolving independently with a different time constant and fed by both the photon flux and the cosmic rays. The method was refined with respect to the technique originally used by Lari et al. 2001 for the reduction of the S1 field only. Thus, a reduction of all ELAIS fields using this improved method seemed appropriate and was carried out.

4. Simulations

Due to the peculiar nature of the data reduction method we make use of, one would like to carefully test its performance on "ideal" data and sources. Besides, systematic effects on flux estimates related to the data reduction method can only be probed by these means. However, due to the corresponding strong peculiarities of our dataset, characterized by several noise features on different time scales, only real data can be taken as representative of instrumental behaviour. Therefore, the effects of additional sources must be somehow simulated on the top of real pixel time histories and data reduction must then be carried out exactly as done for real sources. Quite suitably, as already mentioned, based as it is on a mathematical model, the LARI method offers a straightforward way to model the additional signal produced on real time histories by such additional sources. On this basis, an extensive set of simulations was carried out (and reduced exactly as done with real data) to assess the effects on flux estimates and the overall performance of data reduction. Such simulations allowed to assess the completeness and astrometric/photometric accuracy of our catalogue.

5. Optical/NIR Identifications

Optical and NIR identifications of our sources were carried out on a wide variety of observational data (both taken from data archives and taken by collaborators) to distinguish between stellar and galaxy-like sources, independently assess the astrometric and photometric accuracy and accurately calibrate our photometry in an absolute way on the basis of predicted vs. measured MIR stellar fluxes. Together with available spectroscopy, this allows relevant studies of MIR-selected populations to be carried out on large samples (Gonzalez-Solares et al. 2003; La Franca et al. 2003).

6. The Catalogue

The ELAIS 15 μm Final Analysis catalogue contains about 2000 sources detected with a S/N greater than 5 in the four fields N1, N2, N3 and S1, totaling an area of about 10 deg^2. The catalogue (which will become available at `http://web.pd.astro.it/poe/elais`) makes up the backbone of the "ELAIS Final Catalogue", summarizing the results from the several ELAIS multi-wavelength observational campaigns carried out in the past few years (Rowan-Robinson et al. 2003).

7. Conclusions

A technique for ISO-CAM/PHOT data reduction, the LARI method, was refined and applied to ELAIS 15 μm main fields. The method now allows a substantially more robust and quicker data reduction than originally presented in Lari et al. 2001. Its application, in its new form, to the four fields composing the dataset (including a re-reduction of S1 observations already presented in Lari et al. 2001) has produced a catalogue of about 2000 sources, detected with a S/N ratio greater than 5 over a total area of 10.18 deg^2. Source fluxes span the 0.5 - 100 mJy range, filling the existing gap between the Deep ISO-CAM Surveys and the faint end of IRAS All Sky Survey. The astrometric and photometric accuracy as well as the completeness of the catalogue have been tested through simulations.

References

Cesarsky C.J. et al. 1996, A&A, 315, L32
Gonzalez-Solares E. et al. 2003, in preparation
Kessler M.F. et al. 1996, A&A, 315, L27
La Franca F. et al. 2003, in preparation
Lari C. et al. 2001, MNRAS, 325, 1173
Lari C. et al. 2003, ESA SP 511, in press, astro-ph/0210566
Lemke D. et al. 1996, A&A, 315, L64
Neugebauer G., Habing H. J., van Duinen R. et al. 1984, ApJ, 278, L1
Oliver S., Rowan-Robinson M. et al. 2000, MNRAS, 316, 749
Rowan-Robinson M. et al. 2003, MNRAS, submitted, astro-ph/0308283
Soifer B.T., Houck J.R. and Neugebauer G. 1987, ARA&A, 25, 187
Vaccari M. et al. 2003, in preparation

ELAIS-SOUTH: THE NATURE AND EVOLUTION OF GALAXIES AND AGN IN THE MID-INFRARED

F. La Franca[1], C. Gruppioni[2], I. Matute[1,3], F. Pozzi[2] and the ELAIS consortium

1) Dip. di Fisica, Universita' degli studi Roma Tre, Via della Vasca Navale 84, Roma, Italy

2) INAF, Osservatorio Astronomico di Bologna, Via Ranzani 1, Bologna, Italy

3) MPE, Giessenbachstrasse, Postfach 1312, Garching, Germany

Abstract

We present the results from the spectroscopic identifications of the ELAIS sources in the southern fields S1 and S2. Two main extragalactic classes have been found: z<0.5 star-forming galaxies (~75%), and AGN (~25%). ~20% are dust enshrouded starburst galaxies [e(a) spectra]. All the galaxies appear more dust extincted in the optical than nearby normal galaxies. The evolution of galaxies is fitted by a strong evolution for only a fraction of the whole population. AGN1 evolves according to a pure luminosity evolution, while no evolution is found for the AGN2.

1. Introduction

The extragalactic background light shows that the emission from galaxies at infrared and sub-mm wavelengths is an energetically significant component of the Universe. This emission originates from star-formation activity and active galactic nuclei (AGN). The precise contribution from each type of activity is still debated. In particular, data from deep ISO surveys seem to require strong evolution for galaxies emitting in the infrared wavebands. This result, supported also by the detection of a substantial cosmic Infrared Background (CIRB) in the 140 μm - 1 mm range (see Lagache et al. 1999), has stimulated the development of several evolutionary models for IR galaxies. All these models, can marginally fit the IR/sub-millimeter source counts and the CIRB, but suffer of parameter degeneracy and none of them is based on a local luminosity function (LF) obtained from 15-μm data, being all extrapolated from different IR wavelengths (12, 25, 60 μm). So far, complete spectroscopic samples of 15-μm sources have been obtained only in small fields (i.e. HDF-N and HDF-S), too small and too deep to allow the study of the local LF.

M. Plionis (ed.), Multiwavelength Cosmology, 181-184.

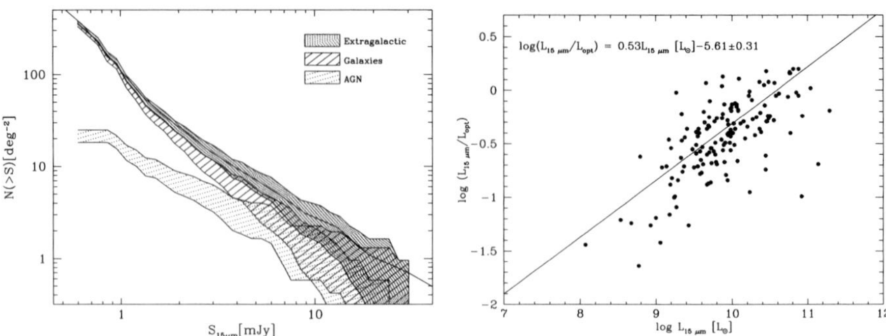

Figure 1. The observed 15-μm source counts of the extragalactic sources (galaxies and AGN) from ELAIS-S1.

Figure 2. L_{15}/L_R as a function of L_{15} from ELAIS-S1.

Here we present an analysis of the 15-μm LF of AGN and galaxies derived from the spectroscopic identifications of the southern fields of the European Large Area *ISO* Survey (ELAIS): the regions S1 and S2. These results are (and will be) published by Pozzi et al. (2003), La Franca et al. (2003), Pozzi et al. (in prep), and Matute et al. (in prep).

2. The sample

The ELAIS (Oliver et al. 2000), with its broad flux range (0.5< $S_{15\mu m}$ <150) and several thousand sources detected over a ~12 deg^2 area (Rowan-Robinson et al. 2003), is the crucial survey to investigate the nature and evolution of the MIR sources in the region where the source counts start diverging from the non-evolution predictions (a few mJy at 15-μm; Elbaz et al. 1999; Gruppioni et al. 2002). The southern field S1 (α(2000) = 00^h 34^m 44.4^s, δ(2000) = -43° 28′ 12″) covers an area of 2°×2°, while S2 (α(2000) = 05^h 02^m 24.5^s, δ(2000) = -30° 36′ 0″) covers an area of 21′×21′. The 15-μm catalogs obtained in the S1 and S2 fields are given by Lari et al. (2001) and Pozzi et al. (2003), respectively. The source counts have been presented by Gruppioni et al. (2002). The results of the optical identifications and spectroscopy analysis for S1 and S2 are given in La Franca et al. (2003) and Pozzi et al. (2003) respectively. Matute et al. (2002) has presented a first estimate of the evolution of AGN1.

S1 contains 462 15-μm sources of which 406 form a complete and highly reliable sub-sample. ~80% sources were optically identified (R≤23.0). Spectroscopic identifications were obtained for ~89% of the optically identified sources. Two main spectroscopic classes have been found: z<0.5 star-forming galaxies (~75%; from absorbed to extreme starbursts: $L_{MIR}\approx 10^8$-10^{11} $L_\odot$), and AGN (both type 1 and 2), which account for ~25% of the sources. About 20% of the extragalactic sources are dust-enshrouded starburst galaxies [e(a)

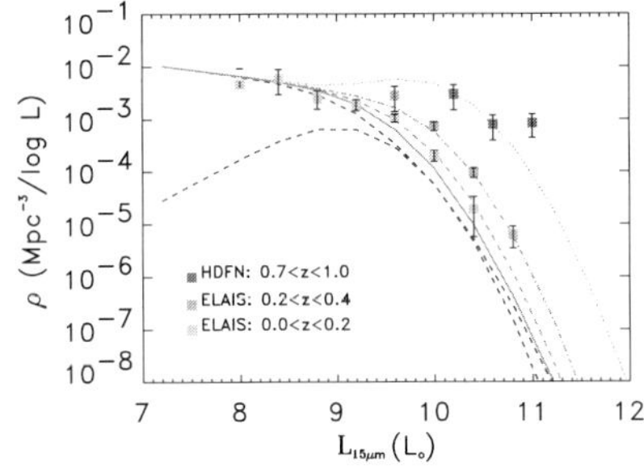

Figure 3. The 15-μm LF for the two galaxy classes, plus the extrapolation to the redshift interval of the HDF-N survey.

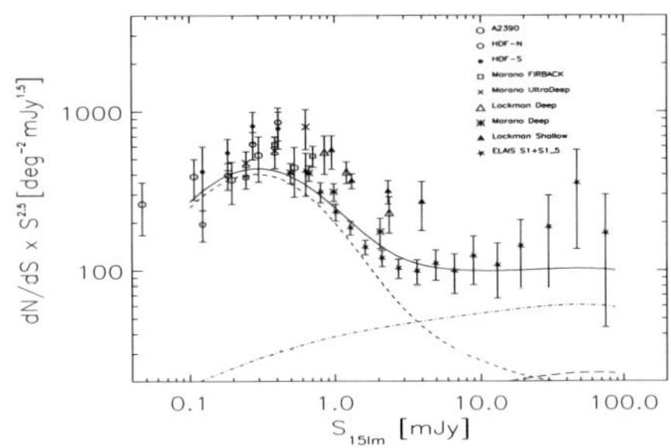

Figure 4. Comparison of model prediction and observed 15-μm source counts.

spectra], and all the starburst galaxies appear more dust extincted in the optical than nearby normal galaxies. S2 contains 43 sources, $\sim$90% sources were optically identified and $\sim$70% of the optically identified sources were spectroscopically identified. The counts for galaxies and AGNs have been derived (Figure 1). All the ELAIS extragalactic sources at fluxes $S_{15\mu m}$>0.6 mJy contribute to $\sim$12% to the CIRB. We found that the average optical-IR SED of galaxies is luminosity dependent: were more luminous galaxies have larger L_{15}/L_{opt} ratio (Figure 2; La Franca et al., 2003).

3. The evolution of galaxies and AGN

To study the 15-μm LF of galaxies and its evolution, we used a high reliable subsample of 153 galaxies from S1 and S2, with redshift z<0.4. We have assumed a modified Schechter LF $\phi(L, z)$ (see Saunders et al. 1990), and two populations for the evolving galaxies: the starbursts with $\log(L_{15\mu m}/L_{\rm R}) \geq -0.4$, and the normal (cirrus) galaxies with $\log(L_{15\mu m}/L_{\rm R}) < -0.4$. The shape of the observed source counts (Euclidean from *IRAS* to a few mJy, followed by a sharp upturn at fainter fluxes) favors the hypothesis of strong evolution for only a fraction of the whole population, with the remaining galaxies giving rise to the Euclidean behavior. We have thus constrained our model by assuming that the cirrus population does not evolve, while allowing the active galaxies (starburst) to evolve both in density and in luminosity, according to $\phi(L, z) = g(z)\phi(L/f(z), 0)$, parameterizing the two evolutions with two power-laws: $g(z) = (1+z)^{K_d}$ and $f(z) = (1+z)^{K_l}$. We found an evolution for the active population of $\sim(1+z)^{3.5}$ both in density and in luminosity up to $z_{break}\sim$1 (Pozzi et al. in prep.). Figure 3 shows the fitted LF, while the

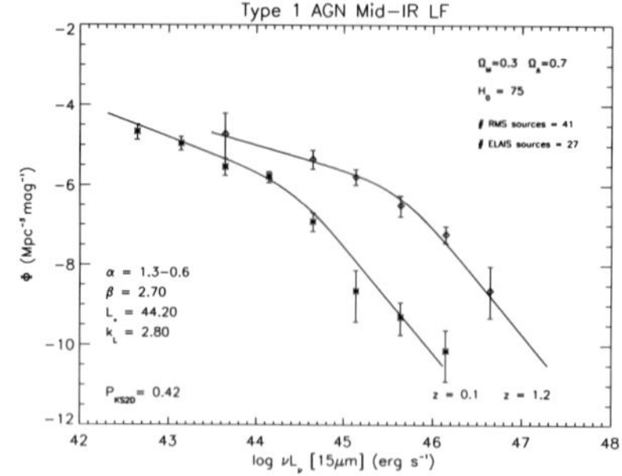

Figure 5. Luminosity function of AGN1.

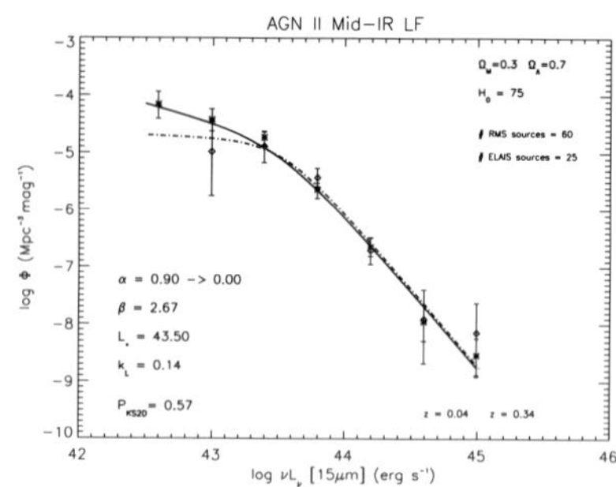

Figure 6. Luminosity function of AGN2.

comparison of model predictions and observed 15-μm source counts is shown in Figure 4.

The analysis of the evolution of AGN was based on 27 AGN1 and 25 AGN2 contained in the S1 and S2 samples. We combined our data with a local sample of 41 AGN1 and 50 AGN2 belonging to the sample of $IRAS$ galaxies with $S_{12\mu m}$>300 mJy, from Rush, Malkan and Spinoglio (RMS; 1993). The adopted shape of the LF was a classical 2 power law evolving according to a pure luminosity evolution (PLE) model $L_{15}(z) = L_{15}(0)(1+z)^{k_L}$ (see Matute et al. 2002). We allowed a flattening of the faint slope α with redshift. In Figure 5 and 6 the LF of AGN1 and AGN2 are shown (errorbars correspond to 1σ confidence level). We found a luminosity evolution for AGN1 of $(1+z)^{2.5}$ up to z=2, while the faint slope flattens from 1.2 at z=0 to 0.4 at z=1.2. No evolution was found for the AGN2 population (Matute et al. in prep).

References

Elbaz, D., et al. 1999, A&A, 351, L37

Gruppioni, C., et al. 1999, MNRAS, 305, 297

Gruppioni, C., Lari, C., Pozzi, F., Zamorani, G., Franceschini, A., Oliver, S., Rowan-Robinson, M. & Serjeant, S. 2002, MNRAS, 335, 831

Gruppioni, C., Pozzi, F., Zamorani, G., Ciliegi, P., Lari, C., Calabrese, E., La Franca, F., Matute, I. 2003, MNRAS, 341, L1

La Franca F., Gruppioni, C., Matute, I., Pozzi, F., et al. 2003, AJ, submitted

Lagache, G., et al. 1999, A&A, 344, 322

Lari, C., et al. 2001, MNRAS, 325, 1173

Matute, I., et al. 2002, MNRAS, 332, L11

Oliver, S., et al. 2000, MNRAS, 316, 749

Pozzi, F., et al. 2003, MNRAS, in press

Rowan-Robinson M., et al. 2003, MNRAS, submitted [astro-ph/0308283]

Rush, B., Malkan, M.A., & Spinoglio, L. 1993, ApJS, 89, 1

Saunders, W., et al. 1990, MNRAS, 242, 318

PROPERTIES OF A LARGE SAMPLE OF ERO'S

Arnouts Stephane
Laboratoire d'Astrophysique de Marseille, Traverse du Siphon, 13376 Marseille, France
Stephane.Arnouts@oamp.fr

Abstract We investigate the properties of a large sample of extremely red objects (~ 500 EROs) based on deep optical and IR datasets from the CDF-South and HDFs. The ERO's surface density is consistent with a monolithic model where ellipticals form between $2 \leq z_{form} \leq 2.5$, while semi-analytic models (SAM) underestimate the density by a factor of 2 to 10. Additional observations (clustering and comoving number density), independent of color modeling, are in good agreement with the expected properties of ellipticals in a hierarchical ΛCDM scenario. This comparison can suggest that SAM provide good predictions for the dark matter haloes assembling while it cannot describe the baryon assembling (star formation) of this extreme population.
Finally, we speculate about the descendants and ancestors of EROs. In the framework of CDM paradigm, the EROs could be the already assembled progenitors of local ellipticals. The brightest population of Lyman Break Galaxies at $z \sim 3$ could go through an ERO stage, while it should not be the case for the bulk of the LBGs (faintest samples) unless they undergo major merging events between $z \sim 3$ and $z \sim 1$.

Keywords: Cosmology: elliptical galaxies – evolution – large-scale structure

1. The surface density of EROs

The extremely red objects have colors consistent with passive ellipticals observed at $z \geq 1$. Therefore they could provide constraints for the two competing models involved in the formation of elliptical galaxies: the hierarchical model where ellipticals form recently through mergers and the monolithic model where ellipticals form at high-z and evolve passively up to now. To investigate some of these issues, we have selected a sample of EROs in the optical-infrared datasets of the Chandra Deep Field South carried out by the ESO Imaging Survey project and the HDF-North and South. The CDFS covers a unique area of ~ 300 sq. arcmin up to $K_{VEGA} \sim 20.5$. The HDFs allow us to enlarge the magnitude baseline up to $K_{VEGA} \sim 22$ in order to probe ellipticals with $L \geq L_*/6.5$ up to $z \sim 1.7$.

In Figure 1, we show the cumulative surface densities of EROs as a function of the K magnitude with $(R-K) \geq 5.3$ (left panel) and $(R-K) \geq 6$ (right panel). Our results agree with other previous estimates. We have mod-

M. Plionis (ed.), Multiwavelength Cosmology, 185-188.

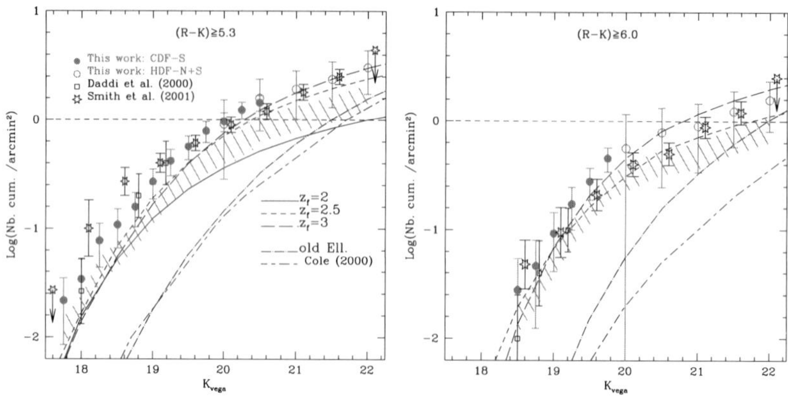

Figure 1. Surface density of EROs with $(R - K) \geq 5.3 - 6$ (right and left panel resp.)

eled the ERO's density by adopting the local IR luminosity function of ellipticals and a pure luminosity evolution with short bursts at formation epochs $2 \leq z_{form} \leq 3$. Since the EROs sample consists of a mix population of ellipticals and dusty star-forming galaxies with still uncertain fraction (see Cimatti's contribution), the hatched regions in Fig 1 show the density of ellipticals assuming 30% to 70% of the whole sample. For both samples, a redshift of formation for ellipticals between $2 \leq z_{form} \leq 2.5$ provide good fits while higher z_{form} seem excluded. We also show the prediction of the semi analytic model from Cole et al. (2000). This model underpredicts the density by a factor of 2 to 10. This result suggests that either hierarchical model underpredicts massive galaxies at $z \geq 1$ or that the semi-analytic recipes for the star formation treatment should be improved. The first point can be tested by measuring the clustering properties of EROs and compared with the elliptical prediction in SAM which is independent of the IR-optical colors.

2. The clustering and comoving density of EROs

In Figure 2(a), we show the correlation lengths (R_0) for different color and magnitude cuts of the ERO's sample which is more than one magnitude deeper than previous works. Note that the R_0 values are sensitive to: (a) the adopted $N(z)$. We use the PLE model with $z_{form} = 2.5$, consistent with Cimatti et al., 2002 (spectroscopic sample at $K \leq 19.2$); (b) the contamination by dusty star-forming galaxies which could slightly decrease the signal.

In Figure 2(b), we show the comoving number densities. The hatched region shows the 30 to 70% fraction of pure ellipticals.

A comparison with local elliptical galaxies from Willmer et al. (1998) shows that EROs have similar clustering amplitudes and comoving densities than local ellipticals. Our results are compared with the elliptical's sample as defined

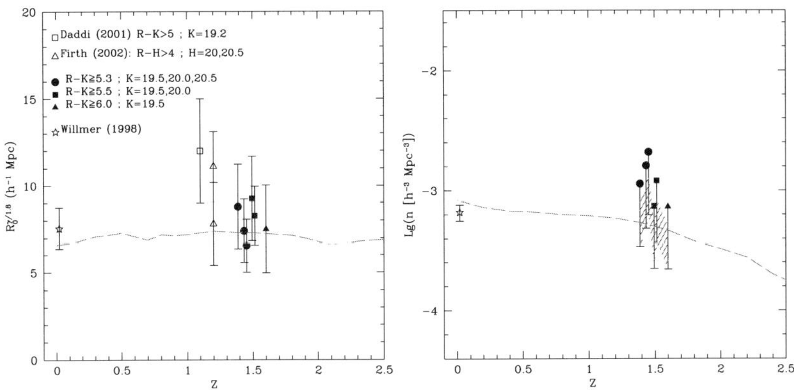

Figure 2. Correlation lengths (left panel) and comoving density of EROs (right panel)

in the SAM from Kauffmann et al. (1999, red long dashed lines). This sample shows no decline in the correlation length with redshift and have similar amplitude than the EROs at $z \sim 1.5$. The old-ERO's comoving number densities also agree with the expected density derived in the SAM from Kauffmann et al (1999). This supports the hypothesis that hierarchical models can predict where and when the ellipticals form while they cannot reproduce the evolution of their stellar content.

3. The ancestors and descendants of EROs

Finally we address the issue of the link between the EROs and other populations at different redshifts. Are EROs the progenitors of present-day ellipticals and/or the descendants of LBGs at $z \sim 3$? To get a hint, we use analytic models in the framework of CDM paradigm to connect the clustering, the comoving number density and the bias at different z as a function of a minimum dark mass threshold (Arnouts et al., 1999; Moustakas & Somerville, 2002). In Figure 3, we plot the properties of the local ellipticals, the EROs ($z \sim 1.5$) and the LBGs from Adelberger et al. (2003, filled dark triangles) and from Foucaud et al. (2003; filled hexagons; see Foucaud's contribution). The minimum DMH ranges are fixed for each population according to their R_0 values. We find $10^{13.0-13.4} M_\odot$ for local ellipticals and $10^{12.6-13.4} M_\odot$ for EROs. For the LBGs we use $10^{11.4-12.4} M_\odot$ corresponding to the two faintest LBG samples. The evolution with z of these three DMH ranges are shown by different hatched regions. The high mass values obtained for the EROs show that EROs should be located in dense regions of the dark matter at $z \geq 1$. Similar masses are found for the local ellipticals supporting the idea that EROs could be the already assembled progenitors of local ellipticals. The brightest population of LBG at $z \sim 3$ occupy similar DMH masses than EROs and local ellipticals and

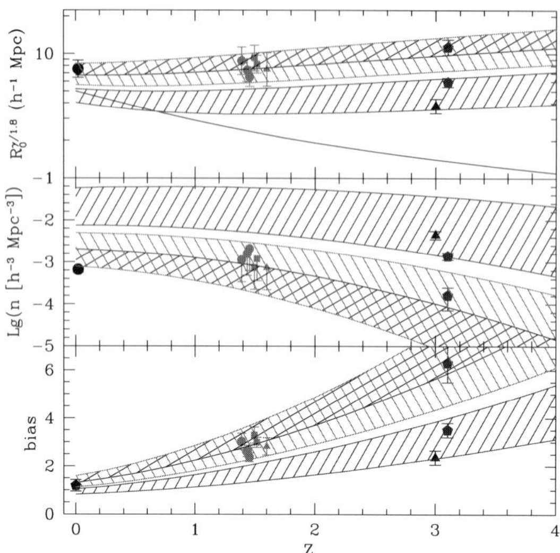

Figure 3 Drawing back and forward the redshift evolution of EROs as compared with local ellipticals and LBGs at $z \sim 3$. The panels show the redshift evolution of the comoving correlation lengths (top), the comoving number density (middle) and the linear bias (bottom). Observations are compared with CDM haloes predictions selected with a fixed minimum mass. The minimum mass ranges are fixed for each population according to their R_0 values (see text).

have a comoving density consistent with the expected density of such a DMH. Therefore the brightest LBGs could go through an ERO stage. In contrast, the bulk of the LBGs (the two faintest samples) occupy smaller DMH and should not follow the same behavior than their bright counterpart unless they undergo major merging events between $z \sim 3$ to $z \sim 1$.

Acknowledgments

I wish to thank my collaborators: C. Benoist, L. Olsen, L. da Costa and EIS team for their support in the analysis of the ERO's sample in the Chandra deep Field South ; E. Vanzella, S. Cristiani, A. Fontana, P. Sarraco, E. Giallongo, L. Moscardini for the analysis of the HDFs; S. Foucaud, HJ Mc Cracken, O. Le Fevre for the analysis of the Lyman break galaxies at z=3.
I thank the organizers for this wonderful conference.

References

Adelberger K., Steidel C., Shapley A. & Pettini M., 2003, ApJ, 584, 45
Arnouts S., Benoist C., da Costa L., Olsen L.F. et al.,2003, submitted to A&A
Cimatti, A.; Daddi, E.; Mignoli, M. et al., 2002, A&A, 381, 68
Cole S., Lacey C., Baugh C. , Frenk C., 2000, MNRAS, 319, 204, (C00)
Daddi E., Cimatti A. and Renzini A., 2000, A&A, 362, L45
Daddi E., Broadhurst T., Zamorani.,Cimatti A. et al.,2001, A&A, 376, 825
Firth A., Somerville R., McMahon R. et al., 2002, MNRAS, 332, 617
Foucaud S., McCracken H.J., Le Fevre O., Arnouts S. et al.,2003, A&A in press
Kauffmann G., Colberg J.M., Diafaro, A. and White S.,1999, MNRAS, 307, 529
Moustakas L.A.and Somerville R.S., 2002, ApJ, 577, 1
Smith G.P., Smail I., Kneib J.P., Czoske O. et al., 2002, MNRAS, 330, 1
Willmer, C., da Costa L., Pellegrini P., 1998, AJ, 115, 869

SIMULATIONS & THEORY

COSMOLOGY AND ASTROPHYSICS WITH CLUSTERS OF GALAXIES (*INVITED*)

Stefano Borgani
Dipartimento di Astronomia, Universit'a di Trieste, via Tiepolo 11, I-34131 Trieste, Italy.
borgani@ts.astro.it

Abstract I discuss the role that X–ray galaxy clusters plays in the determination of cosmological parameter. In particular I focus the discussion on the evolution of the cluster mass function for the determination of the density parameter Ω_m and the normalization of the power spectrum σ_8. Available constraints indicate that Ω_m lies in the range 0.2–0.5, with $\sigma_8 \simeq 0.7$–0.8 for $\Omega_m = 0.3$. I emphasize that the ability of clusters to provide such constraints is determined by the possibility of relating X–ray observable quantities to the cluster mass and, therefore, to our understanding of the physical processes which determine the properties of the ICM. I finally discuss the relevance that hydrodynamical simulations of clusters can play to understand the ICM physics and to calibrate mass estimates from X–ray observational quantities.

Keywords: Cosmology:numerical simulations – galaxies:clusters – hydrodynamics – X–ray:galaxies

1. Introduction

Clusters of galaxies probe the high–density tail of the cosmic density field and their number density is highly sensitive to specific cosmological scenarios (e.g. Press & Schechter 1974). The mass function of galaxy clusters and its evolution is currently used to measure the amplitude of density perturbations on ~ 10 Mpc scales and to place constraints on the value of the matter density parameter Ω_m (e.g. Oukbir & Blanchard 1992; Eke et al. 1998; Borgani et al. 2001). In this way, probing the evolution of the population of galaxy clusters provides a sensitive probe of the cosmological scenario, which is intrinsically and conceptually complementary to that provided by observations of the CMB anisotropies (e.g., Spergel et al. 2003). However, what cosmological models predict is the number density of clusters of a given mass at varying redshifts. On the other hand, the cluster mass is never a directly observable quantity, although several methods exist to estimate it from observations.

M. Plionis (ed.), Multiwavelength Cosmology, 191-198.

The ROSAT All–Sky Survey and deep PSPC and HRI ROSAT pointings have offered during the last decade a unique means to compile extended samples of X–ray clusters from redshift $z \lesssim 0.1$, out to $z \simeq 1.3$, with well defined selection functions (see Rosati, Borgani & Norman 2002, for a recent review). While these samples have been already used to probe the evolution of the cluster population over a fairly large redshift baseline, forthcoming surveys based on XMM–Newton and Chandra data will further contribute to substantially enlarge the statistics of distant clusters.

As of today, the availability of robust relations between cluster masses and X–ray observable quantities represents the main source of uncertainty for using cluster as tools for precision cosmology. For instance, the X–ray luminosity, L_X, is highly sensitive to the details of the gas distribution, to the overall dynamical status of clusters and to the structure of their central cooling regions. At the same time, the presence of unresolved temperature profiles and/or local violation of the condition of hydrostatic equilibrium, may introduce significant uncertainties in the relation between T and the cluster mass M. This calls for the need of understanding in detail the physical processes, which are relevant for establishing the observational properties of clusters.

Among such processes, radiative cooling and heating of the intra–cluster medium (ICM) from supernova and AGN are already known to play important roles in determining the thermodynamical properties of the gas and, therefore, its emissivity and temperature. In this respect, hydrodynamical simulations of clusters represent now invaluable tools to treat such complex physical processes and to shed light on the interplay between the cosmological framework, which defines the pattern of cluster formation, and the astrophysical processes, which determine the X–ray observational properties of the ICM.

In the first part of this contribution, I will discuss how the evolution of the cluster population can be used to place constraints on cosmology, and the level of uncertainty in the determination of cosmological parameter, which is associated to the uncertainty in the knowledge of the ICM properties. In the second part I will show examples of how large hydrodynamical simulations of galaxy clusters and groups can be used to make progress in the description of the ICM physics.

2. Cosmology with galaxy clusters

The mass distribution of dark matter halos undergoing spherical collapse in the framework of hierarchical clustering is described by the Press-Schechter distribution (PS, Press &Schechter 1974). The number of such halos in the mass range $[M, M + dM]$ can be written as $n(M, z)dM = \bar{\rho}\, M\, f(\nu)\, (d\nu/dM)\, dM$ where $\bar{\rho}$ is the cosmic mean density. The function f depends only on the variable $\nu = \delta_c(z)/\sigma_M$. $\delta_c(z)$ is the linear–theory overdensity extrapolated

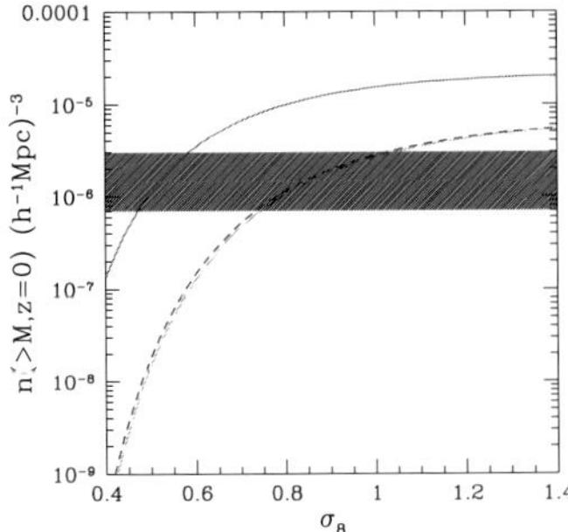

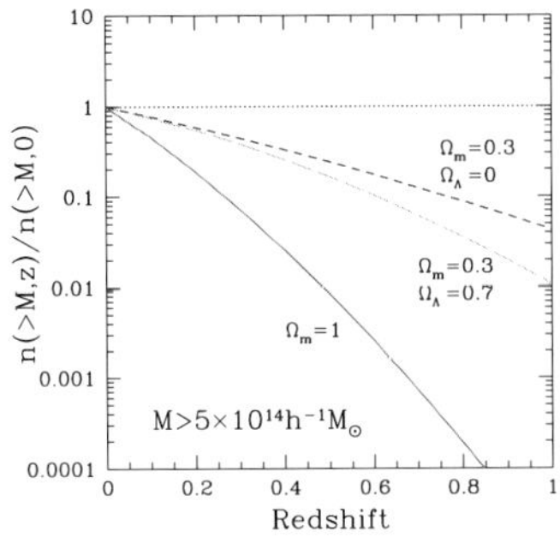

Figure 1. The sensitivity of the cluster mass function to cosmological models. Left: the cumulative mass function at $z = 0$ for $M > 5 \times 10^{14} h^{-1} M_\odot$ for three cosmologies, as a function of σ_8, with shape parameter $\Gamma = 0.2$; solid line: $\Omega_m = 1$; short–dashed line: $\Omega_m = 0.3$, $\Omega_\Lambda = 0.7$; long–dashed line: $\Omega_m = 0.3$, $\Omega_\Lambda = 0$. The shaded area indicates the observational uncertainty in the determination of the local cluster space density. Right: Evolution of $n(>M, z)$ for the same cosmologies and the same mass–limit, with $\sigma_8 = 0.5$ for the $\Omega_m = 1$ case and $\sigma_8 = 0.8$ for the low–density models.

to the present time for a uniform spherical fluctuation collapsing at redshift z, which depends on cosmology through the growth factor for linear density fluctuations ($\delta_c = 1.68(1+z)$ for an Einstein–de Sitter cosmology). The r.m.s. density fluctuation at the mass scale M, σ_M, is connected to the fluctuation power spectrum, $P(k)$, whose normalization is usually expressed in terms of σ_8, the r.m.s. density fluctuation within a top–hat sphere of $8\,h^{-1}$Mpc radius. In their original derivation of the cosmological mass function, Press & Schechter (1974) obtained the expression $f(\nu) = (2\pi)^{-1/2} \exp(-\nu^2/2)$ for Gaussian density fluctuations. Despite its subtle simplicity, the PS mass function has served for more than a decade as a guide to constrain cosmological parameters from the mass function of galaxy clusters. Only with the advent of the last generation of N–body simulations, which are able to span a very large dynamical range, significant deviations of the PS expression from the exact numerical description of gravitational clustering have been noticed (e.g. Jenkins et al. 2001, Evrard et al. 2002; White 2002). Such deviations are interpreted in terms of corrections to the PS approach, e.g. by incorporating the effects of non–spherical collapse (Sheth & Tormen 1999). I show in Figure 1 how the mass function can be used to constrain cosmological models. In the left panel I show that, for a fixed value of the observed cluster mass function, the implied value of σ_8 increases as the density parameter decreases. Determinations of the cluster mass function in the local Universe using a variety of samples and methods indicate that $\sigma_8 \Omega_m^\alpha = 0.4 - 0.6$, where $\alpha \simeq 0.4 - 0.6$, almost independent of the presence of a cosmological constant term providing spatial flatness (e.g. Ikebe et al. 2002; Pierpaoli et al. 2003, and references therein).

The growth rate of the density perturbations depends primarily on Ω_m and, to a lesser extent, on Ω_Λ, at least out to $z \sim 1$, where the evolution of the cluster population is currently studied. Therefore, following the evolution of the cluster space density over a large redshift baseline, one can break the degeneracy between σ_8 and Ω_m. This is shown in the right panel of Figure 1: models with different values of Ω_m, which are normalized to yield the same number density of nearby clusters, predict cumulative mass functions that progressively differ by up to orders of magnitude at increasing redshifts.

A commonly adopted procedure to estimate cluster masses is based on the measurement of the temperature of the ICM. Based on the assumption that gas and dark matter particles share the same dynamics within the cluster potential well, the temperature T and the velocity dispersion σ_v are connected by the relation $k_BT = \beta\mu m_p\sigma_v^2$, where $\beta = 1$ would correspond to the case of a perfectly thermalized gas. If we assume spherical symmetry, hydrostatic equilibrium and isothermality of the gas, the solution of the equation of hydrostatic equilibrium provides the link between the total cluster virial mass, $M_{\rm vir}$, and the ICM temperature: $k_BT \propto \beta^{-1}M_{\rm vir}^{2/3}[\Omega_m\Delta_{vir}(z)]^{1/3}(1+z){\rm keV}$., where $\Delta_{vir}(z)$ is the ratio between the average density within the virial radius and the mean cosmic density at redshift z. The above expression for the mass–temperature relation is fairly consistent with hydrodynamical cluster simulations with $0.9 \lesssim \beta \lesssim 1.3$ (e.g. Bryan & Norman 1998, Frenk et al. 2000). Observational data on the $M_{\rm vir}$–T relation show consistency with the $T \propto M_{\rm vir}^{2/3}$ scaling law, at least for $T \gtrsim 3$ keV clusters, but with a $\sim 40\%$ lower normalization (e.g., Finoguenov et al. 2001). We will discuss in the next section possible reasons for this difference between the observed and the simulated M–T relation. In any case, such uncertainties in the correct value of the normalization of the M–T relation translates into an uncertain determination of σ_8 (for a fixed value of Ω_m): the higher this normalization, the larger the mass corresponding to a given temperature, the larger the value of σ_8 required for the predicted mass function to match the observed X–ray temperature function (XTF; e.g., Pierpaoli et al. 2003; Seljak et al. 2002). In Figure 2 we show the analysis by Pierpaoli et al. (2003), who quantified the change in σ_8 (for $\Omega_m = 0.3$) induced by varying the M–T normalization.

Another method to trace the evolution of the cluster number density is based on the XLF. The advantage of using X-ray luminosity as a tracer of the mass is that L_X is measured for a much larger number of clusters within samples well-defined selection properties. The most recent flux–limited cluster samples contain now a large (~ 100) number of objects, which are homogeneously identified over a broad redshift baseline, out to $z \simeq 1.3$. A useful parameterization for the relation between temperature and bolometric luminosity is $L_{bol} \propto T_X^\alpha(1+z)^A(d_L(z)/d_{L,EdS}(z))^2$, with $d_L(z)$ the luminosity–distance at redshift z for a given cosmology. Independent analyses of nearby clusters

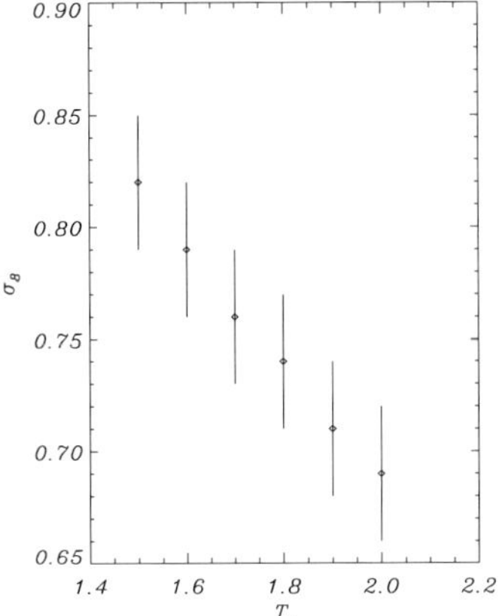

Figure 2 The dependence of σ_8 on the value of the M–T normalization. Here the density parameter is fixed at $\Omega_{\rm m} = 0.3$ (from Pierpaoli et al. 2003).

with $T_X \gtrsim 2$ keV consistently show that $\alpha \simeq 2.5$–3 (e.g. Arnaud & Evrard 1999 and references therein), with no evidence for a strong evolution out to $z \gtrsim 1$ (e.g., Ettori et al. 2003; cf. also Vikhlinin et al. 2002).

In Figure 3 we show the constraints on the σ_8–Ω_m plane obtained from the ROSAT Deep Cluster Survey (Borgani et al. 2001; Rosati et al. 2002). The different panels report the effect of changing in different ways the parameters defining the M–L_X relation, such as the slope α and the evolution A of the L_X–T relation, the normalization β of the M–T relation, and the overall scatter Δ_{M-L_X}. Figure 3 demonstrates that firm conclusions about the value of the matter density parameter Ω_m can be drawn from available samples of X-ray clusters. In keeping with most of the analyses in the literature, based on independent methods, a critical density model cannot be reconciled with data. Specifically, $\Omega_m < 0.5$ at 3σ level even within the full range of current uncertainties in the relation between mass and X-ray luminosity. However, the results shown in Fig. 3 also demonstrate that constraints in the σ_8–Ω_m may change by changing the the M–L_X relation within current uncertainties, by an amount which is at least as large as the statistical uncertainties. This emphasize that the main obstacle toward a precision estimate of cosmological parameter with forthcoming large cluster surveys will lie in the systematics uncertainties in our description of the ICM properties, rather than in the limited statistics of distant clusters.

3. Hydrodynamical simulations of clusters

I present here results from a large cosmological hydrodynamical simulation of concordance ΛCDM model ($\Omega_m = 1 - \Omega_\Lambda = 0.3$, $\Omega_{bar} = 0.04$, $h = 0.7$, $\sigma_8 = 0.8$) within a box of $192\,h^{-1}$Mpc on a side, using 480^3 dark matter and an equal number of gas particles, and a Plummer–equivalent softening for the computation of the gravitational force of $7.5h^{-1}$kpc (Borgani et al. 2003). The run has been realized using GADGET, a massively parallel tree N–body/SPH

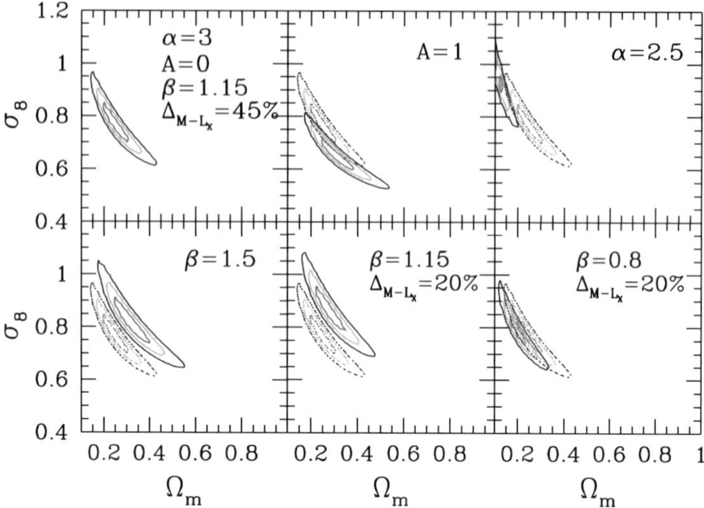

Figure 3. Probability contours in the σ_8–Ω_m plane from the evolution of the X-ray luminosity distribution of RDCS clusters. Different panels refer to different ways of changing the relation between cluster virial mass, M, and X-ray luminosity, L_X, within theoretical and observational uncertainties. The upper left panel shows the analysis corresponding to the choice of a reference parameter set. In each panel, we indicate the parameters which are varied, with the dotted contours always showing the reference analysis.

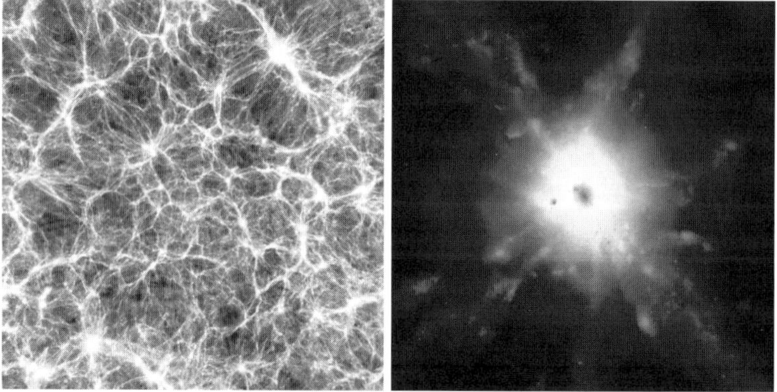

Figure 4. Left panel: map of the gas density over the whole simulation box at $z = 0$, projected through a slice having thickness of 12 h^{-1}Mpc, and containing the most massive cluster found in the simulation (upper right side of the panel). Right panel: zoom into the region of the largest cluster; the cluster is shown out to two virial radii.

code, which treats radiative cooling, star formation from a multiphase ISM and the effect of galactic winds (Springel & Hernquist 2003).

In the left panel of Figure 4, we show a map of the gas distribution at $z = 0$. The panel on the right shows the gas distribution around the most massive cluster found in the simulation and represents a zoom-in of the whole simulation

box by about a factor of 20. The amount of small–scale detail which is visible in the zoom–in of the right panel demonstrates the large dynamic range encompassed by the hydrodynamic treatment of the gas in our simulation.
We show in Figure Figure 5 the comparison between the simulated and the observed M–T relation. The left panel demonstrates that cooling in our simulation is not effective in reducing the M–T normalization to the observed level. In this panel, our results are compared to those by Finoguenov et al. (2001). Making a log–log least square fitting to the relation $\log(M_{500}/M_0) = \alpha \log(T_{500}/\mathrm{keV})$, we obtain $\alpha = 1.59{\pm}0.05$ and $M_0 = (2.5{\pm}0.1)\,10^{13}h^{-1}M_\odot$. Therefore, the normalization of our relation at 1 keV turns out to be higher by about 20 per cent than the observed one. The intrinsic scatter around the best–fitting relation is $\Delta M/M = 0.16$.
A critical issue in this comparison concerns the different procedure used for estimating masses in simulations and in observations. For instance, Finoguenov et al. (2001) estimate masses by applying the equation of hydrostatic equilibrium, assuming a β–model for the gas density profile and a polytropic equation of state of the form $T \propto \rho_{\rm gas}^{\gamma-1}$. In order to verify whether such a procedure leads to a biased estimate of cluster masses, we follow this same procedure to estimate masses of simulated clusters. For each cluster, we fit the gas density profile to the β–model, while temperature and gas–density profiles are fitted to a polytropic equation of state. As shown in the central panel of Fig. 5, the effect of using such a mass estimator is that of lowering the normalization of the M–T relation and to bring it into better agreement with observations.
Allen et al. (2001) used Chandra data to resolve temperature profiles for hot relaxed clusters and avoided the assumption of a β–model for the gas density profile. In this way, the resulting M–T relation from their analysis can be directly compared to the simulation result based on the "true" cluster masses. The resulting best–fitting M_{2500}–T_{2500} relation is plotted as a dashed line in the right panel of Fig. 5 and compared to the results of our simulation. It is quite remarkable that the simulation results now agree with observations on the M–T relation. Still, this result suggests that observed and simulated M–T relations agree with each other when high–quality observational data are used which accurately resolves the surface–brightness profile and the temperature structure of clusters. Thanks to the good statistics offered by our simulation, a reliable calibration of the scatter expected in the M–T relation can be obtained. We suggest that this calibration should be used for the determination of cosmological parameters from the XTF and XLF. This shows the importance of a precise calibration of the relations between theory–predicted mass and X–ray observed quantities, and highlights the important role that large cosmological simulations can play in understanding the systematics involved.

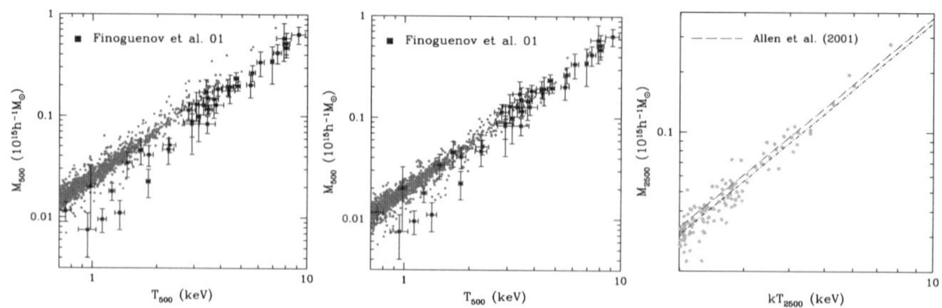

Figure 5. Comparison between the observed and the simulated M–T relation. Left and central panels refer to the relation at $\bar{\rho}/\rho_c = 500$. In the left panel we compare the results from Finoguenov et al. (2001) to the true total masses of simulated clusters. In the central panel, cluster masses are estimated by assuming a β–model and a polytropic equation of state in the solution of the equation of hydrostatic equilibrium. In the right panel the simulation results at $\bar{\rho}/\rho_c = 2500$ are compared to the relation found by Allen et al. (2001) from Chandra for hot relaxed clusters (dashed line); the dot–dashed line is our best–fit to the clusters with $T_{2500} > 2$ keV.

Acknowledgments

I would like to thank all the people who collaborated with me on the results presented in this paper. In alphabetic order, they are: A. Diaferio, K. Dolag, L. Moscardini, G. Murante, C. Norman, E. Pierpaoli, P. Rosati, Volker Springel, G. Tormen, L. Tornatore and P. Tozzi.

References

Allen S.W., Schmidt R.W., Fabian A.C., 2001, MNRAS, 328, L37
Borgani S., et al., 2001, ApJ, 561, 13
Borgani S., et al., 2003, MNRAS, submitted
Bryan G.L., Norman M.L., 1998, ApJ, 495, 80
Eke V.R., et al., 1998, MNRAS, 298, 1145
Ettori S., et al., 2003, A&A, submitted
Evrard A.E., et al., 2002, ApJ, 573, 7
Finoguenov A., Reiprich T.H., Bөhringer H., 2001, A&A, 369, 479
Frenk C.S., et al. 2000, ApJ, 525, 554
Ikebe Y., et al., 2002, A&A, 383, 773
Jenkins A., et al., 2001, MNRAS, 321, 372
Oukbir J., Blanchard A., 1992, A&A, 262, L21
Pierpaoli E., et al., 2003, MNRAS, 342, 163
Press W.H., Schechter P., 1974, ApJ, 187, 425
Rosati P., Borgani S., Norman C., 2002, ARAA, 40, 539
Sheth R.K., Tormen G., 1999, MNRAS, 308, 119
Seljak U., 2002, MNRAS, 337, 769
Spergel D.N., et al., 2003, ApJ, in press (preprint astro–ph/0302209)
Springel V., Hernquist L., 2003, MNRAS, 339, 289
Vikhlinin A., et al., 2002, ApJ, 578, L107
White M., 2002, ApJS, 143, 241

STRUCTURE FORMATION IN DYNAMICAL DARK ENERGY MODELS

A.V. Macciò[1], S.A. Bonometto[1], R, Mainini[1] & A. Klypin[2]
(1) Physics Dep. G. Occhialini, Univ. of Milano–Bicocca & I.N.F.N., Sezione di Milano Piazza della Scienza 2, 20126 Milano, Italy
(2) New Mexico State University, Las Cruces, New Mexico, USA
maccio@mib.infn.it, bonometto@mib.infn.it, mainini@mib.infn.it, aklypin@nmsu.edu

Abstract We perform n–body simulations for models with a DE component. Besides of DE with constant negative $w = p/\rho \geq -1$, we consider DE due to scalar fields, self–interacting through RP or SUGRA potentials. According to our post–linear analysis, at $z = 0$, DM power spectra and halo mass functions do not depend on DE nature. This is welcome, as ΛCDM fits observations. Halo profiles, instead, are denser than ΛCDM. For example, the density at $10\,h^{-1}$kpc of a DE $\sim 10^{13} M_\odot$ halo exceeds ΛCDM by $\sim 40\,\%$. Differences, therefore, are small but, however, DE does not ease the problem with cuspy DM profiles. On the contrary it could ease the discrepancy between ΛCDM and strong lensing data (Bertelmann 1998, 2002). We study also subhalos and find that, at $z = 0$, the number of satellites coincides in all DE models. At higher z, DE models show increasing differences from ΛCDM and among themselves; this is the obvious pattern to distinguish between different DE state equations.

Keywords: Audio quality measurements, perceptual measurement techniques

1. Introduction

Deep survey and CBR data confirm that $\sim 70\,\%$ of the world is Dark Energy (see, e.g., Efstathiou et al 2002, Percival et al 2002, Spergel et al 2003, Tegmark et al 2001, Netterfield et al 2002, Pogosian et al 2003, Kogut et al 2003), as needed to have the accelerated expansion shown by SNIa data (Riess et al 1998, Perlmutter et al 1999). The nature of Dark Energy (DE) is a puzzle. ΛCDM needs a severe fine–tuning of vacuum energy. DE with constant negative $w = p/\rho > -1$ has even less physical motivation. Apparently, the only viable alternative is dynamical DE, a *classical* self–interacting scalar field ϕ (Wetterich 1985). Among potentials $V(\phi)$ with a tracker solution, limiting the impact of initial conditions, Ratra–Peebles (1988, RP hereafter) and SUGRA (Brax & Martin 1999, 2000) potentials bear a particle physics motivation.

M. Plionis (ed.), Multiwavelength Cosmology, 199-202.

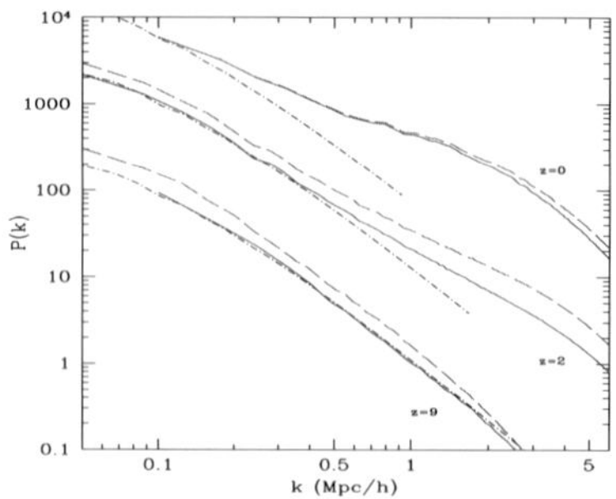

Figure 1. Spectrum evolution for ΛCDM (solid line) and RP (long dashed); the dot–dashed line is the linear prediction for ΛCDM.

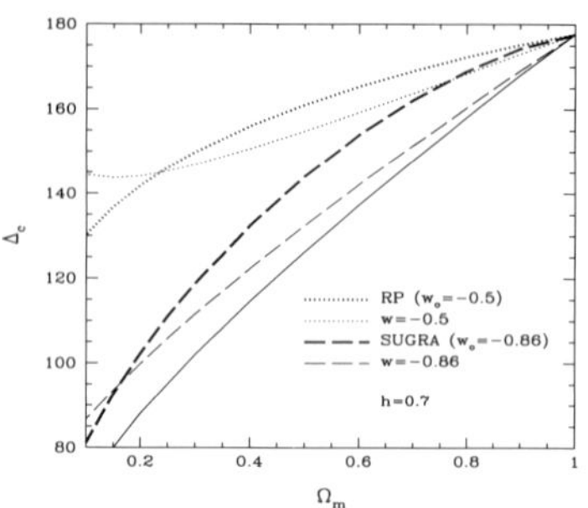

Figure 2. The virial density contrast Δ_c vs. Ω_m at $z = 0$. Models are indicated in the frame

Studying a dynamical DE model requires: (i) a linear treatment, to yield CBR spectra and transfer function; (ii) a post–linear treatment, to yield halo virial density contrasts and mass functions; (iii) a non–linear treatment. Here we report results on (ii) and (iii). We use the n–body program ART, modified to deal with any dependence of Ω_m (matter density parameter) on a (scale factor). Mainini et al (2003b) give analytical fitting formulae for such dependence. Further details are in Mainini, Macciò & Bonometto (2003a) and Klypin et al (2003). RP and SUGRA are parametrized by the energy scale Λ/GeV.

2. Non–linear results

Fig. 1 shows the evolution of the spectrum, as obtained from simulations of ΛCDM and RP models (Λ/GeV$= 10^3$), the most distant models treated. Models were normalized so to obtain the same number of halos at $z = 0$.
Halos were extracted from simulations using the virial density contrasts Δ_c obtained by Mainini et al (2003b), where one can find plots for the dependence $\Delta_c(a)$; here we show Δ_c dependence on Ω_m at $z = 0$ (Fig. 2) Figs. 3 & 4 show the mass function and its evolution in a number of models.
Using ART facilities, a particular halo was magnified in all simulations. Fig. 5 shows that its profile is NFW with a concentration depending on DE nature. Concentration can also be considered on a statistical basis. Fig. 6 shows how halo concentrations depend on the model. Here concentrations are defined as the ratio between the radius r_c at which the density contrast is 110 and the radius r_s in the NFW expression of the radial density.
We also studied how the number of satellites of a halo depends on DE nature. In Fig. 7 we report such dependence. However, also in this case, once care

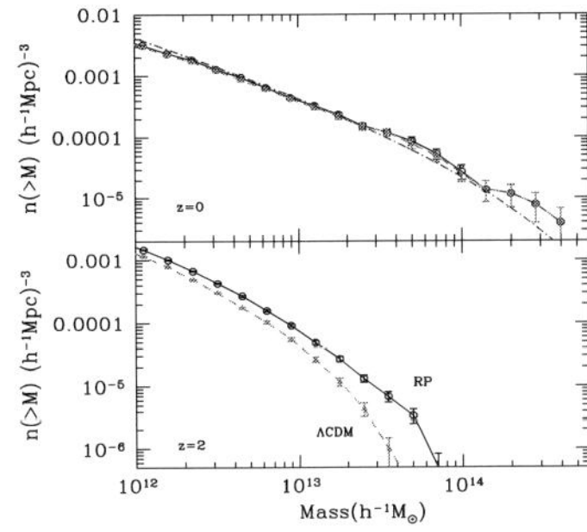

Figure 3. Mass function at $z = 0$ and $z = 2$ for the same models of Fig. 1. Evolution is faster for ΛCDM than for RP

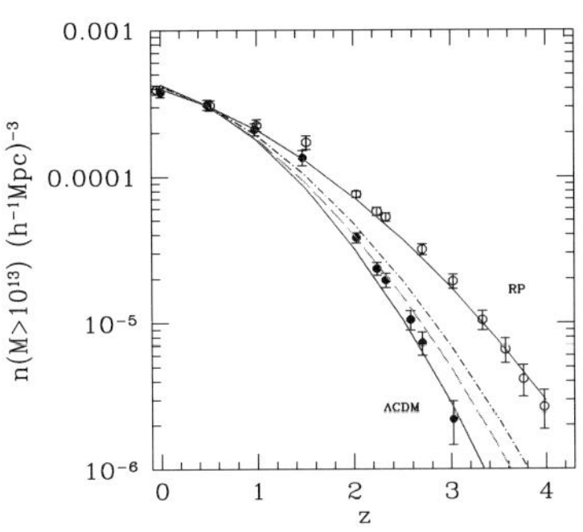

Figure 4. Halo number evolution in various models. Unlabeled curves refer to SUGRA and constant $w = -0.8$

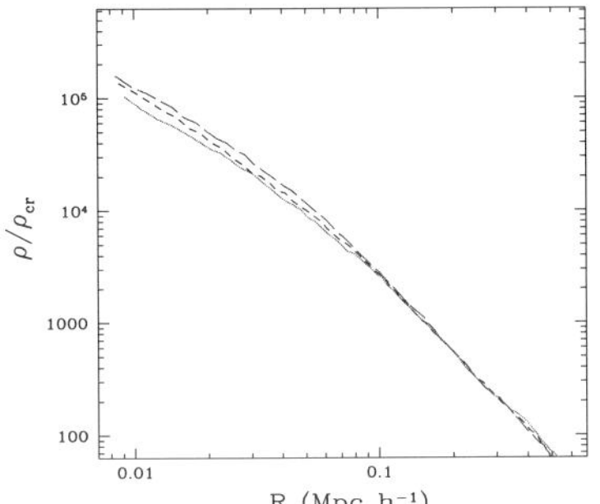

Figure 5. Density profile for a single magnified halo. Solid, short dashed, long dashed lines refer to ΛCDM, SUGRA, RP.

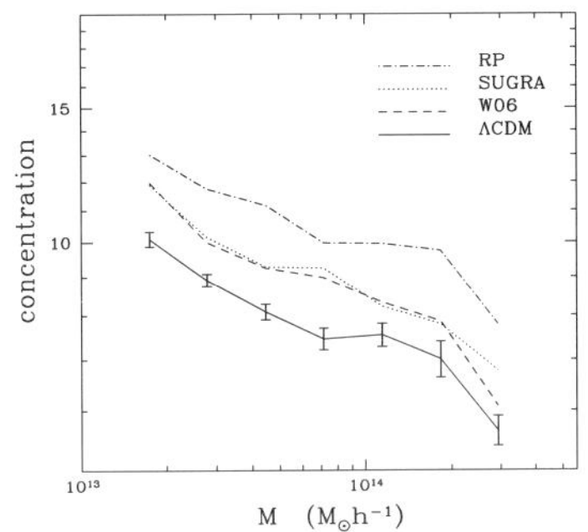

Figure 6. Concentration distribution in various models

is payed to properly normalize numbers to the same central halo velocity, no appreciable dependence on DE nature can be found.

3. Conclusions

In this paper we showed how a simple modification of the program ART permits to perform a wide analysis of dynamical DE models. This task is simplified by the very structure of the program, which uses the scale factor a as *time*–variable and requires only

$$dt/da = H_o^{-1}\sqrt{a\,\Omega_m(a)/\Omega_m(a_o)}$$

(H_o: today's Hubble parameter), to detail the action of forces. Once $\Omega_m(a)$ (the dependence of the matter density parameter on the scale factor) is as-

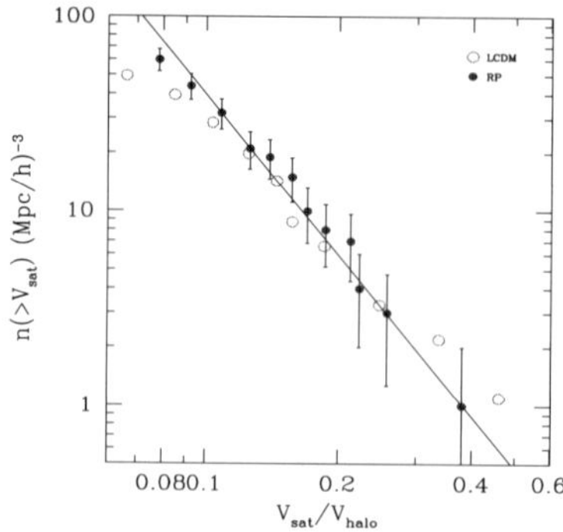

Figure 7. Number of halo satellites

signed, the dynamical problem is then properly defined. Most of the preliminary work was then performed at the post–linear level. This provided us suitable expressions for the virial density contrast, so that halos can be selected in the correct way in all models, and also the required fitting expressions for $\Omega_m(a)$.

Discriminating DE models from ΛCDM essentially requires good data at high redshift. A discrimination at $z = 0$ can be made only using an observable sensitive to the concentration distribution. In principle, such an observable exists and is related to strong lensing (giant halo statistic). Further work in this direction is in progress.

References

Bartelmann M., Huss A., Carlberg J., Jenkins A. & Pearce F. 1998, A&A 330, 1

Bartelmann M., Perrotta F. & Baccigalupi C. 2002, A&A 396, 21

Brax, P. & Martin, J., 1999, Phys.Lett., B468, 40 and 2000, Phys.Rev. D, 61, 103502

Efstathiou, G. et al., 2002, MNRAS, 330, 29

Klypin, A., Macciò, A.V., Mainini R. & Bonometto S.A., 2003, astro–ph/0303304 and ApJ (submitted)

Kogut et al., 2003, astro–ph/0302213

Mainini, R., Macciò, A.V. & Bonometto, S.A., 2003a, NewA 8, 172

Mainini, R., Macciò, A.V., Bonometto, S.A., & Klypin, A., 2003b, astro–ph/0303303 and ApJ (in press)

Netterfield, C. B. et al. 2002, ApJ, 571, 604

Percival W.J. et al., 2002, astro-ph/0206256, MNRAS (in press)

Perlmutter S. et al., 1999, ApJ, 517, 565

Pogosian, D., Bond, J.R., & Contaldi, C. 2003, astro-ph/0301310

Ratra B., Peebles P.J.E., 1988, Phys.Rev.D, 37, 3406

Riess, A.G. et al., 1998, AJ, 116, 1009

Spergel et al. 2003, astro–ph/0302209

Tegmark, M., Zaldarriaga, M., & Hamilton, AJ 2001, Phys.R., D63, 43007

Wetterich C., 1985, Nucl.Phys.B, 302, 668

STUDY OF GALAXY CLUSTER PROPERTIES FROM HIGH-RESOLUTION SPH SIMULATIONS

G. Yepes[1], Y. Ascasibar[1,2], S. Gottlöber[3] and V. Müller[3]

[1] *Grupo Astrofísica, Universidad Autónoma de Madrid*

[2] *Theoretical Physics, Oxford University*

[3] *Astrophysikalisches Institut Potsdam*

Abstract

We present some of the results of an ongoing collaboration to study the dynamical properties of galaxy clusters by means of high resolution adiabatic SPH cosmological simulations. Results from our numerical clusters have been tested against analytical models often used in X-ray observations: β model (isothermal and polytropic) and those based on universal dark matter profiles. We find a universal temperature profile, in agreement with AMR gasdynamical simulations of galaxy clusters. Temperature decreases by a factor 2-3 from the center to virial radius. Therefore, isothermal models (e.g. β model) give a very poor fit to simulated data. Moreover, gas entropy profiles deviate from a power law near the center, which is also in very good agreement with independent AMR simulations. Thus, if future X-ray observations confirm that gas in clusters has an extended isothermal core, then non-adiabatic physics would be required in order to explain it.

1. Introduction

Clusters of galaxies are the largest gravitationally bound structures in the universe. Therefore, they have often been considered as a canonical data set for cosmological tests. During the last two decades, a great effort has been devoted to investigate the mass distribution in CDM haloes by means of numerical N-body simulations. It is now firmly established that dark matter density profiles can be fitted by an universal two-parameter function, valid from galactic to cluster scales. For the gas component, the situation is less clear. The ICM is in the form of a hot diffuse X-ray emitting plasma, where the cooling time (except in the innermost regions) is typically longer than the age of the universe. Adiabatic gasdynamical simulations have therefore been used to study the formation and evolution of galaxy groups and clusters in different cosmologies. The Santa Barbara Cluster Comparison Project (Frenk et al., 1999, SBCCP) showed a clear difference between SPH and Eulerian Adaptive Mesh Refine-

M. Plionis (ed.), Multiwavelength Cosmology, 203-206.

ment (AMR) codes. While (the only one available at that time) Bryan and Norman's AMR code predicted an isentropic gas profile at the center, all the SPH codes used in SBCCP predicted an isothermal gas distribution almost to the virial radius of the Coma-like simulated cluster. As pointed out by several people (e.g. Borgani et al., 2002; Serna et al., 2003), the standard SPH method could suffer from entropy conservation problems. This is particularly accentuated in low mass resolution SPH simulations (see Lewis et al., 2000). A new implementation of SPH has been recently proposed (Springel and Hernquist, 2002) in which entropy conservation is much better fulfilled.
In order to asses the reliability of our numerical results, we did an extensive convergence study in terms of resolution (mass and spatial), as well as numerical technique. For this last purpose, we re-simulated one of our clusters with 3 different numerical codes: Tree-SPH GADGET, both with the standard (Springel et al., 2001) and the entropy conserving SPH implementation (Springel and Hernquist, 2002), as well as the Eulerian AMR code ART (Kravtsov et al., 2002). Radial profiles of gas and dark matter are compared in Figure 1. The agreement between AMR and the entropy version of GADGET is remarkable. The standard SPH GADGET still shows the same trend reported in Frenk et al. (1999), although our mass resolution is 64 times better (512^3 effective particles).

2. Numerical experiments

We have carried out a series of high-resolution gasdynamical simulations of cluster formation in a flat LCDM universe ($\Omega_{\rm m} = 0.3$; $\Omega_\Lambda = 0.7$; $h = 0.7$; $\sigma_8 = 0.9$; $\Omega_{\rm b} = 0.02\,h^{-2}$). Simulations were run with the entropy conserving SPH version of the parallel Tree code GADGET. We have selected 15 clusters extracted from a low-resolution (128^3) volume of $80h^{-1}$ Mpc. Each object has been re-simulated by means of the multiple mass technique (e.g. Klypin et al., 2001). We use 3 levels of mass refinement, reaching an effective resolution of 512^3 CDM particles ($\sim 3 \times 10^8\ h^{-1}$ $\rm M_\odot$). Gas has been added in the highest resolved area only. The gravitational smoothing was set to $\epsilon = 2 - 5\ h^{-1}$ kpc, depending on number of particles within the virial radius (Power et al., 2003). The minimum smoothing length for SPH was fixed to the same value as ϵ. The X-ray temperature of these objects ranges from 1 to 3 keV. For a more extended description of the numerical experiments, the reader is referred to Ascasibar (2003).

3. Results

A detailed discussion of the results from our numerical experiments can be found elsewhere (Ascasibar, 2003; Ascasibar et al., 2003). Here, we will focus on the radial structure of gas and dark matter in clusters.

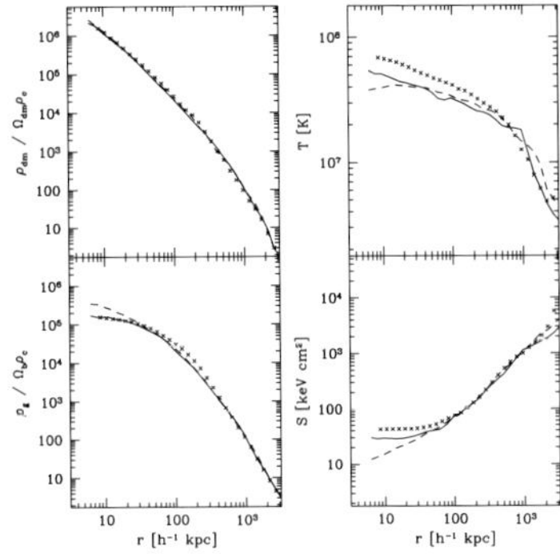

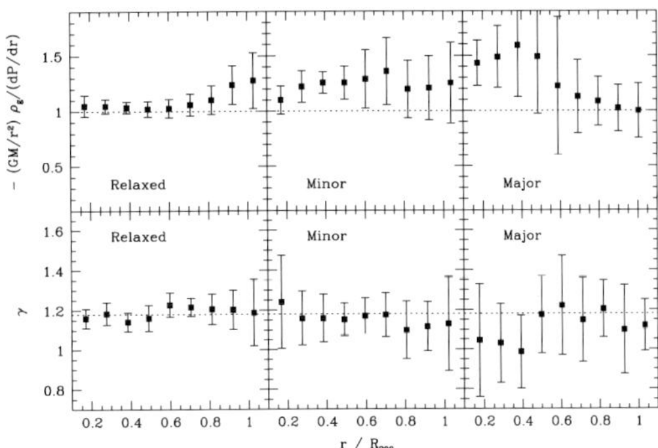

Figure 1. **Left:** Comparison of density, temperature and entropy profiles for a cluster simulated with 3 different numerical hydro codes: Standard SPH GADGET (dashed lines); Entropy-conserving GADGET (solid lines) and eulerian AMR ART code (crosses). **Right:** Testing Hydrostatic equilibrium (upper panel) and polytropic equation of state (lower panel) in our numerical clusters, classified according to their dynamical state.

We have considered four self-consistent analytical models, based on the hypotheses that the hot ICM gas is in hydrostatic equilibrium with the dark matter halo and that it follows a polytropic equation of state. Two of our models assume NFW (Navarro et al., 1997) and MQGSL (Moore et al., 1999) formulae to describe the CDM density profile, whereas the other two assume a β-model for the gas distribution. One is an isothermal version with $\beta = 2/3$ (BM) and the other is a polytropic model with $\gamma = 1.18$ and $\beta = 1$ (PBM). We have first tested the hypothesis of hydrostatic equilibrium (HE) and polytropic equation of state (e.o.s) for the gas in our clusters. In Figure 1 (right) we show the results of this test. H.E. is nicely fulfilled by those clusters that are in a relaxed or minor merger state. The e.o.s for the gas in these clusters can be reasonably approximated by a constant polytropic index of $\gamma \sim 1.2$. Then, we compared the simulated radial distributions of gas and dark matter with each analytical model. By fitting the numerical X-ray surface brightness, the models based on universal CDM profiles are able to estimate the ICM properties within $30 - 40\%$ errors. β-models yield similar estimates for $r \geq 0.1R_{200}$, but the shape of the inferred profiles at smaller radii are severely misleading.

In Figure 2 (left) we plot the spherically averaged temperature profile of our halos, together with the best fit for each analytical model. As can be seen, the β-model gives the poorest fit because gas in simulations is far from being isothermal.

The projected emission-weighted temperature profile is also shown in Figure 2 (right), compared with recent AMR cluster simulations (Loken et al., 2002) and with X-ray observations. Both sets of simulations predict the same uni-

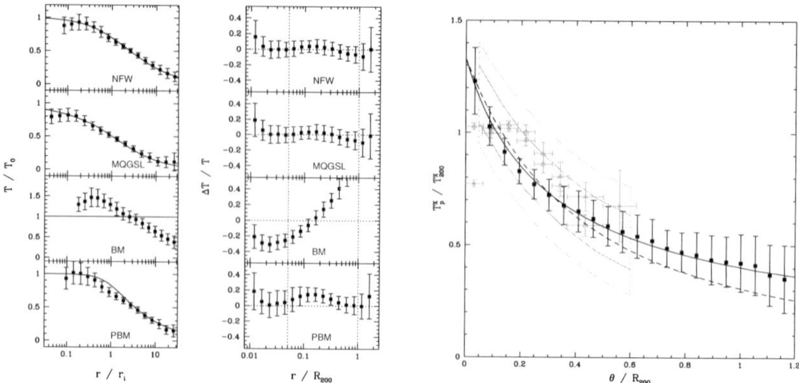

Figure 2. **Left:** 3d Temperature profiles from clusters (points) and the corresponding fit from 4 analytical models (see Ascasibar et al., 2003 for more information). **Right:** Projected emission-weighted temperature profile (black squares with error bars). Dashed line: results from AMR simulations (Loken et al., 2002), solid line fit to our data Loken et al., 2002. Observational data from De Grandi and Molendi, 2002 (triangles) and Markevitch et al., 1998 (enclosed boxes).

versal gas temperature profile for clusters, with no indication of an isothermal core. This is one of the few cases in which SPH and AMR simulations agree so well on one issue. If the existence of a large isothermal core is indeed confirmed by upcoming X-ray observations, it would be an indication that non-adiabatic processes must be considered in numerical simulations of galaxy cluster formation.

References

Ascasibar, Y. (2003). PhD thesis, Univ. Autonoma de Madrid (Spain), (astro-ph/0305250).

Ascasibar, Y., Yepes, G., Müller, V., and Gottlöber, S. (2003). *MNRAS,* (astro-ph/0306264).

Borgani et al. (2002). *MNRAS*, 336:409–424.

De Grandi, S. and Molendi, S. (2002). *ApJ*, 567:163–177.

Frenk et al. (1999). *ApJ*, 525:554–582.

Klypin, A., Kravtsov, A. V., Bullock, J. S., and Primack, J. R. (2001). *ApJ*, 554:903–915.

Kravtsov, A. V., Klypin, A., and Hoffman, Y. (2002). *ApJ*, 571:563–575.

Lewis et al. (2000). *ApJ*, 536:623–644.

Loken et al. (2002). *ApJ*, 579:571–576.

Markevitch, M., Forman, W. R., Sarazin, C. L., and Vikhlinin, A. (1998). *ApJ*, 503:77.

Moore, B., Quinn, T., Governato, F., Stadel, J., and Lake, G. (1999). *MNRAS*, 310:1147–1152.

Navarro, J. F., Frenk, C. S., and White, S. D. M. (1997). *ApJ*, 490:493.

Power et al. (2003). *MNRAS*, 338:14–34.

Serna, A., Domínguez-Tenreiro, R., and Saiz, A. (2003). *MNRAS, submitted.*

Springel, V. and Hernquist, L. (2002). *MNRAS*, 333:649–664.

Springel, V., Yoshida, N., and White, S. D. M. (2001). *New Astronomy*, 6:79–117.

X-RAY CLUSTER PROPERTIES IN SPH SIMULATIONS OF GALAXY CLUSTERS

R. Valdarnini
SISSA, Vie Beirut 2-4 34014, Trieste Italy

Abstract Results from a large set of hydrodynamical SPH simulations of galaxy clusters in a flat LCDM cosmology are used to investigate cluster X-ray properties. The physical modeling of the gas includes radiative cooling, star formation, energy feedback and metal enrichment that follows from the explosions of SNe type II and Ia. The metallicity dependence of the cooling function is also taken into account. It is found that the luminosity-temperature relation of simulated clusters is in good agreement with the data, and the X-ray properties of cool clusters are unaffected by the amount of feedback energy that has heated the intracluster medium (ICM). The fraction of hot gas f_g at the virial radius increases with T_X and the distribution obtained from the simulated cluster sample is consistent with the observational ranges.

Keywords: SPH simulations, heating of the ICM.

1. Introduction

There is a wide observational evidence Allen & Fabian 1998; Markevitch 1998 that the observed cluster X-ray luminosity scales with temperature with a slope which is steeper than that predicted by the self-similar scaling relations ($L_X \propto T_X^3$). This implies that low temperature clusters have central densities lower than expected Ponman, Cannon & Navarro 1999. This break of self-similarity is usually taken as a strong evidence that non-gravitational heating of the ICM has played an important role in the ICM evolution. A popular model which has been considered as a heating mechanism for the ICM is supernovae (SNe) driven-winds White 1991. An alternative view is that radiative cooling and the subsequent galaxy formation can explain the observed $L_X - T_X$ relation because of the removal of low-entropy gas at the cluster cores Bryan 2000. In this proceedings I present preliminary results from a large set of hydrodynamical SPH simulations of galaxy clusters. The physical modeling of the gas includes a number of processes (see later), and the simulations have been performed

M. Plionis (ed.), Multiwavelength Cosmology, 207-210.

in order to investigate the consistency of simulated cluster scaling relations against a number of data.

2. Simulations and results

Hydrodynamical TREESPH simulations have been performed in physical coordinates for a sample of 120 test clusters. A detailed description of the procedure can be found in Valdarnini (2003, V03). The cosmological model is a flat CDM model, with vacuum energy density $\Omega_\Lambda = 0.7$, matter density parameter $\Omega_m = 0.3$ and Hubble constant $h = 0.7 = H_0/100 Km sec^{-1} Mpc^{-1}$. $\Omega_b = 0.019h^{-2}$ is the value of the baryonic density. The clusters are the 120 most massive ones identified at $z = 0$ in a cosmological $N-$body run with a box size of $200h^{-1}$ Mpc. The virial temperatures range from $\sim 6KeV$ down to $\sim 1KeV$. The simulations have $N_g \simeq 70,000$ gas particles. The cooling function of the gas particles also depends on the gas metallicity, and cold gas particles are subject to star formation. Once a star particle is created it will release energy into the surrounding gas through SN explosions of type II and Ia. The feedback energy (10^{51} erg) is returned to the nearest neighbor gas particles, according to its lifetime and IMF. SN explosions also inject enriched material into the ICM, thus increasing its metallicity with time.

Observational variables of the simulated clusters are plotted in Fig. 1 as a function of the cluster temperature T_X. Different symbols are for different redshifts ($z = 0$, $z = 0.06$ and $z = 0.11$). For the sake of clarity, not all the points of the numerical sample are plotted in the figure. Global values $A_{Fe} = M_{Fe}/M_H$ of the iron abundances for the simulated clusters are shown in panel (a). A comparison with real data of a small sample subset (four clusters , see V03) shows a good agreement with the measured values. The values of A_{Fe} appear to increase with T_X, even though the observational evidence of an iron abundance increasing with T_X is statistically weak Mushotzky & Lowenstein 1997.

The fraction of hot gas $f_g = M_g/M_T$ is defined within a given radius as the ratio of the mass of hot gas M_g to the total cluster mass M_T. The fraction f_g has been calculated at a radius enclosing a gas overdensity of $\delta = 500$ relative to the critical density. The values of f_g of the numerical cluster sample are plotted in panel (b). There is a clear tendency for the f_g distributions to increase with cluster temperature. This is in accord with theoretical predictions of the radiative cooling model Bryan 2000, and with numerical simulations Muanwong et al. 2002 ; Dave et al. 2002. A subsample of the f_g distribution at $z = 0$ is found to be statistically consistent (V03) with the data of Arnoud & Evrard (1999). Furthermore, Sanderson et al. (2003) found strong observational evidence for a dependence of f_g with T_X.

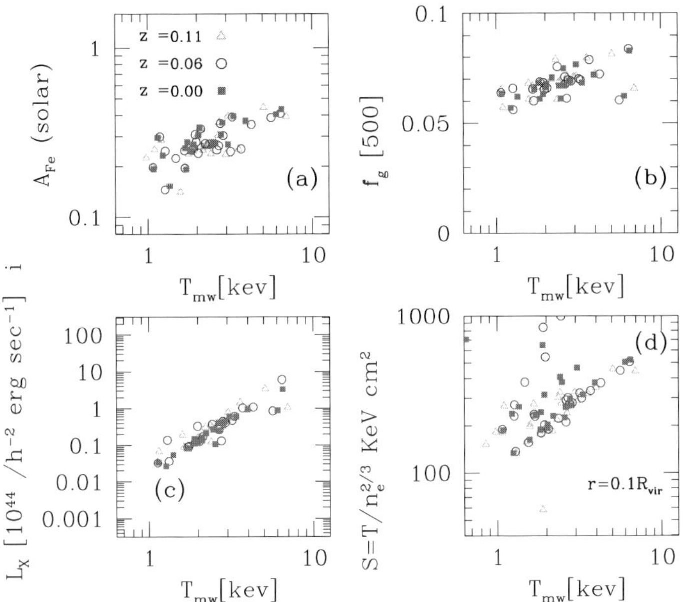

Figure 1. **(a)**: Average iron abundances of the simulated cluster sample at $r = 0.5h^{-1}Mpc$ are plotted versus cluster temperatures. Different symbols refer to different redshitfs. **(b)**: As in panel (a), but for the gas fractions $f_g = M_g(< r)/M_T(< r)$; the fractions are evaluated at the radius within which $\rho/\rho_c = \delta$, with $\delta = 500$. **(c)**: Bolometric X-ray luminosities are plotted as a function of the temperature. **(d)**: Cluster core entropies $S(r) = k_B T(r)/n_e(r)$ at the radius $r = 0.1r_{vir}$ are shown against cluster temperatures.

The values of the bolometric X-ray luminosity L_X are shown in panel (c) as a function of the cluster temperature. Mass-weighted temperatures have been used as unbiased estimators of the spectral temperatures Mathiesen & Evrard 2001. A central region of size $50h^{-1}Kpc$ has been excised Markevitch 1998 in order to remove the contribution to L_X of the central cooling flow. The L_X of the simulations at $z = 0$ are in good agreement with the data (V03) over the entire range of temperatures. An additional run has been performed for the cluster with the lowest temperature, but with a SN feedback energy of 10^{50} erg for both SNII and Ia. The final L_X of the run is very similar to that of standard run. This demonstrates that the final X-ray luminosities of the simulations are not sensitive to the amount of SN feedback energy that has heated the ICM.

The core entropies of the clusters are displayed in panel (d) against the cluster temperatures. The cluster entropy is defined as $S(r) = k_B T(r)/n_e(r)$, where n_e is the electron density. The core entropy is calculated at a radius which is 10% of the cluster virial radius. The result indicates that for low temperatures part of the sample has clusters with core entropies which approach a floor at $\sim$ 100kev cm^{-2}, while for some clusters the decline of entropy with temperature is steeper and close to the self-similar predictions ($S \propto T$). This different behaviour could be due to the different dynamical histories of the clusters. It is worth stressing that for all the plotted quantities there is little evolution below $z = 1$.

To summarize, simulation results give final X-ray properties in broad agreement with the data. These findings support the so-called radiative model, where the X-ray properties of the ICM are driven by the efficiency of galaxy formation rather than by the heating due to non-gravitational processes.

References

Allen S. W., Fabian A. C., 1998, MNRAS, 297, L57

Arnaud M., Evrard A. E., 1999, MNRAS, 305, 631

Bryan G. L., 2000, ApJ, 544, L1

Dave R., Katz N., Weinberg D.H., 2002, ApJ, 579, 23

Mathiesen B., Evrard A. E., 2001, ApJ, 546, 100

Markevitch M., 1998, ApJ, 504, 27

Muanwong O., Thomas P. A., Kay, S. T., Pearce F. R., 2002, astro-ph/0205137

Mushotzky R. F., Lowenstein, M., 1997, ApJ, 481 L63

Ponman T. J., Cannon D. B., Navarro J. F., 1999, Nature, 397, 135

Sanderson A. J. R, Ponman, T. J., Finoguenov A., Lloyd-Davies E. J. & Markevitch, M., 2003, MNRAS, 340, 989

Valdarnini, R. 2003, MNRAS, 339, 1117

White R. E., 1991, ApJ, 367, 69

EVOLUTION OF MAGNETIC FIELDS IN GALAXY CLUSTERS

Klaus Dolag
Dipartimento di Astronomia, Universitą di Padova, Italy

Abstract Cosmological simulations of magnetic fields in galaxy clusters show that remarkable agreement between simulations and observations of Faraday rotation can be achieved assuming that seed fields of $\approx 10^{-9}$G where present at redshifts $\approx$ 15-20. Previous simulations showed that the structure of the seed field is irrelevant for the final intracluster field. On average, the field grows exponentially with decreasing cluster redshift, but merger events cause steep transient increases in the field strength. A set of new, high resolution simulations of clusters will allow us to study the predicted magnetic field in galaxy clusters in even more detail. Using the observed correlation between X-ray surface brightness and Faraday rotation measure we confirm that the magnetic field scales with the intracluster gas density. The observations of low temperature clusters also indicate that the central magnetic field strength is correlated with the cluster temperature, as predicted by the simulations.

1. Introduction

Observations of the so called Faraday Rotation Measure as well as observations of diffuse radio haloes in galaxy clusters consistently show that clusters of galaxies are pervaded by magnetic fields of $\sim \mu$G strength. Coherence of the observed Faraday rotation across large radio sources demonstrates that there is at least a field component that is smooth on cluster scales. The origin of such fields is largely unclear.

We used magneto hydrodynamical simulations of structure formation to study the evolution and the final magnetic field structure, investigating the obtained magnetic field profile and the typical reversal scale. The models generally reproduce the observed rotation measure very well. Specially the obtained correlation between the rotation measure and the x-ray surface brightness, which nicely reproduce the observed one, suggest, that also the magnetic field in real galaxy clusters has a similar structure.

M. Plionis (ed.), Multiwavelength Cosmology, 211-214.

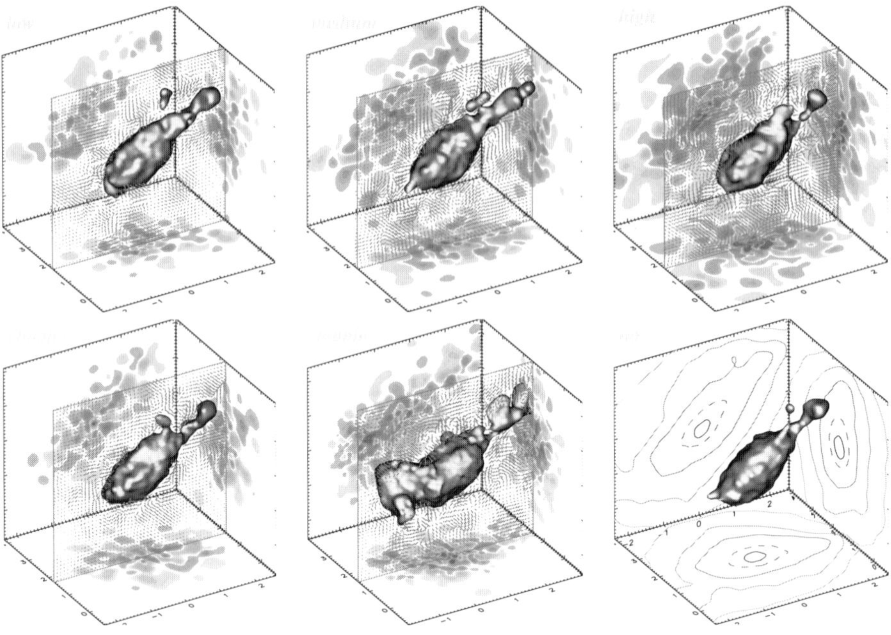

Figure 1. Shown a simulated cluster for different magnetic field configurations as labeled. The gray contours in the boxes are isodensity curves of the intra cluster gas. The patches on the box sides show the Faraday rotation measure. The frame in the middle of the boxes shows a slice through the intracluster magnetic field, with the arrows indicating field strength.

2. Previous simulations

We used the cosmological MHD code described in Dolag, Bartelmann & Lesch (1999,2002) to simulate the formation of magnetized galaxy clusters from an initial density perturbation field. Our main results can be summarized as follows: (i) Initial magnetic field strengths are amplified by approximately three orders of magnitude in cluster cores, one order of magnitude above the expectation from flux conservation and spherical collapse. (ii) Vastly different initial field configurations (homogeneous or chaotic) yield results that cannot significantly be distinguished. (iii) Micro-Gauss fields and Faraday-rotation observations are well reproduced in our simulations starting from initial magnetic fields of $\sim 10^{-9}$ G strength at redshift 15 for both explored cosmologies, namely an EdS and a flat, low density universe. Figure 1 shows one simulated clusters for different initial magnetic field configurations.

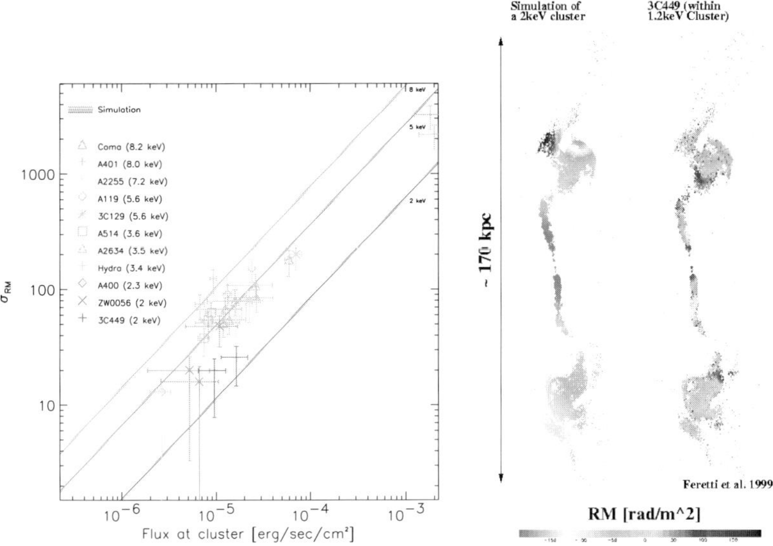

Figure 2. Left Panel shows the Lx-RM correlation for a collection of observations of embedded radio sources in different clusters compared with predictions from the simulations. Data points see references in Dolag 2002. Right panel shows a comparison of observed and synthetic RM embedded in a 1kev cluster.

3. Lx-RM correlation

As the simulations predict the statistics of the rotation measure within our models, we can comparing this with the x-ray emission to provide an opportunity to measure the shape of the magnetic field in galaxy clusters. This mean, we are comparing the two line of sight integrals

$$\int n_{\mathrm{e}}^2\,\sqrt{T}\,\mathrm{d}x \Longleftrightarrow \int n_{\mathrm{e}}\,B_{\|}\,\mathrm{d}x$$

with each other. Neglecting the dependency on the temperature, the slope of the correlation tells us, how the magnetic filed scales with the density inside the cluster. Left part of Figure 2 shows the correlation between the x-ray surface brightness and the RMS in a synthetic observation compared with real observations. Clear to see that both agree. The slope of one translate into a linear relation between magnetic field and intracluster gas density. Specially the low temperature clusters are showing some shift of this correlation with temperature, as predicted by the simulations.

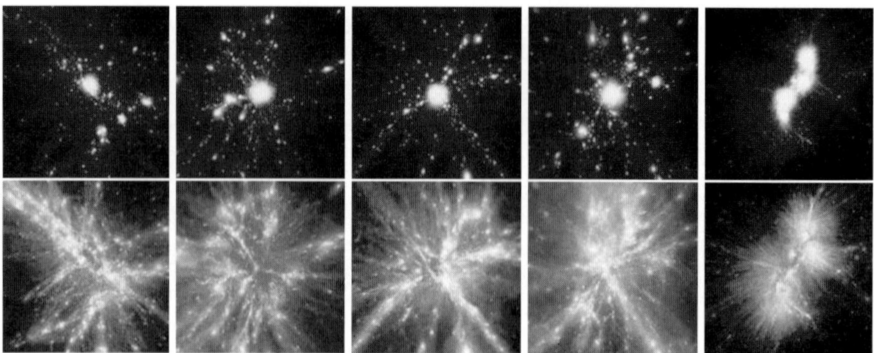

Figure 3. Visualization of the gas density and the temperature of the simulations containing the 5 most massive clusters in our set.

4. Next Generation of Simulations

We created a set of 10 suitable initial conditions for cluster simulations. This was done quite carefully by optimizing the number of particles needed for the objects we want to simulate and the corresponding mass resolution of 1.3×10^9 $M_\odot$ for the dark matter particles. This leads to a very large improvement of the resolution with respect to other simulations done before. Fig. 3) shows a visualization of the final distribution of the gas density (upper row) and temperature distribution (lower row) for the simulations containing the 5 biggest clusters. The most massive cluster in our simulations contains at the end more than 2 Millions gas particles within R_{200}. The initial conditions are also trimmed to have the same mass resolution for all objects: this allows us to better understand the effects of the additional physics onto the dynamics of the objects in our simulations. Two simulations also contain a slightly larger region of space. One is hosting a small supercluster structure, the other one is hosting a long filamentary structure connecting three massive clusters. These two simulations will allow us to accurately study the gas properties between galaxy clusters. See `http://dipastro.pd.astro.it/~cosmo/Clusters/`

The right part of Figure 2 shows first results from one of these high resolution simulations including magnetic fields. Clear to see that these new high resolution simulations are suitable to study not only average properties like radial profile but also detailed structure of predicted rotation measure maps.

References

Dolag, K., Bartelmann, M., Lesch, H. 1999, A&A, 348, 351

Dolag, K., Bartelmann, M., Lesch, H., 2002, A&A, 387, 383

Dolag 2002, Proceedings to 'Matter and Energy in Clusters of Galaxies', Taiwan, Co-Editors S. Bowyer and C.-Y. Hwang, ASP 2002

FIRST STARBURSTS AT HIGH REDSHIFT: FORMATION OF GLOBULAR CLUSTERS

Oleg Y. Gnedin
Space Telescope Science Institute
ognedin@stsci.edu

Abstract Numerical simulations of a Milky Way-size galaxy demonstrate that globular clusters with the properties similar to observed can form naturally at $z > 3$ in the concordance ΛCDM cosmology. The clusters in our model form in the strongly baryon-dominated cores of supergiant molecular clouds. The first clusters form at $z \approx 12$, while the peak formation appears to be at $z \sim 3-5$. The zero-age mass function of globular clusters can be approximated by a power-law $dN/dM \propto M^{-2}$ in agreement with observations of young massive star clusters.

For the first time it has been possible to include the formation of globular clusters self-consistently into the hierarchical galaxy formation model Kravtsov and Gnedin, 2003. The simulations used in our study were performed using the Eulerian gasdynamics+N-body ART code which uses an adaptive mesh refinement approach to achieve high dynamic range Kravtsov et al., 1997. Several physical processes critical to various aspects of galaxy formation are included: star formation; metal enrichment and thermal feedback due to supernovae type II and type Ia; self-consistent advection of metals; metallicity- and density-dependent cooling and UV heating due to the cosmological ionizing background. The simulation follows the early ($z > 3$) stages of the evolution of a Milky Way-type galaxy, $M \approx 10^{12} h^{-1}$ M$_\odot$ at $z = 0$.

Although the resolution achieved in the disk region is very high by cosmological standards, $\Delta x \approx 50$ pc at $z = 4$, it is still insufficient to resolve the formation of stellar clusters. The resolution, however, is sufficient to identify the potential sites for GC formation. The natural candidates are cores of giant molecular clouds in high-redshift galaxies Harris and Pudritz, 1994. Accordingly, we complement the simulations with a physical description of the gas distribution on a subgrid level. It assumes an isothermal structure of the unresolved core and a universal density threshold of cluster star formation, $\rho_{\rm sf} = 10^4$ M$_\odot$ pc^{-3}. The gas at densities above $\rho_{\rm sf}$ is converted into a star

M. Plionis (ed.), Multiwavelength Cosmology, 215-218.

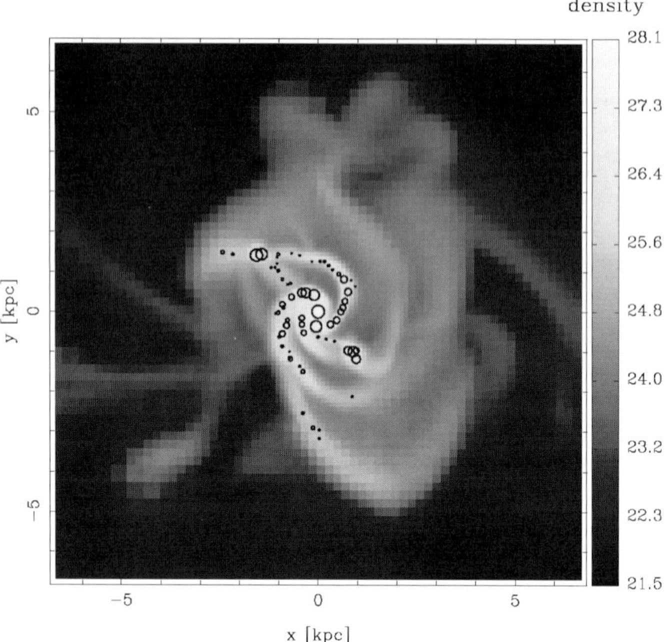

Figure 1. Projected gas density in the most massive disk at $z = 4$. A nearly face-on disk has prominent spiral arms and is in the process of very active accretion and merging. The globular clusters identified at this epoch are shown by circles.

cluster with the efficiency $\epsilon = M_*/M_{\rm core} = 60\%$. The size of the cluster after the gradual loss of the remaining gas is $R_* = R_{\rm core}/\epsilon$.

The first globular clusters formed around $z \approx 12$ and cluster formation continued at least until $z \approx 3$. The preferred epoch of globular cluster formation ($z \sim 3-5$) corresponds to the cosmic time of only $1-2$ Gyr, when the gas supply is abundant in the disks of the progenitor halos and the merger rate of the progenitors is high. All old globular clusters thus appear to have similar ages.

Most globular clusters in our model form in halos of mass $> 10^9$ $M_\odot$. *Within the progenitor systems, globular clusters form in the highest-density regions of the disk.* In the most massive disk in the simulation (Figure 1) the newly formed clusters trace the spiral arms and the nucleus, similarly to the young star clusters observed in merging and interacting galaxies Whitmore et al., 1999.

Note that although the high-redshift globular clusters form in gaseous disks, the subsequent accretion of their parent galaxies would lead to tidal stripping and disruption. The clusters are likely to share the fate of the stripped stars

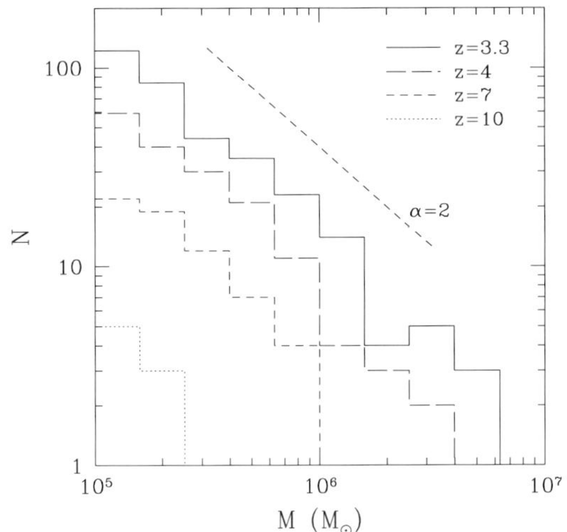

Figure 2. The build-up of the initial mass function of globular clusters at four epochs. The straight dashed line shows a power-law, $N \propto M^{-2}$. The slope α becomes shallower with decreasing redshift and saturates at $\alpha = 2.05 \pm 0.07$ for $z < 4$.

that build up the galactic stellar halo and should have roughly *spherical spatial distribution at* $z = 0$.

The total mass of clusters within a parent galactic halo correlates with the total galaxy mass $M_{\rm h}$: $M_{\rm GC} = 3.2 \times 10^6\,{\rm M}_\odot\,\left(M_{\rm h}/10^{11}\,{\rm M}_\odot\right)^{1.13\pm0.08}$. The global efficiency of cluster formation, $M_{\rm GC}/M_{\rm h}$, therefore depends only weakly on the galaxy mass. The mass of the globular cluster population in a given region also strongly correlates with the local average star formation rate density: $M_{\rm GC} \propto \Sigma_{\rm SFR}^{0.75\pm0.06}$ at $z = 3.3$. A similar correlation, albeit with a significant scatter, was reported for the observed present-day galaxies Larsen, 2002. In our model the correlation arises because both the star formation rate and mass of the globular cluster population are controlled by the same parameter, the amount of gas in the densest regions of the ISM.

The metallicities, which the clusters acquire from the gas in which they form, are remarkably similar to the metal-poor part of the Galactic cluster distribution. Note that at all epochs the dynamical time of the molecular cores is very short ($\sim 10^6$ yrs), which means that the galactic gas is pre-enriched even before the first clusters form. Thus *the oldest globular clusters do not contain the oldest stars in the Galaxy*.

The mass function of the model clusters at all output epochs (Figure 2) is well fit by a power-law $dN/dM \propto M^{-2}$ for $M > 10^5\,{\rm M}_\odot$, in agreement with the

mass function of *young* star clusters in normal spiral and interacting galaxies Zhang and Fall, 1999. The power-law slope at high masses also resembles that of the high-redshift mass function of dark matter halos in the hierarchical CDM cosmology Gnedin, 2003. We find that although the most massive cluster in each galaxy correlates with the parent galaxy mass, $M_{\rm max} \propto M_{\rm h}^{1.3\pm0.1}$, a significant scatter around this relation implies that for a given halo the masses of individual clusters can vary by a factor of three. Therefore, the *overall shape of the mass function of globular clusters in our model does not follow directly from the mass function of their parent halos, but depends also on the mass function of molecular cloud cores within individual galaxies.*
The slope of the mass function steepens with increasing mass. We find a relatively shallow, $\alpha \approx 1.7$, mass function for the small-mass clusters forming in lower-density cores and the steep, $\alpha \approx 2$, mass function for massive globular clusters forming in the densest regions of the disk. The steepening of the luminosity function with increasing luminosity is also observed for young star clusters in starbursting galaxies Whitmore et al., 1999; Larsen, 2002. *Thus over the range of masses typically probed in observations, $M > 10^5$ M$_\odot$, the mass function can be approximated by a power-law because the expected curvature over this range is rather small.*

References

Gnedin, O. Y. (2003). Formation of Globular Clusters: In and Out of Dwarf Galaxies. *in Extragalactic Globular Cluster Systems, ed. M. Kissler-Patig (Springer), p. 224-229;* astro-ph/0210556.

Harris, W. E. and Pudritz, R. E. (1994). Supergiant molecular clouds and the formation of globular cluster systems. *ApJ*, 429:177–191.

Kravtsov, A. V. and Gnedin, O. Y. (2003). Formation of globular clusters in hierarchical cosmology. *ApJ, submitted; astro-ph/0305199.*

Kravtsov, Andrey V., Klypin, Anatoly A., and Khokhlov, Alexei M. (1997). Adaptive Refinement Tree: A new high-resolution N-body code for cosmological simulations. *ApJ Suppl.*, 111:73–94.

Larsen, S. S. (2002). The Luminosity Function of Star Clusters in Spiral Galaxies. *AJ*, 124:1393–1409.

Whitmore, B. C., Zhang, Q., Leitherer, C., Fall, S. M., Schweizer, F., and Miller, B. W. (1999). The Luminosity Function of Young Star Clusters in "the Antennae" Galaxies (NGC 4038-4039). *AJ*, 118:1551–1576.

Zhang, Q. and Fall, S. M. (1999). The Mass Function of Young Star Clusters in the "Antennae" Galaxies. *ApJ Lett.*, 527:L81–L84.

EXPECTED PROPERTIES OF PRIMEVAL GALAXIES AND CONFRONTATION WITH OBSERVATIONS

Daniel Schaerer
Geneva Observatory, 51, Ch. des Maillettes, CH-1290 Sauverny, Switzerland
daniel.schaerer@obs.unige.ch

1. Properties of low metallicity and PopIII starbursts – predictions and confrontation with observations

Based on existing or new stellar tracks and appropriate non-LTE model atmospheres we have recently undertaken a systematic study of the expected rest-frame EUV, UV to optical properties of starbursts of zero and higher metallicity (Schaerer 2002, 2003, hereafter S02, S03). These studies provide predicted SEDs including stellar and nebular emission (continuum and lines), detailed ionizing properties, calibrations for star formation indicators (S03), and also predicted metal production rates for various elements (S02). These should be useful for a variety of studies concerning high redshift galaxies, cosmological reionisation studies and others.

Among the main observational characteristics found for PopIII and metal-poor starbursts are:

1) Continuum emission is expected to be dominated by nebular (bound-free, free-free and two photon) emission even in the rest-frame UV domain (S02). This leads to a flatter intrinsic SED compared to normal galaxies.

2) The maximum Lyα emission (i.e. neglecting possible radiation transfer effects through the galaxian ISM and IGM and dust) of integrated stellar populations increases quite strongly with decreasing metallicity. This is illustrated in Fig. 1a considering also various cases of the IMF at low metallicity.

3) Strong He II recombination lines (He II $\lambda1640$, He II $\lambda4686$,...) are a quite unique signature due to hot massive main sequence stars of PopIII/very low metallicity (cf. Tumlinson & Shull 2000). Significant He II emission is, however, only expected at metallicities below $\lesssim 10^{-5}$ solar (S03). This is e.g. illustrated by plotting the ratio of He^+ to H ionizing photons, $Q(\mathrm{He}^+)/Q(\mathrm{H})$, computed for a SF population at equilibrium, i.e. for constant SF (Fig. 1b). These findings are also confirmed by the study of Panagia, Stiavelli, and collaborators (see these proceedings).

M. Plionis (ed.), Multiwavelength Cosmology, 219-222.

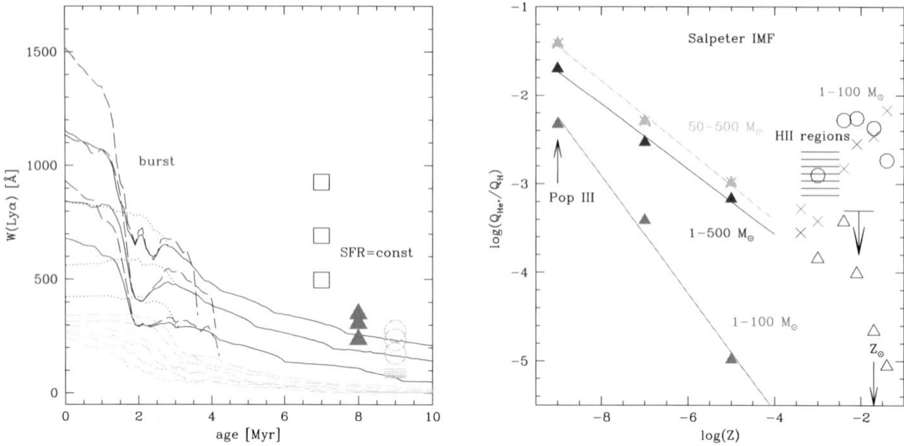

Figure 1. **a–left:** Predicted Lyα equivalent widths for bursts of different metallicities and IMFs. **b–right:** Predicted hardness $Q(\mathrm{He}^+)/Q(\mathrm{H})$ as a function of metallicity for starbursts between PopIII and normal metallicities. Figures taken from Schaerer (2003) – see there for a complete explanation of the figures

In short, the following three characteristics can be identified as fairly clear signatures of very metal-poor or PopIII starbursts ("primeval galaxies"): Strong Lyα emission, narrow nebular He II emission, and very weak or absent metal lines. We now discuss more precisely these "criteria" and confront them with existing observations.

1.1 Strong Lyα emission

If Lyα equivalent widths $\gtrsim$ 500 A are measured, this would, according to Fig. 1a, require stellar populations with metallicities $Z \lesssim 10^{-5}$ (i.e. [Z/H] $\lesssim -3.3$ in solar units) and/or an extreme IMF. Such large $W(\mathrm{Ly}\alpha)$ cannot be explained by stellar photoionisation in galaxies with a Salpeter like IMF and higher metallicities. Alternative explanations include AGNs.

In fact a good fraction of the Lyα emitters found by the LALA survey seems to show such large equivalent widths (Malhotra & Rhoads 2002). However, it must be remembered that the equivalent widths are determined from narrow-band (NB) and R-band imaging, which could lead to important uncertainties, as e.g. noted by comparing similar data from NB imaging and spectroscopy of SUBARU data of Ouchi et al. (2003 and private comm.). Spectroscopic follow-up of 5 LALA sources seem, however, to confirm their previous equivalent widths measurements from imaging (Rhoads et al. 2003). In principle it is thus possible that (some of) the strong Lyα emitters of the LALA are such primeval galaxies. Whether such large numbers of objects at relatively low redshifts (e.g. at $z = 4.5$ Malhotra & Rhoads 2002) would truly be expected seems surprising (cf. Scannapieco et al. 2003). No doubt detailed follow-up work on these objects will be of great interest.

1.2 Nebular He II emission

As mentioned above the detection of nebular He II emission lines above a certain level (e.g. $W(\text{He II } \lambda1640) \gtrsim 10$ A or He II $\lambda1640/\text{H}\beta \gtrsim 0.05$, see S03) would be a very strong case for a very metal-poor if not metal-free stellar population if shown to originate from stellar photoionisation (as opposed e.g. to AGN activity).

Tumlinson et al. (2001) have speculated whether such objects could already have been found "accidentally" at $z \lesssim 5$, confused with Lyα emitters whose redshift is identified by a single line measurement. Their simulations (in rough agreement with our more detailed results) show that this should then correspond to PopIII objects with high star formation rates comparable to those determined for Lyman break galaxies. This may again be questionable on grounds of the fairly small probability of existence of PopIII galaxies at such "low" redshifts. However, the current observations do not allow to prove or disprove their speculation.

During this conference my attention was brought to the discovery of an unusual $z = 3.357$ lensed galaxy showing narrow He II emission lines (among others) indicative of a fairly high excitation (see Fosbury et al. 2003). These authors speculate that this might be ionised by a very hard stellar spectrum possibly due to an extremely metal-poor stellar cluster. However, this seems quite unlikely for the following reasons. First numerous metal lines are seen (including in the UV spectrum) indicating an ISM metallicity of $\sim$ 1/20 solar, and [O III] λ5007/H$\beta \sim 7.5$ is typical of metal-poor H II galaxies. Furthermore, no case is known where the stellar metallicity is lower than the nebular one. This would require quite exotic formation and mixing scenarios, in contradiction with our current knowledge of SF galaxies in the nearby Universe. Finally, an alternative explanation (obscured AGN) exists capable of reproducing the observed emission line properties (Binette et al. 2003). Although this peculiar object is certainly of great interest it appears unlikely that it is related to an extremely metal-poor or even PopIII starburst.

1.3 Weak/absent metal lines

Establishing the weakness or absence of metal lines will ultimately be important to prove the extremely low or zero metallicity case. From simple photoionisation modeling Panagia (these proceedings and 2003) propose e.g. that [O III] λ5007/H$\beta \lesssim 0.1$ (0.01) should indicate a metallicity $\lesssim 10^{-3}$ (10^{-4}) solar, approximately corresponding to the expected metallicity of second generation stars (see also Scannapieco et al.). Above redshift $z \gtrsim 4$, a measurement of [O III] λ5007 will require sensitive space instruments like those planned for the JWST. Up to redshift $z \lesssim 9$ such observations should be feasible with the NIRSpec multi-object spectrograph (e.g. Panagia 2003). Beyond that, explorations will be very difficult and time consuming as single object spectroscopy (MIRI) will only be available at the corresponding wavelengths ($\lambda > 5 \mu m$).

2. Into the future...

From the above it is evident that no genuine PopIII object or extremely metal-poor galaxy ($Z/Z_{\odot} \lesssim 10^{-3...-4}$) has been found so far. Currently the best candidates are the high Lyα equivalent widths objects from the LALA survey whose properties remain puzzling. However, given their fairly large number and relatively low redshift ($z \sim 4.5$) it would at first be surprising if many would truly correspond to this category.

Other independent options to search for such "primeval" galaxies and higher z objects should be explored. The feasibility of such studies has been addressed e.g. in Schaerer & Pelló (2001) and Pelló & Schaerer (2002). Possible avenues include among others the use of particular broad-band selection criteria and ultra-deep near-IR imaging to find new candidates. Preliminary results from such studies are discussed by Richard et al. (2003) in these proceedings. There is little doubt that the great progress made over recent years on the exploration of the high redshift Universe will continue and possibly even truly "primeval" galaxies or PopIII objects will be found in this decade and before the JWST will be available. We look forward to exciting times!

References

Binette, L., et al., 2003, A&A, 405, 975
Fosbury, R.A.E., et al., 2003, ApJ, in press (astro-ph/0307162)
Malhotra, S., Rhoads, J.E., 2002, ApJ, 565, L71
Ouchi, M., et al., 2003, ApJ, 582, 60
Panagia, N., 2003, astro-ph/0309417
Pelló , R., Schaerer, D., 2002, in "Science with the GTC", astro-ph/0203203
Rhoads, J.E., et al., 2003, AJ, 125, 1006
Richard, J., et al., 2003, astro-ph/0308543
Schaerer, D., 2002, A&A, 382, 28 (S02)
Schaerer, D., 2003, A&A, 397, 527 (S03)
Schaerer, D., Pelló, R., 2001, in "Scientific Drivers for ESO Future VLT/VLTI Instrumentation", astro-ph/01072740
Scannapieco, E., et al., 2003, ApJ, 589, 35
Tumlinson, J., Shull, J.M., 2000, ApJ, 528, L65
Tumlinson, J., Giroux, M.L., Shull, J.M., 2001, ApJ, 550, L1

VOID HIERARCHY AND COSMIC STRUCTURE

Rien van de Weygaert[1] & Ravi Sheth[2,3]
[1] *Kapteyn Institute, Univ. Groningen, P.O. Box 800, 9700 AV Groningen, The Netherlands*
[2] *Dept. of Physics and Astronomy, University of Pittsburgh, 3941 O'Hara St., PA 15260, U.S.A.*
[3] *NASA/Fermilab Astrophysics Group, Batavia, IL 60510-0500, U.S.A.*
rks12@pitt.edu, weygaert@astro.rug.nl

Abstract Within the context of hierarchical scenarios of gravitational structure formation we describe how an evolving hierarchy of voids evolves on the basis of *two* processes, the *void-in-void* process and the *void-in-cloud* process. The related analytical formulation in terms of a *two-barrier* excursion problem leads to a self-similarly evolving peaked void size distribution.

1. Introduction: Excursions

Hierarchical scenarios of structure formation have been very successful in understanding the formation histories of gravitationally bound virialized haloes. Particularly compelling has been the formulation of a formalism in which the collapse and virialization of overdense dark matter halos within the context of hierarchical clustering can be treated on a fully analytical basis. This approach was originally proposed by Press & Schechter (1974), which found a particularly useful and versatile formulation and modification in the the *excursion set formalism* (Bond et al. 1991).
It is based on the assumption that for a structure to reach a particular nonlinear evolutionary stage, such as complete gravitational collapse, the sole condition is that its *linearly extrapolated primordial density* should attain a certain value. The canonic example is that of a spherical tophat overdensity collapsing once it reaches the collapse barrier $\delta_c \approx 1.69$. The successive contributions to the local density by perturbations on a (mass) resolution scale S_m may be represented in terms of a density perturbation random walk, the cumulative of all density fluctuations at a resolution scale smaller than S_m. By identifying the largest scale at which the density passes through the barrier δ_c it is possible (1) to infer at any cosmic epoch the mass spectrum of collapsed halos and (2) to reconstruct the merging history of each halo (see Fig. 2, top right).
In this study we demonstrate that also the formation and evolution of foamlike patterns as a result of the gravitational growth of primordial density perturba-

M. Plionis (ed.), Multiwavelength Cosmology, 223-226.

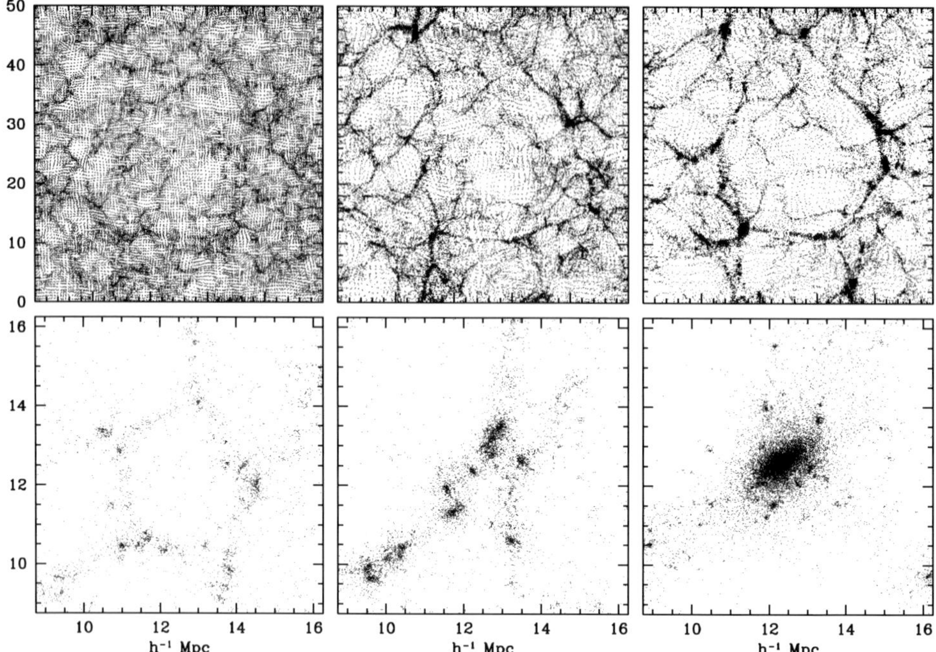

Figure 1. Illustration of the two essential "void hierarchy modes": (top) the *void-in-void* process (top), with a void growing through the merging of two or more subvoids; (bottom) the *void-in-cloud* process: a void demolished through the gravitational collapse of embedding region.

tions is liable to a successful description by the excursion set analysis. This is accomplished by resorting to a complementary view of clustering evolution in which we focus on the evolution of the *voids* in the Megaparsec galaxy and matter distribution, spatially *the* dominant component. An extensive description of this work can be found in Sheth & Van de Weygaert (2003, MNRAS, submitted).

2. Void Evolution: the two processes

Primordial underdensities are the progenitors of voids. Because underdensities are regions of suppressed gravitational attraction, representing an effective repulsive gravity, matter flows out of their interior and moves outward to the edges of these expanding *voids*. Voids *expand*, become increasingly *empty* and develop an increasingly *spherical* shape (Icke 1984). Matter from the void's interior piles up near the edge: usually a ridge forms around the void's rim and at a characteristic moment the void's interior shells take over the outer ones. At this *shell-crossing* epoch the void reaches maturity and becomes a nonlinear object expanding self-similarly, the implication being that the majority of observed voids is at or near this stage (Blumenthal et al. 1992). As voids develop from underdensities in the primordial cosmic density field, the interaction with internal substructure and external surrounding structures translates into a continuing process of hierarchical void evolution (Dubinski et al. 1993, Van de Weygaert & Van Kampen 1993).

The evolution of voids resembles that of dark halos in that large voids form from mergers of smaller voids that have matured at an earlier cosmic time (Fig. 1, top row). However, in contrast to dark halos, the fate of voids is ruled by two processes. Crucial is the realization that the evolving void hierarchy does not only involve the *void-in-void* process but also an additional aspect, the *void-in-cloud* process. Small voids may not only merge into larger encompassing underdensities, they may also disappear through collapse when embedded within a larger scale overdensity (Fig. 1, bottom row). In terms of the *excursion set approach*, it means that the *one-barrier* problem for the halo population has to be extended to a more complex *two-barrier problem.* Voids not only should ascertain themselves of having decreased their density below the *void barrier* δ_v of the *shell-crossing* transition. For their survival and sheer existence it is crucial that they take into account whether they are not situated within a collapsing overdensity on a larger scale which crossed the collapse barrier δ_c. They should follow a random walk path like type "3" in Fig. 2 (bottom right), rather than the *void-in-cloud* path "4". The repercussions of this are far-reaching and it leads to a major modification of the void properties and distribution.

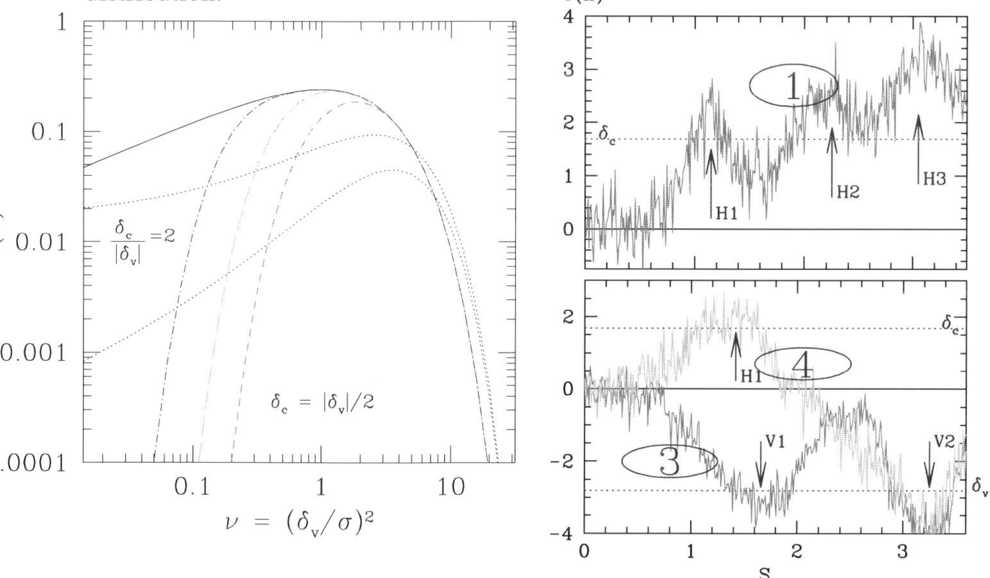

Figure 2. **Left**: Void Size Probability Function for various values of the *void-and-cloud* parameter δ_c/δ_v: 1/2 (short dashed), 1 (dashed), 2 (dot-dashed), and ∞ (solid). For reference, the dotted curves are predictions by the "adaptive peaks model" of Appel & Jones (1990). **Right**: Excursion Set Random Walks, local density $\delta(\mathbf{x})$ as function of mass resolution S_m. Clearly, the fluctuations are larger as resolution S_m increases. (Top) Halo formation. Horizontal dashed bar is the collapse barrier δ_c. (Bottom) Fate of void V2, illustrated by 2 possible random walks: curve "4" is a *void-in-cloud* process, curve "3" a *void-in-void* process.

3. Void Distribution: Universal, Peaked and Self-Similar

Analytically, the resulting expression follows by evaluating the fraction of walks which first cross δ_v at S, and which do not cross δ_c until after they have crossed δ_v (Sheth & Van de Weygaert 2003). An insightful approximation of this distribution, in terms of the "self-similar" void density $\nu \equiv \delta_v^2/S$ is

$$\nu f(\nu) \approx \sqrt{\frac{\nu}{2\pi}}\, \exp\left(-\frac{\nu}{2}\right)\, \exp\left(-\frac{|\delta_v|}{\delta_c}\frac{D^2}{4\nu} - 2\frac{D^4}{\nu^2}\right), \qquad (1)$$

in which the $D \equiv (|\delta_v|/(\delta_c - \delta_v)$ parameterizes the the relative impact of void and halo evolution on the hierarchically evolving population of voids.

The resulting distributions, for various values of D, are shown in Figure 2 (left). The void size distribution is clearly *peaked about a characteristic size*: the *void-in-cloud* mechanism is responsible for the demise of a sizeable population of small voids. The halo mass distribution diverges towards small scale masses, so that in terms of numbers the halo population is dominated by small mass objects. The void population, on the other hand, is "void" of small voids and has a sharp small-scale cut-off.

Four additional major observations readily follow from this analysis: (•) At any one cosmic epoch we may identify a *characteristic void size* which also explains why in the present-day foamlike spatial galaxy distribution voids of $\sim 20 - 30h^{-1}$Mpc are the predominant feature; (•) The *void distribution evolves self-similarly* and the *characteristic void size increases with time*: the larger voids present at late times formed from mergers of smaller voids which constituted the dominating features at earlier epochs (Fig. 1, top panels); (•) Volume integration shows that at any given time *the population of voids approximately fill space*, apparently squeezing the migrating high-density matter in between them; (•) As the size of the major share of voids will be in the order of that of the characteristic void size this observation implies that the *cosmic matter distribution* resembles a *foamlike packing of spherical voids of approximately similar size and excess expansion rate*.

In all, a slight extension and elaboration on the original extension formalism enables the framing of an analytical theory explaining how the characteristic observed weblike Megaparsec scale galaxy distribution, and the equivalent frothy spatial matter distribution seen to form in computer simulations of cosmic structure formation, are natural products of a hierarchical process of gravitational clustering. A continuously evolving hierarchy of voids produces a dynamical foamlike pattern whose characteristic dimension grows continuously along with the evolution of cosmic structure, a Universe which at any one cosmic epoch is filled with bubbles whose size corresponds to the scale just reaching *maturity*.

THE MERGING HISTORY OF MASSIVE BLACK HOLES

Marta Volonteri,[1] Francesco Haardt,[2] Piero Madau[1] & Alberto Sesana[2]

[1] *University of California, Santa Cruz, CA 95064, USA*

[2] *Universita dell'Insubria, Como, Italy*

Abstract We investigate a hierarchical structure formation scenario describing the evolution of a Super Massive Black Holes (SMBHs) population. The seeds of the local SMBHs are assumed to be 'pregalactic' black holes, remnants of the first POPIII stars. As these pregalactic holes become incorporated through a series of mergers into larger and larger halos, they sink to the center owing to dynamical friction, accrete a fraction of the gas in the merger remnant to become supermassive, form a binary system, and eventually coalesce. A simple model in which the damage done to a stellar cusps by decaying BH pairs is cumulative is able to reproduce the observed scaling relation between galaxy luminosity and core size. An accretion model connecting quasar activity with major mergers and the observed BH mass-velocity dispersion correlation reproduces remarkably well the observed luminosity function of optically-selected quasars in the redshift range 1<z<5. We finally asses the potential observability of the gravitational wave background generated by the cosmic evolution of SMBH binaries by the planned space-born interferometer LISA.

Keywords: cosmology: theory – black holes – galaxies: evolution – quasars: general

1. Introduction

In Volonteri et al. 2003 (Paper I) we have assessed a model for the assembly of SMBHS at the center of galaxies that trace their hierarchical build-up far up in the dark halo 'merger tree'. We have assumed that the first 'seed' black holes (BHs) had intermediate masses, $m_{seed} \approx 150\,m_{\odot}$, and formed in (mini)halos collapsing at $z \sim 20$ from high-σ density fluctuations (cfr. Madau and Rees 2001). These pregalactic holes evolve in a hierarchical fashion, following the merger history of their host halos. During a merger event BHs approach each other owing to dynamical friction, and form a binary system. Stellar dynamical processes drive the binary to harden and eventually coalesce.
The merger history of dark matter halos and associated black holes is followed through Monte Carlo realizations of the merger hierarchy (merger trees) which

M. Plionis (ed.), Multiwavelength Cosmology, 227-230.

allow to track the evolution of SMBH binaries along cosmic time, and analyze how their fate is influenced by the environment (e.g stellar density cusps).

2. Accretion history

The first stars must have formed out of metal-free gas, with the lack of an efficient cooling mechanism possibly leading to a very top-heavy initial stellar mass function. If stars form above 260 $m_\odot$, after 2 Myr they would collapse to massive BHs containing at least half of the initial stellar mass. The mass density parameter of our '3.5-σ' pregalactic holes is $\Omega_{seed} \geq 2 \times 10^{-9}\,h$. This is much smaller than the density parameter of the supermassive variety found in the nuclei of most nearby galaxies, $\Omega_{\rm SMBH} \approx 2 \times 10^{-6}$, Clearly, if SMBHs form out of very rare Pop III BHs, the *present-day mass density of SMBHs must have been accumulated during cosmic history via gas accretion, with BH-BH mergers playing a secondary role*. This is increasingly less true, of course, if the seed holes are more numerous and populate the 2- or 3-σ peaks instead, or halos with smaller masses at $z > 20$ (Madau and Rees 2001).
To avoid introducing additional parameters to our model, as well as uncertainties linked to gas cooling, star formation, and supernova feedback, we adopt a simple prescription for the mass accreted by a SMBH during each major merger assuming that in every accretion episode the BH accretes a mass proportional to the observed correlation between stellar velocity dispersion and SMBH mass. The normalization factor is of order unity and is fixed in order to reproduce both the stellar velocity dispersion and SMBH mass relation observed locally and the optical LF of quasars in the redshift range 1<z<5.

3. Dynamical evolution of BH binaries

During the merger of two halo+BH systems of comparable masses, dynamical friction drags in the satellite hole toward the center of the newly merged system, leading to the formation of a bound BH binary in the violently relaxed stellar core (Begelman et al. 1980). The subsequent evolution of the binary is determined by the initial central stellar distribution. As the binary separation decays, the effectiveness of dynamical friction slowly declines and the BH pair then hardens via three-body interactions, i.e., by capturing and ejecting at much higher velocities the stars passing close to the binary (Quinlan1996). The hardening of the binary modifies the stellar density profile, removing mass interior to the binary orbit, depleting the galaxy core of stars, and slowing down further hardening. If the hardening continues sufficiently far, gravitational radiation losses finally take over, and the two BHs coalesce in less than a Hubble time. The merger timescale is computed adopting a simple semi-analytical scheme that qualitatively reproduces the evolution observed in N-body simulations.

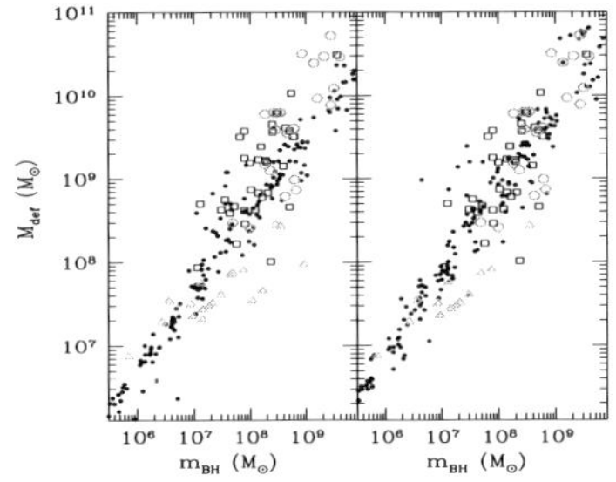

Figure 1. Mass deficit produced at $z = 0$ by shrinking SMBHs in our merger tree as a function of nuclear SMBH mass (filled dots). Left panel: cusp regeneration case. Right panel: core preservation case. Galaxies without a core are shown as vertical arrows at an arbitrary mass deficit of $10^6 m_\odot$. Empty squares, circles and triangles: mass deficit inferred in a sample of local galaxies by Milosavljevic et al. (2002) and Faber et al. (1997).

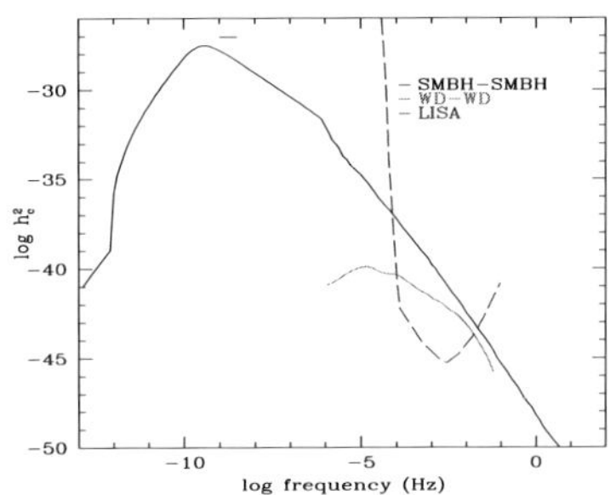

Figure 2. Gravitational background from the ensemble of cosmic SMBH binaries. The characteristic strain squared is plotted vs. the wave frequency (solid line). Dashed line: LISA sensitivity window for one year observation. Thick dash at $f \simeq 10^{-9}$: current limits from pulsar timing experiments (Lommen(2002)). Lower solid line: the spectrum expected from the Galactic population of white dwarf binaries from Farmer and Phinney2003.

If was first proposed by Ebisuzaki et al. (1991) that the heating of the surrounding stars by a decaying SMBH pair would create a low-density core out of a preexisting cuspy stellar profile. The ability of SMBH binaries in shaping the central structure of galaxies depends also on how galaxy mergers affect the inner stellar density profiles, i.e. on whether cores are preserved or steep cusps are regenerated during major mergers. To bracket the uncertainties and explore different scenarios we run two different sets of Monte Carlo realizations (Volonteri et al. 2003, Paper II). In the first (' cusp regeneration') we assume, that the stellar cusp $\propto r^{-2}$ is promptly regenerated after every major merger event. In the second (' core preservation') the effect of the hierarchy of SMBH binary interactions is instead cumulative. The simple models for core creation described above yields core radii that scale almost linearly with galaxy mass, $r_c \propto M_0^{0.8 \div 0.9}$ in the range $10^{12} < M_0 < 4 \times 10^{13} m_\odot$. A similarly scaling relation was observed by Faber et al. (1997) in a sample of local galaxies. A better test of our model predictions against galaxy data is provided by the 'mass deficit', i.e. the mass in stars that must be added to the observed cores to produce a stellar r^{-2} cusp (see fig. 1). The cusp regeneration model clearly underestimates the mass deficit observed in massive 'core' galaxies, while the core preservation one produces a correlation between 'mass deficit' and the mass of nuclear SMBHs with a normalization and slope that are comparable to the observed relation.

4. Gravitational waves from SMBH binaries

We have computed the gravitational wave (GW) background (in terms of the characteristic strain spectrum) due to the cosmic population of SMBHs evolving along the lines discussed in the previous sections. Details and a complete discussion can be found in Sesana et al.2003.
Using a linearized GR theory of GW emission and propagation, framed in a cosmological context, we find that the broad band GW spectrum can be divided into three main different regimes (fig. 2): for frequencies $\lesssim 10^{-9}$ Hz, the background is shaped by SMBH binaries in the three-body interactions regime (orbital decay driven by stellar scattering); in the intermediate band, $10^{-9}\lesssim f \lesssim 10^{-6}$ Hz, GW emission itself accounts for most of the potential energy losses, and the strain has the "standard" $f^{-2/3}$ behaviour; finally, for $f \gtrsim 10^{-6}$, the GW spectrum is formed by the convolution of the emission at the last stable orbit from individual binaries. In the first two regimes, the strain is dominated by rare low redshift events ($z \lesssim 2$) involving BHs genuinely supermassive. At larger frequencies the contribute from smaller and smaller masses starts dominating.
We have considered the possible future observability of the GW background by the planned LISA interferometer, showing that one year observation would suffice to detect the GW background due to SMBH binaries. Such background should largely overcome that due to galactic WD binaries. In the LISA window ($10^{-4}\lesssim f \lesssim 0.1$ Hz), the main GW sources are BH binaries in the mass range $10^3 \lesssim M \lesssim 10^7$ $M_\odot$, with a relevant contribution from $3 \lesssim z \lesssim 20$. In the nHz regime, probed by pulsar timing experiments, the amplitude of the characteristic strain from coalescing SMBH binaries is close to current experimental limits (Lommen 2002, see also Wyithe and Loeb 2003).

References

Volonteri, M., Haardt, F., and Madau, P. ApJ, 582, 559 (2003) (Paper I)

Madau, P., and Rees, M. J. ApJ, 551, L27 (2001)

Begelman, M. C., Blandford, R. D., Rees, M. J. Nature, 287, 307 (1980)

Quinlan, G. D. NewA, 1, 35 (1996)

Ebisuzaki, T., Makino J., and Okumura S. K. Nature, 354, 212 (1991)

Volonteri, M., Madau, P., and Haardt, F. ApJ, 593, 661 (2003)

Faber, S. M., et al. AJ, 114, 1771 (1997)

Milosavljevic, M., Merritt, D., Rest, A., van den Bosch, F. C. MNRAS, 331, L51 (2002)

Sesana, A., Haardt, F., Volonteri, M. and Madau, P., in prep. (2003)

Farmer, A.J., and Phinney, E.S. MNRAS. submitted (2003) (astro-ph/0304393)

Lommen, A.N., Proceedings of the 270. WE-Heraeus Seminar on Neutron Stars, Pulsars and Supernova Remnants, MPE report 278, eds W. Becker, H. Lesch & J. Truemper, p.114 (2002)

Wyithe, J.S.B., and Loeb, A., ApJ, in press (2003)

CRITIQUE OF TRACKING QUINTESSENCE

Sidney Bludman
Deutsches Elektronen-Synchrotron DESY,Hamburg
University of Pennsylvania, Philadelphia

Abstract Recent observations show that quintessence, if it exists, must now be crawling, even if it once was tracking. This means that any quintessence field stopped tracking relatively early, that the quintessence potential now has appreciable curvature, so that the equation of state $w_Q(z)$ is still reducing for accessible red-shifts. The observational bound on w_Q rejects constant w_Q and inverse-power potentials, but allows the SUGRA potential for a range of initial curvature values.

1. The dark energy density is now exactly or nearly static

The universe is flat, presently dominated by smooth energy, and the cosmological expansion has been accelerating since red-shift $z \sim 0.7$. While programs to test for quintessence and to reconstruct its potential are underway, we already know that radiation/matter equality took place at red-shift $z_{eq} = 3454^{+385}_{-392}$, that $h \equiv H_0/100 = 0.72 \pm 0.05$, and we live at a time when

$$\Omega_{Q0} = 0.71 \pm 0.07, \qquad \bar{w_Q} < -0.78\ (95\%\ CL) \tag{1}$$

1, 2. (The supernova observations fit an average

$$\bar{w_Q}(N) = \int_0^N w_Q(N')\, dN'/N.$$

Because these observations average over a small range in z, in which the quintessence field and $w_Q(z)$ change relatively little, the fitted $\bar{w_Q}$ bound is also an approximate bound on w_{Q0}.) The background density, $\rho_B = (11.67a + 0.003378)/a^4\ meV^4$, is now $\rho_{B0} = 11.67\,meV^4$ and was $\rho_{Bi} = 0.003378\,GeV^4$ at fiducial red-shift $z = 10^{12}$. Where necessary, we will fix $h^2 = 1/2$, so that the present critical density and smooth energy density are $\rho_{cr0} = 40.5\,meV^4$, $\rho_{Q0} = 28.8\,meV^4$.

We will show that these observational constraints already allow only crawling quintessence (Section 2) or potentials that are now fast-rolling(Section 3). Our

M. Plionis (ed.), Multiwavelength Cosmology, 231-236.

treatment transcends the many earlier treatments of the tracking condition 3, 4, 5, 6, 7 and of inverse power potentials 8, 9 by calculating the post-tracker behavior in the present quintessence-dominated era, by considering the range of initial conditions that would lead to tracking, and by treating the numerical problems encountered in cosmological dynamics, particularly in the freezing and tracking eras. But first we review the attractor condition, in order to show how the basin of attraction shrinks for potentials satisfying the observational constraints.

Canonical quintessence models the smooth energy dynamically by a spatially homogeneous light classical scalar field, with canonical kinetic energy $K = \dot{\phi}^2/2$, minimal gravitational coupling, zero true cosmological constant, rolling down its self-potential $V(\phi)$. With this canonical form, the scalar field equation of motion

$$\ddot{\phi} + 3H\dot{\phi} + dV/d\phi = 0, \tag{2}$$

has the first integral

$$V(N) = \rho_Q - d\rho_Q/6dN = \rho_Q(1 - w_Q)/2, \tag{3}$$

where the energy density and pressure are

$$\rho_Q = \dot{\phi}^2/2 + V(\phi), \qquad P/c^2 = \dot{\phi}^2/2 - V(\phi), \tag{4}$$

and the quintessence barotropic index $\gamma_Q(N) \equiv -d\ln\rho_Q/3dN = ((\rho + P/c^2)/\rho)_Q \equiv 1 + w_Q$ lies between 0 and 2. Because the scalar field does not cluster even on supercluster scales, its Compton wavelength is $\gg 300\ Mpc$, so that its mass $m_Q = \sqrt{d^2V/d\phi^2} \lesssim 1.5E - 36\ meV$ must be incredibly small.

From the energy integral, $(\varkappa d\phi/dN)^2 = 6x^2$, where $x^2 \equiv \dot{\phi}^2/2\rho = 3\gamma_Q\Omega_Q$ is the quintessence kinetic energy fraction of the total energy density and $\varkappa = \sqrt{8\pi G} \equiv 1/M_P = 1/2.44E18\ GeV$ is the reduced Planck mass. The kinetic/potential energy ratio $K/V = (1 + w_Q)/(1 - w_Q$ changes at the rate $\Delta(N) - 1 \equiv d\ln((1 + w_Q)/(1 - w_Q))/6dN$. For the *roll*, we obtain

$$\lambda \equiv -d\ln V/\varkappa d\phi = \sqrt{3\gamma_Q/\Omega_Q} \cdot \Delta, \tag{5}$$

and

$$\varkappa d\phi/dN = \sqrt{3\gamma_Q\Omega_Q}, \qquad dw_Q/dN = 3(1 - w_Q^2)(\Delta - 1). \tag{6}$$

The total barotropic index is $\gamma = \gamma_B\Omega_B + \gamma_Q\Omega_Q = 1 + w_B\Omega_B + w_Q\Omega_Q$, where the dimensionless ratios, $\Omega_Q,\ \Omega_B \equiv \rho_B/(\rho + \rho_Q)$, are the energy density fractions in quintessence and in background, and $\gamma_Q,\ \gamma_B \equiv -d\ln\rho_B/3dN$ are their corresponding barotropic indices. Integrating the first equation (6), would give an implicit relation between ϕ and $V(\phi)$, so that, the potential can be reconstructed from the equation of state $w_Q(N)$.

Table 1. Potentials Described by Roll $\lambda = -d\ln V/\varkappa d\phi$ and Curvature $1/\beta(\phi) \equiv d^2 \ln V/(\varkappa d\phi)^2/\lambda^2$.

$V(\phi)$	$\lambda(\phi)$	$\beta(\phi)$	**NAME**
$\exp -\lambda\varkappa\phi$	$\lambda = const$	∞	exponential
$1/\sinh^\alpha(\tilde{\alpha}\varkappa\phi)$	$(\alpha\tilde{\alpha})\coth(\tilde{\alpha}\varkappa\phi)$	$\alpha\cosh^2(\tilde{\alpha}\varkappa\phi)$	const w_Q
$\phi^{-\alpha}$	$\alpha/\varkappa\phi$	$const = \alpha$	inverse power
$const$	0	0	cosmological const
$\phi^{-\alpha}\cdot\exp(\varkappa\phi)^2/2$	$\alpha/\varkappa\phi - \varkappa\phi$	$(\alpha - \varkappa^2\phi^2)^2/(\alpha + \varkappa^2\phi^2)$	SUGRA

Besides λ, the potential depends on the curvature $\eta \equiv \varkappa^2 d^2V/d\phi^2$. When the curvature is small, $\ddot{\phi}$ is negligible in the equation of motion (2). When the potential is flat ($\epsilon \equiv \lambda^2/2 \ll 1$), the kinetic energy $\dot{\phi}^2/2$ is negligible in the quintessence energy (3). During ordinary inflation, both these conditions hold (*slow roll approximation*): the expansion is dominated by the cosmological drag and the field is nearly frozen. In quintessence, on the other hand, the acceleration began only recently, so that the roll and curvature may still be appreciable. This invalidates the slow roll approximation for quintessence, so that the dynamical equations need to be integrated numerically.
Unless the quintessence field undergoes a phase transition, the quintessence field rolls monotonically towards a minimum at $\phi = \infty$ or at infinity: either way, the potentials we consider are always convex with curvature $1/\beta(\phi) = d^2 \ln V/\varkappa^2 d\phi^2/\lambda^2 = \Gamma - 1$, where $\Gamma \equiv V d^2V/d\phi^2/(dV/d\phi)^2 \equiv \eta/2\epsilon$. The constant values $\beta = 0, \alpha, \infty$ define the cosmological constant, inverse α power, and exponential potentials. The bottom row in Table 1 is the more realistic SUGRA potential, in which $\beta(\phi)$ varies for $\phi \gtrsim M_P$.

In place of the phase variables ϕ, $w_Q \equiv (P/\rho)_Q$, we may use $x(N)^2$, $y(N)^2 \equiv V_Q/\rho$, for which the equations of motion are 4, 5,6,7

$$dx/dN = -3x + \lambda\sqrt{3/2}y^2 + 3x\gamma/2 \tag{7}$$

$$dy/dN = -\lambda\sqrt{3/2}xy + 3y\gamma/2 \tag{8}$$

$$d\lambda/dN = -\sqrt{6}\lambda^2 x/\beta \quad \text{or} \quad d(1/\lambda)/dN = \sqrt{6}x/\beta. \tag{9}$$

Thus $\gamma = \gamma_Q\Omega_Q + \gamma_B\Omega_B = 2x^2 + \gamma_B(1 - x^2 - y^2)$, is a time-dependent function of the scalar field $\phi(N)$, and $x^2 + y^2 = \Omega_Q$, $2x^2 = \Omega_Q\gamma_Q, x^2/y^2 = K/V = (1 + w_Q)/(1 - w_Q)$, $d\ln(x^2/y^2)/dN = 6(\Delta - 1)$. The three-element system (6-8) is autonomous, except for the slow change in $\gamma_B(N)$ from 4/3 to 1, while gradually going from the radiation-dominated to the matter-dominated universe, around red-shift $z_{eq} = 3454$.
The magnitude of V must be chosen so that the value $V_0 = \rho_{cr0}y_0^2 = \rho_{cr0}(1 - w_{Q0})/2$, is reached at present. For example, inverse power potentials, require mass scale $M = (V_0\phi_0^\alpha)^{1/(4+\alpha)}$, listed in the penultimate column of Table

1. For $\alpha < 0.2$, this mass scale is close to observed neutrino masses and to the present radiation temperature, possibly suggesting some role for the neutrino mass mechanism or for the matter/radiation transition, in bringing about quintessence dominance. For larger α values, this mass scale can be considerably larger, suggesting the larger scales we know in particle physics.

While the evolution of a homogeneous scalar field depends only on its equation of state $w_Q = P/\rho_Q c^2$, the evolution of its fluctuations depends also on the effective sound speed $c_s^2 = (dP/d\dot{\phi})/(d\rho_Q/d\dot{\phi})$. With the linear form for the kinetic energy $K = \dot{\phi}^2/2$ that canonical quintessence assumes, $c_s^2 = c^2$ and $dw_Q/dz > 0$. Non-linear forms for the kinetic energy, like k-essence, would allow violations of the Weak Energy Condition and give different sound speed and structure evolution. Despite this difference in sign of dw_Q/dz, k-essence is hardly distinguishable from quintessence, unless $c_s^2 \approx 0$ since the surface of last scattering 10. We will consider only canonical quintessence evolution with the Friedmann expansion rate $H^2 = \varkappa^2 \rho/3$.

2. Acceptable inverse power potentials require fine-tuned initial conditions

For the simple inverse power law potentials $V(\phi) = M^{4+\alpha}/\phi^\alpha$, the curvature $\beta = \alpha = constant$, so that the equation of motion (2) has exactly scaling solutions in both the radiation- and the matter-dominated eras. As quintessence becomes appreciable, these trajectories curve towards the $x = 0$ (asymptotic de Sitter) axis. Such inverse power potentials arise naturally in supersymmetric condensate models for QCD or instanton SUSY-breaking, but require appreciable quantum corrections when $\phi \gtrsim M_P$.

The tracker is a *fixed* attractor for an exponential, but a slow-moving (*instantaneous*) attractor for inverse power potentials. The second column in Table 2 gives the range in tracker values w_{Qtr}, from $\gamma_B\alpha/(\alpha+2) - 1 = (\alpha - 6)/3(\alpha+2)$, during the radiation-dominated era, to $-2/(\alpha+2)$, during the matter-dominated era. The third column gives the initial Ω_Q at $z = 10^{12}$, for an attractor solution to evolve to the present $\Omega_{Q0} = 0.71$. The next five columns, between the two vertical double bars, summarize the post-tracker, present values for inverse power potentials. The penultimate column tabulates the scalar mass scales needed, in order to reach $\Omega_{Q0} = 0.71$. For small α, the mass scale needed is small, of the order of the present radiation temperature, and $\phi_0 < M_P$, so that quantum corrections are still small. For large α, quintessence would have dominated early enough to interfere with structure formation in the matter-dominated era and quantum corrections would now need to be applied.

For an inverse power potential, $\lambda = \lambda_0(yH/y_0H_0)^{2/\alpha}$ in terms of present values of y, H, λ. Integrating equations (7,8) then gives, in the last col-

Table 2. Tracker and Present ($\Omega_{Q0} = 0.71$) Solutions for Inverse Power Potentials.

α	w_{Qtr}	logΩ_{Qtri}	w_{Q0}	$(K/V)_0$	λ_0	$\varkappa\phi_0$	M	log Ω_{Qi}
6	0..-0.25	-10.9	-0.41	0.42	1.52	3.95	5 PeV	-40..-2
1	-0.55..-0.67	-29.3	-0.76	0.13	0.91	1.10	2 keV	-40..-2
0.5	-0.73..-0.80	-35.2	-0.86	0.07	0.69	0.73	5 eV	-42..-14
0.1	-0.94..-0.95	-41.9	-0.97	0.02	0.35	0.28	12 meV	-42..-32
0	-1	-44.1	-1	0	0	-	2.5 meV	-44.1

Table 3. Quintessence Potentials Which Track Early, But Slow Roll Now.

Quintessence Potential $V(\phi)$	**Theoretical Origin**
$M^4[\cos(\phi/f) + 1]$	String, M-theory pseudo Goldstone light axion
$M^{4+\alpha}\phi^{-\alpha} \cdot \exp\frac{1}{2}(\varkappa\phi)^{\beta}/2$	Supergravity (SUGRA)
$M_P^4[A + (\varkappa\phi - \varkappa\phi_m)^{\alpha}]\exp(-\lambda\varkappa\phi)$	Exponential modified by prefactor; M-theory

umn, the range of initial Ω_{Qi} that would flow onto the presently-allowed values $\Omega_{Q0} = 0.71,\ w_Q \leq -0.78$.

Fast-rolling $\alpha = 6$ trajectories, out of a broad basin of attraction, would track by now to $w_{Q0} = -0.41$, which is excluded observationally. As α decreases, w_{Q0} decreases, but the basin of attraction shrinks. For $\alpha < 0.5$, the presently-tracking range in initial values of $\log\Omega_{Qi}$ is already more than fourteen orders of magnitude narrower than the initial range of the good trackers originally proposed 3 for $\alpha = 6$. For a cosmological constant ($\alpha = 0$), the present value is realized, of course, only if it is initially fixed at its present value $\rho_Q = 28.8\ meV^4$.

For inverse power potentials, the observations that w_Q is already close to the cosmological constant value -1, thus requires $\alpha < 1$, so that the energy was *always* potential dominated ($x << y$). These relatively slow-roll trajectories never track, but crawl towards their present values, only because they were initially tuned close to these values.

3. Conclusions

The observations $\bar{w_Q} < -0.78$ will allow phase trajectories that are presently insensitive to initial conditions, only if the curvature, $1/\beta(\phi)$, increases rapidly near the present epoch $\varkappa\phi \sim 1$. For example, the SUGRA models have minima at $\varkappa\phi = \sqrt{\alpha}$. Far below this minimum, they behave and track like inverse power potentials, but near the minimum, the curvature increases rapidly. This makes the SUGRA phase trajectories scale in the background-dominated era, but then curve down to lower w_{Q0} values, in marginal agreement with observations, for a large range in α values.

This needed post-tracking behavior is illustrated by the popular potentials listed in Table 3. These potentials require fine-tuning of their parameters, but may yet be distinguished from true cosmological constants by combined supernova and CBR observations in the next decades. Canonical quintessence thus requires fine-tuning of initial conditions and/or the functional form of the quintessence potential.

We have not considered models with a true cosmological constant, quantum corrections to classical quintessence, non-linear kinetic energies, nor large extra dimension or brane cosmology models, for which the Friedmann equation is modified at very early times. Within canonical quintessence, two consequences follow from the attractor behavior of the equations of motion:

(1) For an early attractor to produce a smooth energy, that is now insensitive to initial conditions, canonical quintessence requires a potential curvature $\beta(\phi)$ that is now changing rapidly, for a $w_Q(z)$ that is now still diminishing. Otherwise, dynamical vacuum energy appears hardly distinguishable, theoretically and phenomenologically, from the small cosmological constant it was designed to explain or from a time-dependent cosmological constant.

(2) Unless the smooth energy disappears in the future, the future de Sitter attractor is inevitable, with two consequences: (i) For reasons canonical quintessence does not explain, we live in a transient, special time, after matter-domination and large scale structure formation; (ii) The future event horizon will make some future events causally disconnected, so that galaxies outside the Local Supercluster and indeed asymptotic particles and an S-matrix will ultimately become unobservable. Thus cosmological observations already stress the special time when we live and may ultimately require modifying basic quantum mechanical concepts

References

D. N. Spergel *et al*, submitted to Astro. Ph. (2003).

J. L. Tonry *et al*, arXiv:astro-ph/0305008 (2003).

P. J. Steinhardt, L. Wang and I. Zlatev, Phys. Rev. **D59** 1999, 123504.

P. Ferreira and M. Joyce, Phys. Rev. Lett. **79** (1997), 4740; Phys. Rev. **D57** (1998), 6022.

A. R. Liddle and J. Scherer, Phys. Rev. **D59** 1998, 02359.

E. J. Copeland, A. Liddle and D. Wands, Phys. Rev. **D57** 2001, 4686.

S. C. C. Ng, N. J. Nunes and F. Rosati, Phys. Rev. **D64** 2001, 083510.

B. Ratra and P. J. E. Peebles, Phys. Rev. **D37** (1988), 3406.

C. Wetterich, Nucl. Phys. **B302** (1988), 668.

V. Barger and D. Marfata, Phys. Lett **B498** (2001), 67.

X-RAY WAVELENGTHS

COSMOLOGICAL CONSTRAINTS FROM X-RAY OBSERVATIONS OF GALAXY CLUSTERS (*INVITED*)

Steven W. Allen
Institute of Astronomy
Madingley Road, Cambridge CB3 0HA
United Kingdom
swa@ast.cam.ac.uk

Abstract Chandra X-ray observations of rich, dynamically relaxed galaxy clusters allow the properties of the X-ray gas and the total gravitating mass to be determined precisely. By combining Chandra results on the X-ray gas mass fractions in clusters with independent measurements of the Hubble constant and the mean baryonic matter density of the universe, we obtain a tight constraint on the mean matter density of the universe, $\Omega_{\rm m}$, and an interesting constraint on the dark energy density, Ω_Λ. Using these results, together with the observed local X-ray luminosity function of the most X-ray luminous galaxy clusters, a mass-luminosity relation determined from Chandra and ROSAT X-ray data and weak gravitational lensing observations, and the mass function predicted by numerical simulations, we also obtain a precise constraint on the normalization of mass fluctuations in the nearby universe, σ_8.

Keywords: X-rays: galaxies: clusters – gravitational lensing — cosmological parameters

1. Introduction

Accurate measurements of the masses of clusters of galaxies are of profound importance to cosmological studies. Galaxy clusters are the largest gravitationally-bound objects known and their spatial distribution, redshift evolution and matter content provide sensitive probes of cosmology.

Currently, our two best methods for measuring the masses of clusters use X-ray observations and studies of gravitational lensing by clusters. X-ray mass measurements are based on the assumption that the hot, X-ray emitting ($\sim 10^8$ K) gas in clusters is in hydrostatic equilibrium. The distribution of mass can be determined once the temperature and density profiles of the gas are known. The launch of the Chandra X-ray Observatory and XMM-Newton have revolutionized X-ray studies, permitting the first detailed, spatially-resolved X-ray

M. Plionis (ed.), Multiwavelength Cosmology, 239-246.

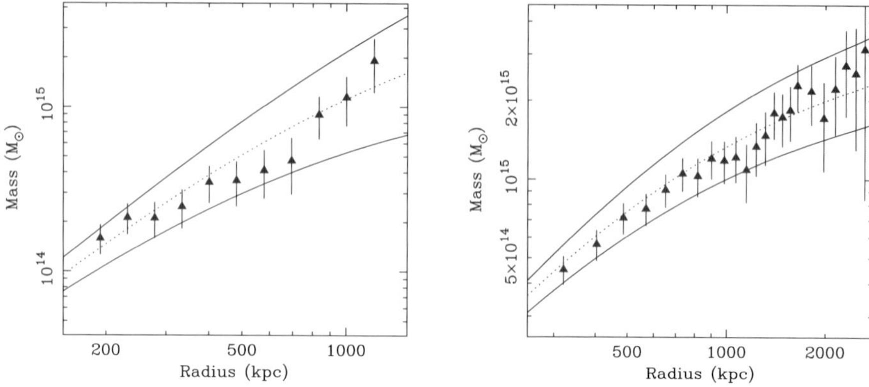

Figure 1. A comparison of the mass measurements obtained from Chandra X-ray observations (solid lines) and wide field weak lensing studies (triangles) of two of the dynamically relaxed clusters in our sample: Abell 2390 (left; Allen et al. 2001) and RXJ1347.5-1145 (right; Allen et al. 2002b). Error bars are 68 per cent confidence limits.

spectroscopy of clusters and opening the door to precise mass measurements for dynamically relaxed systems from X-ray data. Gravitational lensing studies, in contrast, offer a method for measuring the masses of clusters that is essentially free from assumptions about the dynamical state of the gravitating matter, allowing these techniques to be applied to dynamically active systems. The last decade has seen rapid progress in lensing studies, to the point where obtaining weak lensing mass measurements for the largest clusters has become a relatively straightforward task (*e.g.* Dahle et al. 2002). The combination of X-ray and gravitational lensing observations provides the best way to obtain robust measurements of the masses of clusters, as well as a powerful probe of the dynamical states of these systems.

2. X-ray and lensing mass measurements

Fig. 1 shows a comparison of the Chandra X-ray and weak lensing results on the projected mass profiles (luminous plus dark matter) for two of the dynamically relaxed clusters in our sample: Abell 2390 (left) and RXJ1347.5-1145 (right) – the most X-ray luminous cluster known. We find that for both clusters the mass profiles can be parameterized using the 'universal' form determined from simulations by Navarro, Frenk & White (1997), with concentration parameters and scale radii within the ranges expected for clusters of these masses. The agreement between the independent lensing and X-ray mass measurements is good in both cases, confirming the validity of the hydrostatic assumption used in the X-ray analysis and ruling out significant non-thermal pressure support on large spatial scales.

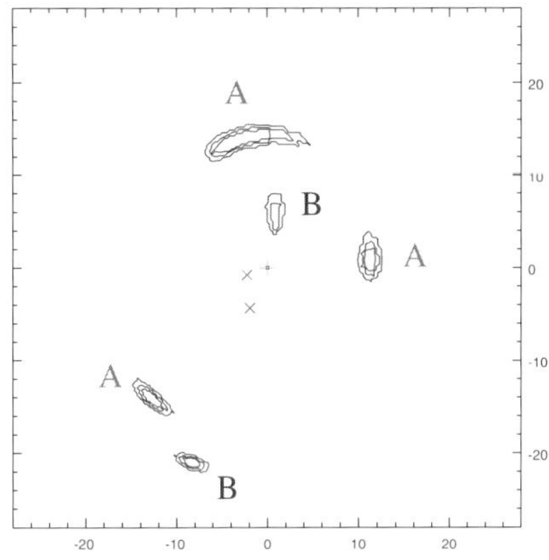

Figure 2. The observed (left) and predicted (right) gravitational arc geometry in MS2137.3-2353 (Allen & Schmidt, in preparation). The source positions in the model are denoted by crosses. The central position for the dominant mass clump in the cluster is marked with a plus sign. Arc redshifts are measured by Sand et al. (2003).

Fig. 2 shows the observed (left) and predicted (right) gravitational arc geometry in MS2137.3-2353 (Allen & Schmidt, in preparation). We find that a simple NFW mass model, based on the best-fitting Chandra X-ray mass model, with an ellipticity and orientation matching that of the cD galaxy, provides a good description for the overall arc geometry in the cluster. Similar results are obtained for other clusters *e.g.* Abell 1835 (Schmidt et al. 2001) and RXJ1347.5-1145 (Allen et al. 2002).

3. Cosmological constraints from the X-ray gas mass fraction in dynamically relaxed clusters

The matter content of rich clusters of galaxies is thought to provide an almost fair sample of the matter content of the universe as a whole (White et al. 1993). The observed ratio of the baryonic to total mass in clusters should therefore closely match the ratio of the cosmological parameters $\Omega_{\rm b}/\Omega_{\rm m}$, where $\Omega_{\rm b}$ and $\Omega_{\rm m}$ are the mean baryon and total mass densities of the Universe in units of the critical density. The combination of robust measurements of the baryonic mass fraction in clusters with accurate determinations of $\Omega_{\rm b}$ from cosmic nucleosynthesis calculations (constrained by the observed abundances of light elements at high redshifts) can therefore be used to determine $\Omega_{\rm m}$.

The baryonic mass content of rich clusters of galaxies is dominated by the X-ray emitting intracluster gas, the mass of which exceeds the mass of optically luminous material by a factor ~ 6; the baryonic mass fraction in stars, deter-

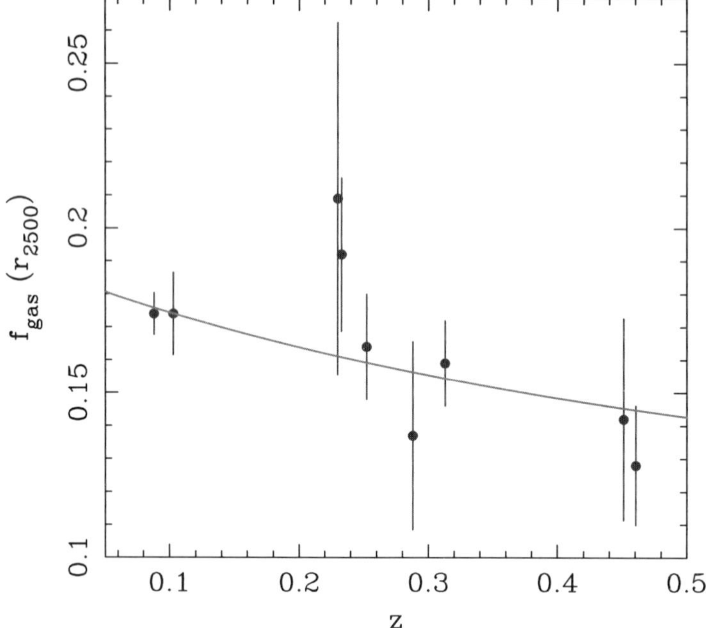

Figure 3. The apparent redshift variation of the X-ray gas mass fraction measured within r_{2500} for a reference SCDM ($h = H_0/100 \text{km s}^{-1} \text{Mpc}^{-1} = 0.5$) cosmology. The curve shows the predicted $f_{\text{gas}}(z)$ behaviour for the best-fitting cosmology. Note that our analysis includes data for one extra cluster at $z = 0.54$, not shown in the figure (Allen et al. , in preparation).

mined from detailed studies of nearby clusters, scales as $0.19h^{0.5}f_{\text{gas}}$, where f_{gas} is the fraction of the total mass in X-ray emitting gas (*e.g.* Fukugita, Hogan & Peebles 1998.) Since the X-ray emissivity of the gas is proportional to the square of its density, the gas mass profile can be precisely determined from the X-ray data. With the advent of accurate measurements of Ω_{b} (*e.g.* Kirkman et al. 2003 and references therein) and a precise determination of the Hubble Constant, H_0 (Freedman et al. 2001), the dominant uncertainty in determining Ω_{m} from the baryonic mass fraction in clusters had lain in the measurements of the total (luminous plus dark) matter distributions in the clusters. However, the new Chandra data and gravitational lensing studies have now reduced the uncertainties in the baryonic mass fraction measurements for dynamically relaxed clusters to $\sim$ 10 per cent, an accuracy comparable to the current Ω_{b} and H_0 results.

Expanding on these simple arguments, we can also constrain the dark energy density if we examine the apparent variation of the observed f_{gas} values with redshift. Such methods were first proposed by Sasaki (1996) and Pen (1997). The key point is that when measuring the X-ray gas mass fraction from the X-ray data, one needs to adopt a reference cosmology. The measured f_{gas} value for each cluster will depend upon the assumed angular diameter distance to

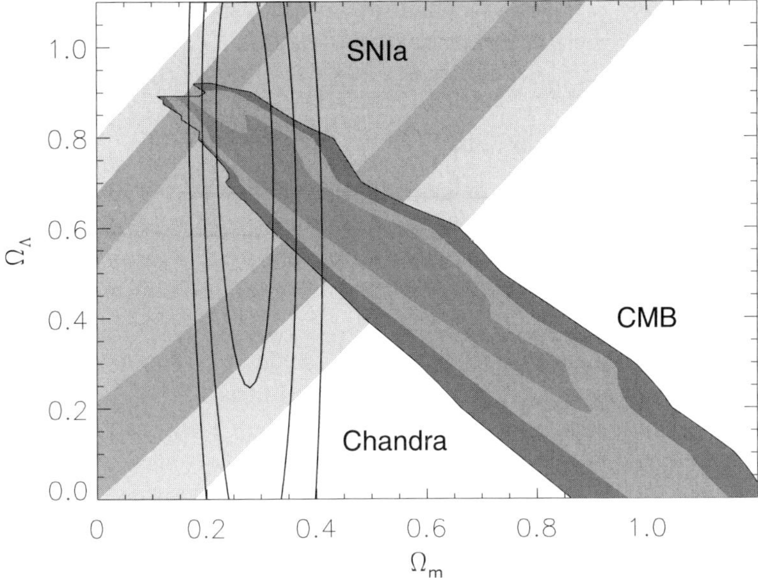

Figure 4. The joint 1, 2 and 3 σ confidence contours on Ω_m and Ω_Λ determined from the current Chandra f_{gas}(z) data (bold contours), and independent analyses of pre-WMAP CMB data and the properties of distant supernovae (adapted from Jaffe et al. 2001).

the source as $f_{gas} \propto D_A^{1.5}$. Thus, although we expect (based on theory and numerical simulations) the observed f_{gas} values to be invariant with redshift, this will only *appear* to be the case when the assumed cosmology matches the true, underlying cosmology.

Fig. 3 shows the measured f_{gas} values as a function of redshift for the current, published sample of 9 clusters for an assumed Einstein-de-Sitter (SCDM: $\Omega_m = 1.0, \Omega_\Lambda = 0.0$) cosmology. The results indicate an apparent drop in f_{gas} as the redshift increases. In order to obtain our constraints on cosmological parameters, we have fitted the data with a model which accounts for the expected apparent variation in the $f_{gas}(z)$ values as the true, underlying cosmology is varied. In detail, the fitted model function is

$$f_{gas}^{mod}(z) = \frac{b\,\Omega_b}{\left(1+0.19\sqrt{h}\right)\Omega_m}\left[\frac{h}{0.5}\,\frac{D_A^{\Omega_m=1,\Omega_\Lambda=0}(z)}{D_A^{\Omega_m,\,\Omega_\Lambda}(z)}\right]^{1.5} \qquad (1)$$

which depends on Ω_m, Ω_Λ, Ω_b, h and a bias factor b. The ratio $(h/0.5)^{1.5}$ accounts for the change in the Hubble constant between the considered model and default SCDM cosmology ($h = 0.5$), and the ratio of the angular diameter distances accounts for deviations in the geometry of the Universe from the Einstein-de-Sitter case. The bias parameter b is motivated by gasdynamical

simulations which suggest that the baryon fraction in clusters is slightly lower than for the universe as a whole. We include a Gaussian prior $b = 0.83 \pm 0.04$, appropriate for such hot, massive clusters (Eke, Navarro & Frenk 1998). We also include Gaussian priors on the Hubble constant, $h = 0.72 \pm 0.08$, the final result from the Hubble Key Project (Freedman et al. 2001), and $\Omega_b h^2 = 0.0214 \pm 0.0020$, from cosmic nucleosynthesis calculations constrained by the observed abundances of light elements at high redshifts (Kirkman et al. 2003). The best-fit parameters and marginalized 1σ error bars, based on data for the present sample of 10 X-ray luminous, dynamically relaxed clusters are $\Omega_m = 0.27 \pm 0.03$ and $\Omega_\Lambda = 0.78 \pm 0.33$.
Fig. 4 shows how the constraints from the Chandra f_{gas} data compare with those obtained by Jaffe et al. (2001) from studies of pre-WMAP cosmic microwave background (CMB) data (incorporating the COBE Differential Microwave Radiometer, BOOMERANG-98 and MAXIMA-1 data of Bennett et al. 1996, de Bernardis et al. 2000 and Hanany et al. 2000, respectively) and the properties of distant supernovae (incorporating the data of Riess et al. 1998 and Perlmutter et al. 1999). The agreement between the results obtained from the independent methods is striking: all three data sets are consistent at the 1σ confidence level, and favour a model with $\Omega_m = 0.3$ and $\Omega_\Lambda = 0.7 - 0.8$. Our results are also consistent with those obtained from more recent CMB data (Spergel et al. 2003, Allen et al. , in preparation) and analyses of the 2dF Galaxy Redshift Survey data (Percival et al. 2002).

4. The normalization of mass fluctuations in the Universe

Combining Chandra results on the masses of dynamically relaxed clusters with wide field weak lensing studies of more dynamically active systems and ROSAT measurements of their X-ray luminosities, we can create a mass-luminosity relation. Fitting these data with a power law model of the form

$$\log_{10}\left[\frac{E(z)\,M_{200}}{h_{50}^{-1}\mathrm{M}_\odot}\right] = \alpha\log_{10}\left[\frac{L}{E(z)\,10^{44}\,h_{50}^{-2}\,\mathrm{erg\,s^{-1}}}\right] + \log_{10}\left[\frac{M_0}{h_{50}^{-1}\mathrm{M}_\odot}\right], \quad (2)$$

where $E(z)$ accounts for the cosmological scaling of the critical density, we obtain best-fitting values and 68 per cent confidence limits from bootstrap simulations of $\log_{10}(M_0/h_{50}^{-1}\,M_\odot) = 14.29^{+0.20}_{-0.23}$ and $\alpha = 0.76^{+0.16}_{-0.13}$. The observed slope is in good agreement with the predicted slope of the mass–*bolometric* luminosity relation from simple, ‘adiabatic’ models of gravitational collapse, $\alpha = 0.75$.

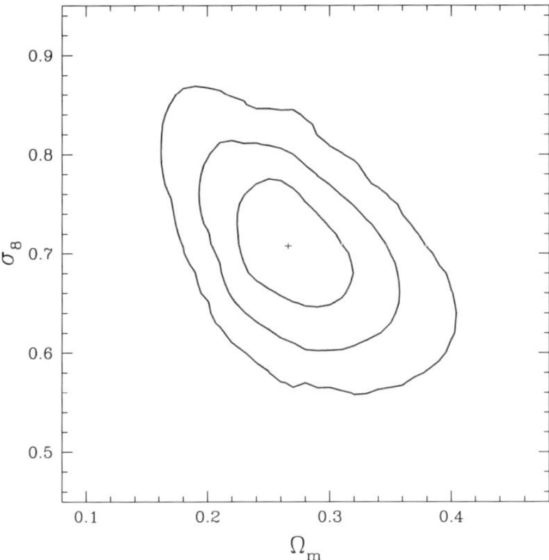

Figure 5. The joint 1, 2 and 3 sigma confidence contours on σ_8 and Ω_m from the present study of X-ray and lensing cluster data. A flat ΛCDM cosmology is assumed.

We can combine the observed mass-luminosity relation with the predicted mass function of clusters from the Hubble Volume simulations of Evrard et al. (2002) to predict the X-ray luminosity function (XLF) of clusters (the number density as a function of luminosity) as a function of cosmological parameters: σ_8 – the rms variation of the linearly-evolved density field on $8h^{-1}$Mpc scales – and Ω_m. These predictions can then be compared with the observed XLF, determined from the ROSAT All-Sky Survey (Ebeling et al. 2000; BÃűhringer et al. 2002), to constrain these parameters.

Fig 5. shows the results on σ_8 and Ω_m (for an assumed flat ΛCDM cosmology) obtained by this method, having also folded in the constraints on Ω_m from the X-ray gas mass fraction data discussed in Section 3. We obtain the best-fit results $\sigma_8 = 0.71 \pm 0.04$ and $\Omega_m = 0.27 \pm 0.03$ (marginalized 68 per cent confidence limits). Our results are in good agreement with those obtained by other groups using similar methods (*e.g.* Reiprich & BÃűhringer 2002; Schuecker et al. 2003) as well as independent analyses of pre-WMAP CMB+2dF data (*e.g.* Lahav et al. 2002), post-WMAP CMB+2dF data (*e.g.* Allen, Schmidt & Bridle 2003), measurements of evolution in the cluster XLF (*e.g.* Borgani et al. 2001), and some recent findings based on studies of cosmic shear (*e.g.* Jarvis et al. 2003, Brown et al. 2003). However, our value for σ_8 is lower than those obtained by some other studies of cosmic shear (*e.g.* van Waerbeke et al. 2002; Refregier et al. 2002), and the reported detection of excess power at high multipoles in the CMB power spectrum (Bond et al. 2003).

The constraints on cosmological parameters and, in particular, the constraints on dark energy, will continue to improve as further high-quality X-ray and gravitational lensing data are gathered.

Acknowledgments

I thank my collaborators Robert Schmidt, Andy Fabian and Harald Ebeling and the conference organizers for a highly enjoyable meeting.

References

Allen, S.W., Ettori, S., Fabian, A.C. 2001, MNRAS, 324, 877
Allen, S.W., Schmidt, R.W., Fabian, A.C. 2002a, MNRAS, 334, L11
Allen, S.W., Schmidt, R.W., Fabian, A.C. 2002b, MNRAS, 335, 256
Allen, S.W., Schmidt, R.W., Fabian, A.C., Ebeling H., 2003, 342, 287
Bennett, C. et al. , 1996, ApJ, 464, L1
BÃűhringer H. et al. , 2002, ApJ, 566, 93
Bond J.R. et al. , 2002, ApJ, submitted (astro-ph/0205386)
Borgani S. et al. , 2001, ApJ, 561, 13
Brown M.L. et al. , 2003, MNRAS, 341, 100
Dahle H., Kaiser N., Irgens R.J., Lilje P.B., Maddox S., 2002, ApJS, 139, 313
de Bernardis P. et al. , 2000, Nature, 404, 955
Ebeling H., Edge A.C., Allen S.W., Crawford C.S., Fabian A.C., Huchra J.P., 2000, MNRAS, 318, 333
Eke V.R., Navarro J.F., Frenk C.S., 1998, ApJ, 503, 569
Evrard A.E. et al. , 2002, ApJ, 573, 7
Freedman W. et al. , 2001, ApJ, 553, 47
Fukugita, M., Hogan, C.J., Peebles, P.J.E. 1998, ApJ, 503, 518
Hanany, S. et al. , 2000, ApJ, 545, L5
Jaffe A.H. et al. , 2001, Phys. Rev. Lett., 86, 3475
Jarvis M. et al. , 2003, AJ, 125, 1014
Kirkman D. et al. , 2003, ApJS, submitted (astrio-ph/0302006)
Lahav O. et al. , 2002, MNRAS, 333, 961
Navarro, J.F., Frenk, C.S., White, S.D.M. 1997, ApJ, 490, 493
Pen, U. 1997, New Ast., 2, 309
Percival W.J. et al. , 2002, MNRAS, 337, 1068
Perlmutter, S. et al. , 1999, ApJ, 517, 565
Refregier A., Rhodes J., Groth E.J., 2002, ApJL, 572, L131
Riess, A.G. et al. , 1998, AJ, 116, 1009
Sand D.J., Treu T., Ellis R.S., 2002, ApJ, 547, L129
Sasaki, S. 1996, PASJ, 48, L119
Schmidt R.W., Allen S.W., Fabian A.C., 2001, MNRAS, 327, 1075
Schuecker P., Böhringer H., Collins C.A., Guzzo L., 2003, A&A, 398, 867
Spergel D. et al. , 2003, ApJS, 148, 175
Van Waerbeke, L, Mellier, Y., Pelló, R., Pen U.-L., McCracken H.J., Jain B., 2002, A&A, 393, 369
White, S.D.M., Navarro J.F., Evrard A.E., Frenk C.S., 1993, Nature, 366, 429

ON THE INTRACLUSTER MEDIUM IN COOLING FLOW & NON-COOLING FLOW CLUSTERS*

Arif Babul, Ian G. McCarthy & Greg B. Poole
Department of Physics and Astronomy
University of Victoria, 3800 Finnerty Road
Victoria, BC, Canada. V8P 1A1
babul@uvic.ca

Abstract Recent X-ray observations have highlighted clusters that lack entropy cores. At first glance, these results appear to invalidate the preheated ICM models. We show that a self-consistent preheating model, which factors in the effects of radiative cooling, is in excellent agreement with the observations. Moreover, the model naturally explains the intrinsic scatter in the L-T relation, with "cooling flow" and "non-cooling flow" systems corresponding to mildly and strongly preheated systems, respectively. We discuss why preheating ought to be favoured over merging as a mechanism for the origin of "non-cooling flow" clusters.

Keywords: Intracluster Medium, Sunyaev-Zel'dovich Effect, X-rays, Entropy Profiles

1. Introduction

Correlations between the various X-ray and Sunyaev-Zel'dovich Effect (SZE) properties of galaxy clusters offer important clues into the physical processes that have impacted the intracluster medium (ICM). Observed scaling relations have been shown to deviate significiantly from expectations based on numerical simulations and analytic models that only take into account the influence of gravity on the ICM. Such discrepancies have prompted considerations of additional, previously unexamined, gas physics. One model, the preheating model, explores the possibility that the nascent ICM is heated by galactic winds and/or AGN outflows from galaxies existing at the time. Even in its simplest avatar, the model scaling relations, be they SZE v. SZE, SZE v. X-ray or X-ray v. X-ray, are in remarkable agreement with the observations (cf. Babul et al. 2002; McCarthy et al. 2002; McCarthy et al. 2003). However, recent X-ray data from *XMM-Newton* and *Chandra* suggests a potential problem with the

*Research supported by NSERC (Canada) Discovery Grant Award

M. Plionis (ed.), Multiwavelength Cosmology, 247-250.

model. Preheating of the ICM sets an entropy floor that manifests itself as a central core-like structure in the entropy profile. A number of observed profiles show no such cores. Although we recognize that the current set of published *Chandra* and *XMM-Newton* cluster results are biased in favour of "massive cooling flow" systems or active mergers, the very existence of systems with power-law-like entropy profiles needs addressing.

Here, we briefly report on our efforts to understand this particular issue, and its implications for the preheated model. Our investigations involve a re-examination of the assumptions underlying the theoretical model as well as of the observational evidence for and against the model.

2. Re-examining the Theoretical Model

Most preheating models are incomplete in that they do not take into account radiative cooling (cf. Babul et al. 2002). While preheating lowers the efficiency of cooling, it does not mitigate it entirely and over a Hubble time, the effects of cooling can be significant. Several recent studies (Voit et al 2002; Dave et al. 2002), have highlighted the potentially important role of cooling though not necessarily in the context of a preheated model. Traditionally, cooling has been difficult to model. However, we have developed a fast, efficient scheme for doing so. The scheme can factor in the effects of not only preheating and radiative cooling (due to both line and continuum emission) on the ICM over cosmological timescales, but potentially also those due to other processes such as conduction. The scheme is currently being tested against detailed hydrodynamic simulations and the initial results are very encouraging. A detailed description of this scheme will be forthcoming. Here we present some early results for the preheated+cooling model.

3. Cluster Entropy Profiles: Theory and Observations

Figure 1 shows the effects of varying levels of preheating+cooling on cluster entropy profiles. Radiative cooling is very efficient in clusters subjected to low levels of preheating. It causes the central entropy core to disappear on a relatively short timescale and drives the entropy profile into the $r^{1.1}$ power-law form reminiscent of the observed profiles of "cooling flow" (CF) clusters. In contrast, cooling is much less efficient in clusters with highly preheated ICM. We have assembled a preliminary collection of observed entropy profiles in order to see how these compare to the model results. We find, suprisingly, that the observed entropy profiles are not all self-similar and power-law-like, but span the entire range of shapes seen in Figure 1.

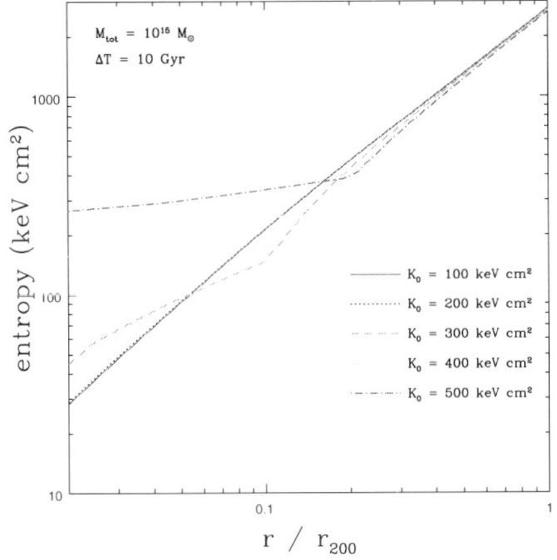

Figure 1. Entropy profiles of a $10^{15}M_\odot$ cluster preheated to different values of $K_\circ$ and observed after 10 Gyrs. The efficacy of radiative cooling increases with decreasing $K_\circ$. In clusters with low $K_\circ$ values, cooling rapidly erases the central core and drives the profile into a $r^{1.1}$ power-law, reminiscent of the observed "cooling flow" clusters profiles. Intriguingly, entropy profiles of actual clusters span the entire range of profile shapes shown.

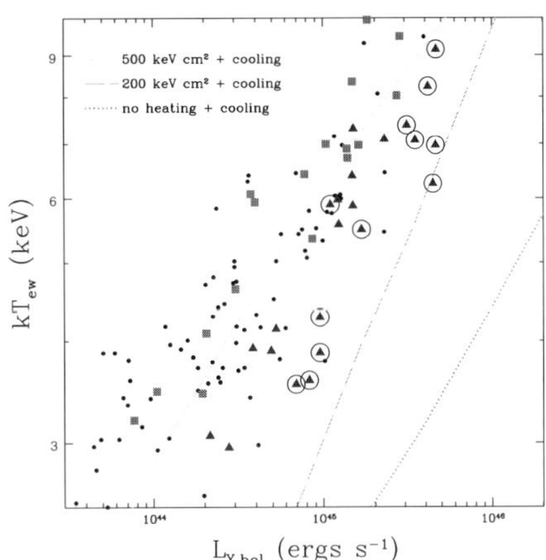

Figure 2. The $L_X - T_X$ relation for $z < 0.2$ clusters constructed using the actual observed values, as opposed to "cooling flow corrected" quantities. Within this locus, we identify the locations of known "cooling flow" and "non-cooling flow" clusters (see text for details). We also show the $L_X - T_X$ results for the preheated+cooling model. We identify NCF and mCF systems with high and low $K_\circ$ values, respectively.

4. Reconstructing the Scaling Relations

Figure 2 shows $L_X - T_X$ relation for preheated+cooling clusters subjected to varying levels of preheating and observed 10 Gyrs later. Also plotted are the (L_X, T_X) for $z < 0.2$ clusters from Horner's ASCA Cluster Catalogue (Horner et al. 2001). Unlike the data typically used in the construction of such plots, this set is *not* "cooling flow corrected". This is important. The theoretical values incorporate the effects of cooling; therefore, a fair comparison with the observations requires that we use data of comparable character.

Both the corrected and the uncorrected data exhibit the same correlations, though the latter has larger scatter. This is well known. However, the scatter is not random. Based on features in the temperature and X-ray SB profiles, we have classified clusters as non-cooling flow clusters (NCF - squares), ordinary cooling flow clusters (CF - triangles), or massive cooling flow clusters (mCF - circled triangles). Although most clusters remain unclassified, it is readily apparent that NCF systems lie close to the upper-left edge of the band while the mCF clusters delineate the opposite (lower-right) edge.

Comparing the preheated+cooling model predictions against the observations, we find the two in excellent agreement. However, in order to account for the breadth and structure within the observed $L_X - T_X$ band, we have to abandon our previous *ad hoc* assumption of uniform energy injection across the entire cluster population. Within our framework, the NCFs correspond to strongly preheated systems ($K_\circ \sim$ 400–500 Kev cm^2) while the mCFs correspond to mildly preheated systems ($K_\circ \sim$ 100–200 Kev cm^2).

5. Non-Cooling Flow Clusters: Products of Preheating?

The assertion that NCF systems are strongly preheated clusters runs counter to the prevailing view. In the latter, NCFs are identified as clusters whose cool dense gas cores, the source of the excess central X-ray emission characteristic of the CF clusters, have been disrupted by major mergers. Images of NCFs with disturbed X-ray morphologies are often used to support this scenario. However, there are numerous CF systems that also appear to be in the throes of on-going mergers. Perseus is one such example (Churazov et al. 2003). The ubiquity of mergers argues against them being the cause of the differences between CFs and NCFs. To test our hypothesis, we have carried out a series of numerical simulation experiments. One distinguishing feature of our study is that our simulations include radiative cooling. Preliminary results suggests that even for nearly head-on 3:1 mergers, variations in the X-ray observables of the primary cluster are extremely short-lived. In particular, we find that if the primary starts out as a CF-like system, by the time the merger remnant has been assimilated, it will have regained its CF-like character. Motl et al. 2003 too get similar results. These findings argue that there ought *not* to be any dynamically relaxed NCF systems. But there are: eg. A1413, A1651, A2319, A3158. That such systems exist at all further indicates an alternate origin for the CF/NCF clusters.

To reiterate, a self-consistent model of the ICM that factors in radiative processes and allows for cluster-to-cluster variations in the level of initial preheating not only is able to account for the existence and the properties of CF and NCF clusters, but also of those that are between these two extremes.

We wish to thank our collaborators, M. Balogh, G. Holder, M. Fardal, T. Quinn, and D. Horner for their contributions to the work described here.

References

Babul, A., Balogh, M. L., Lewis, G. F., & Poole, G. B. 2002, MNRAS, 330, 329

McCarthy, I. G., Babul, A., & Balogh, M. L. 2002, ApJ, 573, 515

McCarthy, I. G., Holder, G. P., Babul, A., & Balogh, M. L. 2003, ApJ, 591, 526

Voit, M. G., Bryan, G. L., Balogh, M. L., & Bower, R. G. 2002, ApJ, 576, 601

Dave, R., Katz, N., & Weinberg, D. H. 2002, ApJ, 579, 23

Horner, D. J. 2001, PhD thesis, University of Maryland

Churazov, E., Forman, W., Jones, C., & Böhringer, H. 2003, ApJ, 590, 225

Motl, P.M., Burns, J.O., Loken, C., Norman, M.L., Bryan, G., 2003, astro-ph/0302427

COSMOLOGICAL CONSTRAINTS FROM THE EVOLUTION OF THE CLUSTER BARYON MASS FUNCTION

A. Vikhlinin, A. Voevodkin, C. R. Mullis, L. VanSpeybroeck, H. Quintana, A. Hornstrup
Harvard-Smithsonian Center for Astrophysics, 60 Garden St, Cambridge MA, 02138, USA
vikhlinin@cfa.harvard.edu

Abstract We determine the evolution of the number density of the galaxy clusters between $z = 0$ and $z = 0.5$. Our method uses the cluster baryon mass as a proxy for the total mass, thereby avoiding any uncertainties of the $M_{\rm tot} - T$ or $M_{\rm tot} - L_X$ relations. Instead, we rely on a well-founded assumption that the $M_b/M_{\rm tot}$ ratio is a universal quantity. Taking advantage of direct and accurate *Chandra* measurements of the gas masses for distant clusters, we find strong evolution of the baryon mass function between $z > 0.4$ and the present. The observed evolution defines a narrow band in the $\Omega_m - \Lambda$ plane which intersects with constraints from the Cosmic Microwave Background and supernovae Ia near $\Omega_m = 0.3$ and $\Lambda = 0.7$. Assuming the flat Universe, cluster evolution favors the values of the dark energy equation of state parameter, $w \approx -1$.

1. Introduction

The growth of large scale structure in the near cosmological past is a sensitive cosmological probe. One of the strongest manifestations of the growth of density perturbations is the formation of virialized objects such as clusters of galaxies. Therefore, the evolution of the cluster mass function provides a sensitive cosmological test.

Most of the difficulties with using cluster evolution as a cosmological probe are on the observational side. The foremost problem is the necessity to relate cluster masses with other observables. The most widely used observable is the temperature of the intracluster medium which should scale with the virial mass as $M \propto T^{3/2}$. To normalize this relation, virial masses still have to be measured in a representative sample of clusters. Unfortunately, theoretical prediction for the number density of clusters with a given temperature is exponentially sensitive to the normalization of the $M - T$ relation. Given that $\pm 50\%$ is a fair estimate of the present systematic uncertainties in the $M - T$

M. Plionis (ed.), Multiwavelength Cosmology, 251-254.

normalization, the impact of this relation on cosmological parameter determination can be very significant.

We apply the evolutionary test to a new sample of distant, $z > 0.4$, clusters derived from the 160 deg^2 *ROSAT* survey (Vikhlinin et al. 1998, Mullis et al. 2003), relying on the evolution of the cluster scaling relations as measured by *Chandra*, and using a modeling technique which does not rely upon the total mass measurements to normalize the $M - T$ or $M - L$ relations and therefore bypasses many of the uncertainties mentioned above.

The proposed method relies on the assumption that the baryon fraction within the virial radius in clusters should be close to the average value in the Universe, Ω_b/Ω_m (White et al. 1993). The baryon mass in clusters is relatively easily measured to the virial radius. To first order, the baryon and total mass are related simply as $M_b = M\,\Omega_b/\Omega_m$, which results in a trivial relation between the cumulative total mass function, $F(M)$, and the baryon mass function, $F_b(M_b)$, $F_b(M_b) = F\,(\Omega_m/\Omega_b\,M_b)$. The average baryon density in the Universe is accurately known, e.g., from the BBN (Burles, Nollet & Turner 2001). Therefore, we have everything needed to convert the theoretical model for the total mass function, $F(M)$, to the prediction for a directly measurable baryon mass function — to compute $F(M)$ one must choose a value of Ω_m, which fixes the scaling between the total and baryon mass.

A detailed description of the datasets used and the technical aspects of the application of this method to model the cluster data are given in Voevodkin & Vikhlinin (2003) and Vikhlinin et al. (2003). The observed baryon mass functions at low redshift ($z < 0.1$) and at high redshift ($0.4 < z < 0.8$, $z_{\rm med} = 0.55$) are shown in Fig. 1. The most notable feature is a factor of ~ 10 decrease in the number density of clusters for the fixed mass at high redshifts.

2. Results

Voevodkin & Vikhlinin (2003) present the modeling of the baryon mass function at low redshift derived from a sample of 50 brightest X-ray clusters from the HIFLUGCS catalog (Reiprich & Böhringer 2001) based on the ROSAT All-Sky Survey. This analysis allows to constrain the amplitude of the linear density perturbations at zero redshift, $\sigma_8 = 0.72 \pm 0.04$ and the matter power spectrum shape parameter $\Omega_M h = 0.13 \pm 0.07$, in good agreement with a number of independent methods. The combined constraints on σ_8 and the shape parameter are shown in Fig. 2.

The evolution in the number density of clusters allows a determination of the growth factor of the density perturbations, $D(z)$ between $z = 0$ and $z = 0.55$. The results of fitting the data from Fig. 1 are shown in Fig. 3. The measurement of $D(z)$ is slightly sensitive to the density parameter. Assuming $\Omega_M = 0.2$, we find $D(0.55) = 0.76 \pm 0.045$; for $\Omega_M = 0.3$, $D(0.55) = 0.73 \pm 0.03$. The

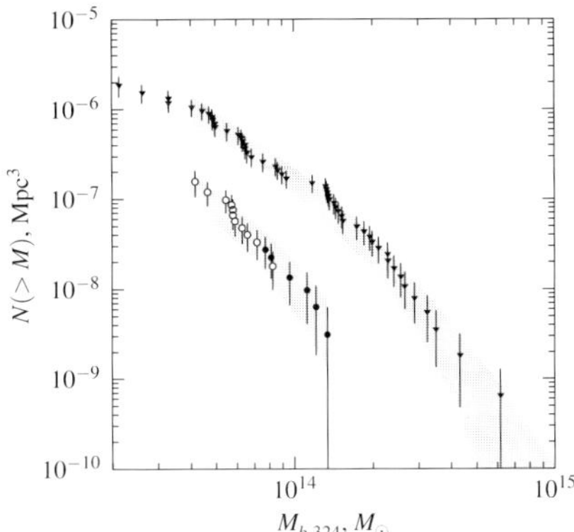

Figure 1. The baryon mass function for the 160 $\deg^2$ survey sample in the redshift interval $0.4 < z < 0.8$. The local mass function from Voevodkin & Vikhlinin (2003) is shown by triangles. The grey band shows a 68% uncertainty interval for the mass functions, including both the Poisson noise and mass measurement uncertainties. The high-redshift mass function is computed for $\Omega_m = 0.3$, $\Lambda = 0.7$; the local mass function is cosmology-independent.

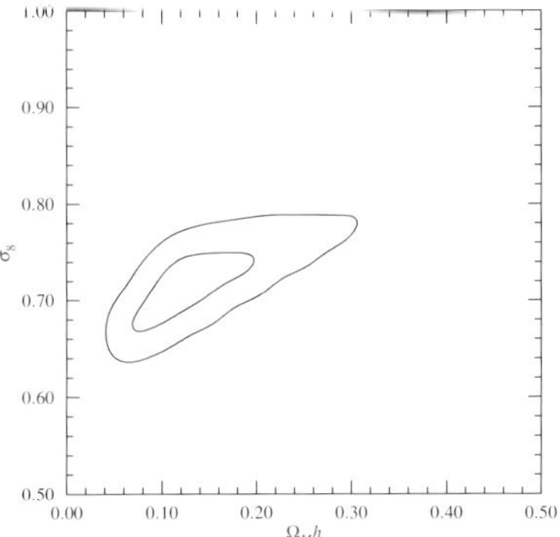

Figure 2. Confidence intervals (68% and 95%) on σ_8 and $\Omega_M h$ from the analysis of the low-redshift baryon mass function, assuming $h = 0.71$, $n = 1$, $\Omega_b h^2 = 0.0224$. See Voevodkin & Vikhlinin (2003) for a discussion of the dependence of these constraints on h, n, and Ω_b.

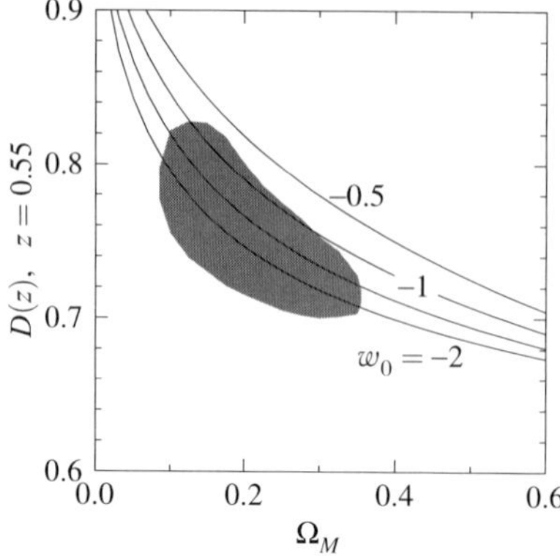

Figure 3. Growth factor of the linear density perturbations at $z = 0.55$ as determined from observed evolution of the cluster baryon mass function. Solid lines show the growth factor in the flat cosmology for several values of the dark energy equation of state parameter, w.

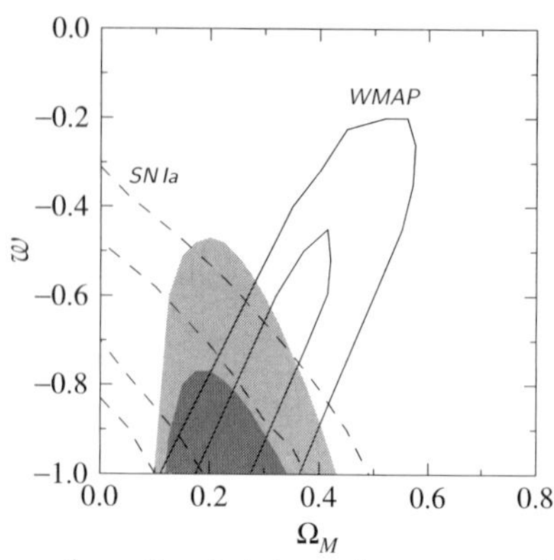

Figure 4. Constraints on the w parameter from the observed cluster evolution (68% and 95% confidence regions) in comparison with the constraints from Sn Ia and WMAP.

values of $\Omega_M > 0.35$ are inconsistent with the shallow slope of the observed mass function (Voevodkin & Vikhlinin 2003).

The growth factor of the density perturbations in the CDM model is determined by the density parameters of matter, dark energy, and by the effective equation of state for the dark energy. In Vikhlinin et al. (2003), we effectively converted the measurement of $D(z)$ into constraints on Ω_M and Ω_Λ assuming that the dark energy is represented by the cosmological constant. This results in a narrow band, $\Omega_m + 0.23\Lambda = 0.41 \pm 0.10$, which intersects with the Sn Ia and CMB results near $\Omega_M = 0.3$ and $\Lambda = 0.7$. Assuming that the Universe is spatially flat, the constraints on $D(z)$ can be converted into constraints on Ω_M and the dark energy equation of state parameter, w (e.g., Linder & Jenkins 2003, cf. solid lines in Fig. 3). The results are presented in Fig. 4. The observed cluster evolution favors $w < 0.8$, in agreement with a combination of the Sn Ia and WMAP results.

References

Burles, S., Nollett, K. M., & Turner, M. S. 2001, ApJ, 552, L1

Linder, E. & Jenkins, A. 2003, MNRAS, in press, astro-ph/0305286

Mullis, C. R. et al. 2003, ApJ, 594, 154

Reiprich, T. H. & Böhringer, H. 2002, ApJ, 567, 716

Vikhlinin, A., McNamara, B. R., Forman, W., Jones, C., Quintana, H., & Hornstrup, A. 1998, ApJ, 502, 558

Vikhlinin, A. et al. 2003, ApJ, 590, 15

Voevodkin, A. & Vikhlinin, A. 2003, ApJ submitted, astro-ph/0305549

White, S. D. M., Navarro, J. F., Evrard, A. E., & Frenk, C. S. 1993, Nature, 366, 429

X-RAY OBSERVATIONS OF THE MOST MASSIVE DLS SHEAR-SELECTED GALAXY CLUSTERS

John P. Hughes,[1] Ian Dell'Antonio,[2] Joseph Hennawi,[3] Vera Margoniner,[4] Sandor Molnar,[1] Dara Norman,[5] David Spergel,[3] J. Anthony Tyson,[4] Gillian Wilson,[6] and David Wittman[4]
[1]*Rutgers University,* [2]*Brown University,* [3]*Princeton University,* [4]*Bell Labs, Lucent Technologies,* [5]*CTIO/NSF AAPF,* [6]*SIRTF Science Center*

Abstract We report on preliminary results of our X-ray survey of the most massive clusters currently identified from the Deep Lens Survey (DLS). The DLS cluster sample is selected based on weak lensing shear, which makes it possible for the first time to study clusters in a baryon-independent way. In this article we present X-ray properties of a subset of the shear-selected cluster sample.

Keywords: dark matter, gravitational lensing, large-scale structure of universe, X-rays: galaxies: clusters

1. Introduction

Nonbaryonic dark matter is apparently the dominant component of galaxy clusters, yet all large samples of clusters to date are selected on the basis of emission from the trace baryons they contain: visible light from galaxies or X-rays from hot intracluster gas. Now, for the first time, we have a direct survey of mass in the Universe that is unbiased with respect to baryons, the Deep Lens Survey (DLS). The DLS is a deep, wide area, multicolor (BVRz') imaging survey being carried out at the NOAO 4-m telescopes. The survey was designed to detect large scale structures in the Universe through weak lensing shear, i.e., distortions to the shapes of distant background galaxies caused by gravitational lensing of massive foreground objects. The DLS team has already shown that shear-selection is effective at finding new galaxy clusters: Wittman et al. (2001) report the discovery of CL J2346+0045, the first galaxy cluster identified by its gravitational effect rather than its radiation.

For this project, 12 square degrees of the DLS data (the maximum sky area available at the time) were processed through the weak lens shear pipeline (Wittman et al. 2003), revealing mass concentrations over a wide range of redshifts. These mass clusters were rank-ordered by their shear signal and the top

M. Plionis (ed.), Multiwavelength Cosmology, 255-258.

candidates were proposed for observation by the *Chandra X-Ray Observatory* in cycle 4. Seven targets were awarded; an additional candidate was available through the *Chandra* archive. One of the principle goals of the follow-up X-ray observations is to confirm that the DLS shear-selected clusters are associated with true virialized, collapsed structures. The basic X-ray information (luminosity, size, morphology, extent of central concentration, and gas temperature) obtained on the clusters will allow us to assess the effect of shear-selection on the L_X–T_X relation, the cluster temperature function, and the relation of these to cluster mass. The full DLS X-ray cluster sample will also allow us to quantify the false-positive rate of "aligned filaments," i.e., line-of-sight projections that appear as spurious mass concentrations in weak lensing shear maps (e.g., White, van Waerbeke, & Mackey 2002).

2. X-ray Observations

We have confidently detected extended X-ray emission from at least five of the eight DLS clusters in the *Chandra* cycle 4 sample. Although the other three targets have been observed, our analysis is not yet complete and we do not comment on them further here.

Figure 1 shows maps of the projected mass (left panels) and X-ray surface brightness (right panels) over the 0.5–2 keV band for several of the DLS mass clusters. In the X-ray images serendipitous point sources have been removed and circles denote the locations of extended X-ray sources detected at signal-to-noise ratios greater than 3. The two highest-ranked shear-selected clusters are strong X-ray sources, with multiple subclusters associated with each system. This can be seen clearly in the X-ray image of DLS cluster 2 (top panel of Figure 1). In the optical images there are a large number of bright galaxies that are presumably cluster members, however no published redshifts are available. The fourth ranked cluster (middle panels) is unusual in having only a single X-ray component. It is centered on a galaxy at a redshift of $z = 0.19$. The last ranked cluster in the *Chandra* cycle 4 sample (bottom panels) shows two significant extended X-ray sources. The southwestern component was confirmed by the DLS team as a massive cluster at $z = 0.68$ (Wittman et al. 2003) and even shows a giant arc from strong lensing. There is no redshift available for the northern X-ray concentration.

In Table 1 we present selected numerical results for the five X-ray clusters associated with shear peaks. In each case only values for the X-ray cluster component with the highest flux are given. Redshift information for clusters 2 and 7 is not yet available; we give very preliminary luminosity and mass estimates based on approximate redshifts from the magnitudes of the member galaxies. Photometric redshifts for all X-ray clusters are in the process of being determined from our imaging data. We used the X-ray luminosity-temperature

Table 1. X-ray Properties of DLS Mass Clusters.

Cluster	z	F_X (0.5–2 keV) (erg s^{-1} cm^{-2})	$L_{\rm bol}$ (erg s^{-1})	M_{200} ($10^{14}\,M_\odot$)
DLS Clu 1	0.298	6.7×10^{-13}	8.5×10^{44}	5.7
DLS Clu 2	$\sim$0.2	2.9×10^{-13}	1.1×10^{44}	1.9
DLS Clu 4	0.1894	1.4×10^{-14}	4.3×10^{42}	0.4
DLS Clu 7	$\sim$0.4	1.9×10^{-14}	3.1×10^{43}	1.0
DLS Clu 8	0.68	2.4×10^{-14}	1.7×10^{44}	2.4

relation (Arnaud & Evrard 1999) to estimate cluster temperatures and then the mass-temperature relation (Evrard, Metzler, & Navarro 1996) to determine the mass within a density contrast of $\delta_c = 200$. These relations are derived from or calibrated against low-redshift clusters; at this point we have not made any adjustments for the redshift range of our sample. Still the mass estimates are consistent with simulations by our group that show an expected mass range for DLS shear-selected clusters extending from roughly $5 \times 10^{13}\,M_\odot$ to $10^{15}\,M_\odot$ with a peak near $3 \times 10^{14}\,M_\odot$. An additional direct measurement of the mass can be made from the shear maps; this work is in progress.

3. Conclusions

The DLS is clearly discovering true three-dimensional clusters of mass, hot X-ray gas, and galaxies. We find a wide range of X-ray luminosity for clusters with similar weak lensing shear. Nearly every shear peak contains multiple X-ray clusters, while the high mass clusters we detect are particularly complex with up to 4 or 5 individually resolved subcomponents. Work in progress on this unique galaxy cluster sample includes measuring X-ray temperatures from *Chandra* and *XMM-Newton* spectra, deriving redshifts and velocity dispersions from ground-based spectroscopy, more detailed mass determinations, and additional numerical simulations.

Acknowledgments

This work was partially supported by *Chandra* grant GO3-4173A and NASA LTSA grant NAG5-3432 to JPH.

References

Arnaud, M, & Evrard, AE 1999, MNRAS, 305, 631
Evrard, AE, Metzler, CA, & Navarro, JF 1996, ApJ, 469, 494
White, M, van Waerbeke, L, & Mackey J 2002, ApJ, 575, 640
Wittman, D., Tyson, JA, Margoniner, VE, Cohen, JG, & Dell'Antonio, IP 2001, ApJ, 557, L89
Wittman, D., Margoniner, VE, Tyson, JA, Cohen, JG, & Dell'Antonio, IP 2003, ApJ, submitted

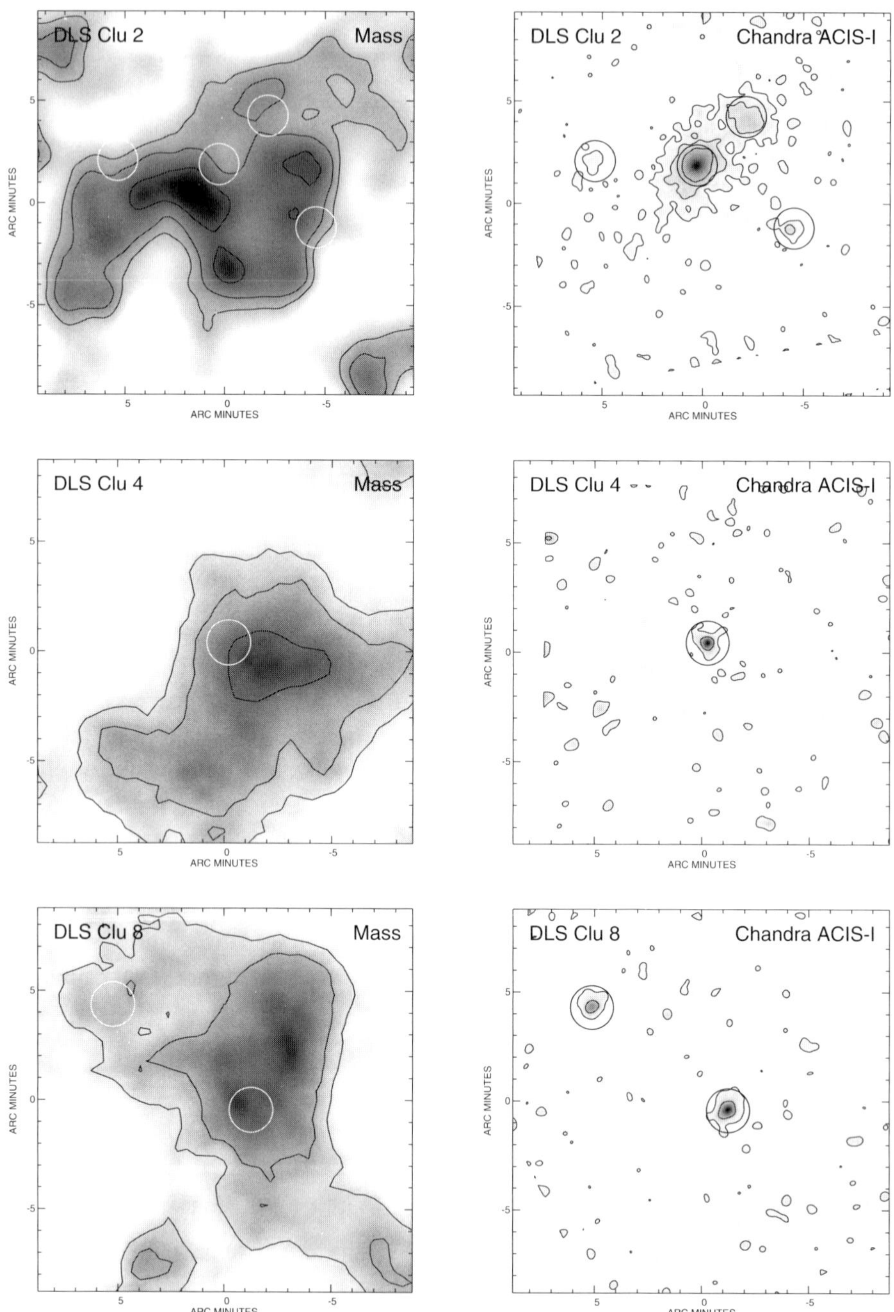

Figure 1. Projected mass (left) and X-ray surface brightness (right) of DLS mass clusters no. 2 (top), 4 (middle), and 8 (bottom).

COSMOLOGY WITH XMM SHARC CLUSTERS

A.Blanchard, S.C. Vauclair & the Ω project collaboration
LAOMP, UPS, CNRS, UMR5572, 14 Av Ed.Belin
31 400 Toulouse, France
alain.blanchard@ast.obs-mip.fr

Abstract We present the preliminary results of the Ω project, a large XMM program devoted to observe distant SHARC clusters. For the first time a measurement of the $L - T$ evolution with XMM is presented. We found clear evidence for a positive evolution of the $L - T$ relation, in agreement with some previous analysis based on ASCA and Chandra observations. Its cosmological implication is also discussed based on a new analysis of different X-ray surveys : EMSS, RDCS, MACS, SHARC, 160 deg^2. Its found that a high matter density model fit remarkably well all these surveys, in agreement with all existing previous analyzes following the same strategy. Concordance models produce far more faint clusters than observed counts. This failure could be the indication of a deviation from the expected scaling of the $M - T$ relation with redshift.

Keywords: Cosmological parameters,X-ray clusters

1. Introduction

The XMM-Ω project (Bartlett et al. 2001) was conducted in order to provide an accurate estimation of the possible evolution of the luminosity–temperature relation at high redshift for clusters of medium luminosity which constitute the bulk of X-ray selected samples, allowing to remove a major source of degeneracy in the determination of Ω_M from cluster abundance evolution.

2. Observed evolution of the $L - T$ relation of X-ray clusters

D.Lumb et al. (2003) present the results of the X-ray measurements of 8 distant SHARC clusters with redshifts between 0.45 and 0.62. By comparing to various local $L - T$ relations, clear evidence for evolution in the $L - T$ relation has been found. The possible evolution has been modeled in the following

M. Plionis (ed.), Multiwavelength Cosmology, 259-262.

way:

$$L_{\mathrm{x}} = L_6(0) \left(\frac{T}{6keV} \right)^{\alpha(0)} (1+z)^{\Gamma} \tag{1}$$

where $L_6(0) \left(\frac{T}{6keV}\right)^{\alpha(0)}$ is the local $L-T$ relation. Γ is found to be of the order of 0.6 in an Einstein-de Sitter cosmology. This results is entirely consistent with previous analyzes (Sadat et al., 1998; Vikhlinin et al, 2002).

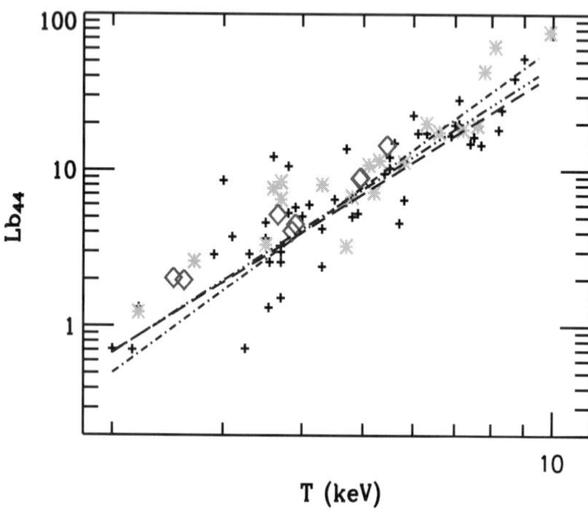

Figure 1. Temperature–luminosity of X-ray clusters: crosses are local clusters from a flux selected sample, stars are from Chandra, diamonds are SHARC clusters from the XMM Ω project.

3. Cosmological interpretation

The evolution of the abundance of X-rays clusters is known to be a powerful cosmological test (Oukbir & Blanchard, 1992). Indeed the evolution of the number of clusters of a given mass is a sensitive function of the cosmological density of the Universe, very weakly depending on other quantities when properly normalized (Blanchard & Bartlett, 1998).

Attempts to apply this test have been performed but still from a very limited number of clusters (typically 10 at redshift 0.35) (Henry, 1997; Viana and Liddle, 1998; Eke et al., 1998; Blanchard et al., 2000). On the other hand, number counts allow one to use samples comprising much more clusters. Indeed using simultaneously different existing surveys: EMSS, SHARC, RDCS, MACs NEP and 160 deg^2 one can use information provided by more than 300 clusters with $z > 0.3$ (not necessarily independent). In order to model clusters number counts, for which temperatures are not known, it is necessary to have a good

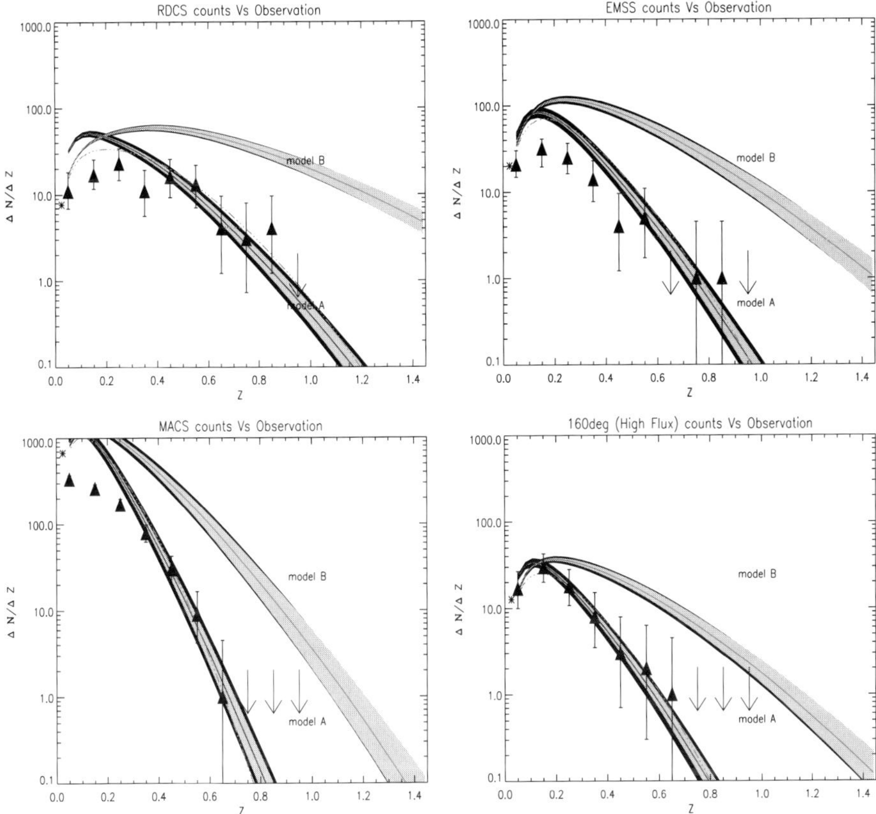

Figure 2. Theoretical number counts in bins of redshift ($\Delta z = 0.1$) for the different surveys: RDCS, EMSS, MACS and 160deg^2 (high flux, corresponding to fluxes $f_x > 2\,10^{-13}$ erg/s/cm^2). Observed numbers are triangles with 95% confidence interval on the density assuming poissonian statistics (arrows are 95% upper limits). The upper curves are the predictions in the concordance model (model A). The lower curves are for critical universe (model B). Different systematics have been investigated (see Vauclair et al. in this volume). Grey areas are uncertainties from uncertainties on σ_8 and on $L-T$ evolution. The 3-dotted-dashed lines show the predicted counts in the concordance model using $M-T$ relation violating the standard scaling with redshift.

knowledge of the $L-T$ relation over the redshift range which is investigated, which information has been provided by XMM and Chandra. Number counts can then be computed:

$$
\begin{aligned}
N(>f_x, z, \Delta z) = & \quad \Omega \int_{z-\Delta z}^{z+\Delta z} \frac{\partial N}{\partial z}(L_x > 4\pi D_l^2 f_x) dz \\
= & \quad \Omega \int_{z-\Delta z}^{z+\Delta z} N(>T(z)) dV(z)
\end{aligned}
$$

$$= \quad \Omega \int_{z-\Delta z}^{z+\Delta z} \int_{M(z)}^{+\infty} N(M,z) dM dV(z) \qquad (2)$$

where $T(z)$ is the temperature threshold corresponding to the flux f_x as given by the observations, being therefore independent of the cosmological model. Several ingredients are needed: the local abundance of clusters as given by the temperature distribution function ($N(T)$), the mass-temperature relation and its evolution, the mass function and the knowledge of the dispersion. Uncertainties in these quantities result in uncertainties in the modeling. However, Vauclair et al. in this issue showed that these sources of -systematics- uncertainties are comparable to statistical uncertainties which are actually very small. Figure 2 illustrates the counts obtained with a standard mass temperature relation: $T = 4\mathrm{keV} M_{15}^{2/3}(1+z)$, the SMT mass function (Sheth et al., 2001), and the $L - T$ relation observed by XMM with its uncertainty. These counts were computed for different existing surveys to which they can be compared.

4. Conclusion

The major results obtained with the Ω project are the first XMM measurement of the evolution of the luminosity-temperature with redshift. A positive evolution has been detected, in agreement with previous results including those obtained by Chandra (Vikhlinin et al., 2002). The second important result is that this evolving $L - T$ produced counts in the concordance model which are inconsistent with the observed counts in all existing published surveys. This could be the sign of a high density universe or a deviation from the expected scaling of the $M - T$ relation with redshift.

References

Bartlett, J. et al., 2001, proceedings of the XXI rencontre de Moriond, astro-ph/0106098

Blanchard, A. & Bartlett, J. 1998, A&A 332, 49L

Blanchard, A., Sadat, R., Bartlett, J. & Le Dour, M. 2000, A&A 362, 809

Eke, V.R., Cole, S., Frenk, C. & Henry, P.J, 1998, MNRAS 298, 1145

Henry, J. P. 1997, ApJ 489, L1

Lumb, D. et al., 2003, A&A, in preparation.

Markevitch, A. 1998, ApJ 504, 27

Oukbir, J. & Blanchard, A. 1992, A&A 262, L21

Sadat, R., Blanchard, A. & Oukbir, J. 1998, A&A 329, 21

Sheth, R. K., Mo, H. J., Tormen, G. 2001, MNRAS, 323 1

Viana, T. P. & Liddle, A. R. 1999, MNRAS, 303, 535

Vikhlinin, A. et al., 2002, ApJL, 578, 107

CONSTRAINTS ON THE DARK MATTER SELF-INTERACTION CROSS-SECTION FROM THE MERGING CLUSTER 1E 0657–56

M. Markevitch[1], A.H. Gonzalez[2], D. Clowe[3], and A. Vikhlinin[1]
[1] *Harvard-Smithsonian Center for Astrophysics, 60 Garden St., Cambridge, MA 02138, USA*
[2] *Department of Astronomy, University of Florida*
[3] *Institut für Astrophysik und Extraterrestrische Forschung der Universitat Bonn*

Abstract We compare new maps of the hot gas, dark matter, and galaxies for 1E 0657–56, a cluster with a rare, high-velocity merger occurring nearly in the plane of the sky. The X-ray observations reveal a prominent bow shock and a bullet-like gas subcluster just exiting the collision site. The optical image shows that the gas bullet lags behind the subcluster galaxies; the weak-lensing mass map reveals a dark matter clump lying ahead of the collisional gas bullet, but coincident with the effectively collisionless galaxies. From these observations, one can directly constrain the cross-section of the dark matter self-interaction. That the dark matter is not fluid-like can be seen directly from the maps; more quantitative limits can be derived from four simple independent arguments. Our most sensitive constraint, $\sigma/m < 1\ \mathrm{cm}^2\ \mathrm{g}^{-1}$, comes from the consistency of the subcluster mass-to-light ratio with the main cluster (and universal) value, which rules out a large mass loss due to dark matter particle collisions.

1. Introduction

Chandra observations of 1E 0657–56, one of the hottest and most luminous galaxy clusters known, revealed a fast-moving, bullet-like, cool subcluster just exiting the core of the main cluster (Markevitch et al. 2002). An optical image shows a galaxy subcluster just ahead of this gas "bullet". A weak-lensing mass map (Clowe et al. 2003, hereafter C03) reveals a dark matter clump coincident with the centroid of the galaxy subcluster (Fig. 1). A combination of these three maps provides gives us a unique possibility to place a direct constraint on the dark matter self-interaction cross-section. Apart from the obvious basic interest for the still unknown nature of dark matter, the possibility of it having a nonzero collisional cross-section has far-reaching astrophysical implications (e.g., Spergel & Steinhardt 2000; Dave et al. 2001; Gnedin & Ostriker 2001; Yoshida et al. 2000).

M. Plionis (ed.), Multiwavelength Cosmology, 263-266.

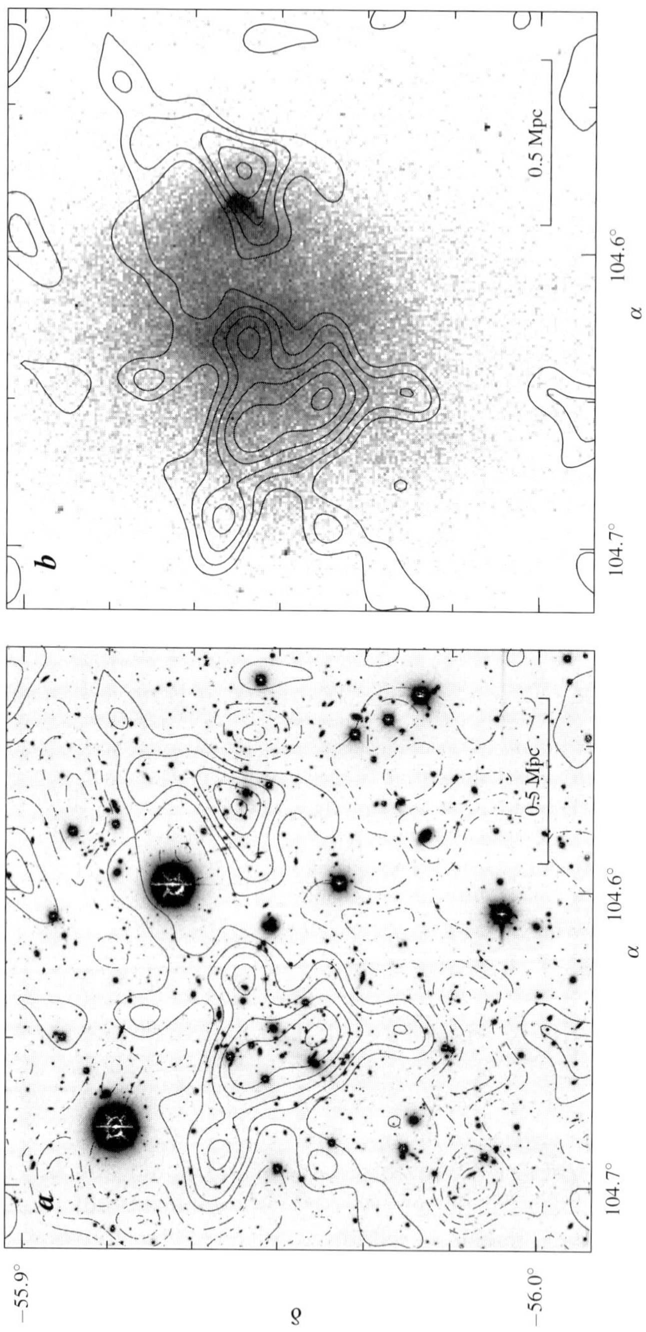

Figure 1. (*a*) Overlay of the weak lensing mass contours on the optical image of 1E 0657–56. Dashed contours are negative (relative to an arbitrary zero level). The subcluster's DM peak is coincident within uncertainties with the centroid of the galaxy concentration. (*b*) Overlay of the mass contours on the X-ray image (only the upper 5 of those in panel *a* are shown for clarity). The gas bullet lags behind the DM subcluster.

2. Collisional cross-section estimates

The dark matter collisional cross-section per unit mass, σ/m (where m is the DM particle mass), can be constrained from the 1E 0657–56 data by at least four independent methods, details of which are given in Markevitch et al. (2003, hereafter M03). They are based on the observed gas–dark matter offset, the absence of an offset between the dark matter and the galaxies, the high subcluster velocity, and the subcluster survival. For these estimates, we will use the X-ray-measured subcluster velocity ($v \simeq 4500$ km s^{-1}) and the weak-lensing measurements of the main cluster's and the subcluster's masses.

2.1 The gas — dark matter offset

The most remarkable feature in Fig. 1*b* is a 23^{+7}_{-11} arcsec offset between the subcluster's DM centroid and the gas bullet (C03). This offset means that the scattering depth of the dark matter subcluster w.r.t. collisions with the flow of dark matter particles cannot be much greater than 1. Otherwise the DM subcluster would behave as a clump of fluid, experiencing stripping and drag deceleration, similar to that of the gas bullet, and there would be no offset between the gas and dark matter. The subcluster's scattering depth is $\tau_s = \Sigma_s\, \sigma/m$, where Σ_s is the DM mass surface density of the subcluster. The measured surface density averaged over the face of the subcluster is $\Sigma_s \simeq 0.1$ g cm^{-2}. Requiring $\tau_s < 1$, we obtain $\sigma/m < 10$ cm^2 g^{-1}.

2.2 The high velocity of the subcluster

The observed velocity of the subcluster, $v = 4500$ km s^{-1}, is in good agreement with the expected free-fall velocity v_{ff} onto the main cluster. For our main cluster's mass profile, a small subcluster falling from a large distance should acquire $v_{ff} \approx 4400$ km s^{-1} at core passage (and even less at the present position). Such an agreement strongly suggests that the subcluster could not have lost much of its momentum to drag forces. For our masses and velocities, each DM particle collision would result in an average momentum loss by the subcluster $\bar{p} \simeq 0.1\, mv$ (M03). Having reached its present location, the subcluster would lose $v - v_{ff} = \Sigma_m\, \bar{p}\, \sigma/m^2$, where $\Sigma_m \simeq 0.3$ g cm^{-2} is the main cluster mass column. Conservatively disregarding dynamic friction and requiring, say. $v - v_{ff} < 1000$ km s^{-1}, we get $\sigma/m < 7$ cm^2 g^{-1}.

2.3 No offset between the dark matter and the galaxies

Another remarkable feature in Fig. 1*a* is the coincidence of the subcluster dark matter centroid with that of the galaxy distribution. Suppose the DM particles are experiencing frequent collisions and a resulting displacement. The effectively collisionless galaxies need time to redistribute over the subcluster and come into equilibrium with a DM gravitational potential that is lagging behind. This timescale happens to be greater than the time it takes for the subcluster to pass through most of the main cluster's observed mass column Σ_m. If, on this timescale, *most* of the subcluster's DM particles experience another collision, there would be a persistent offset between the galaxy and DM peaks, just like that observed between the DM and the X-ray gas. For each DM particle presently in the subcluster, the collision probability has been $\tau_{\rm m} = \Sigma_{\rm m}\,\sigma/m$. Simply requiring that $\tau_{\rm m} < 1$, we obtain $\sigma/m < 3\ {\rm cm}^2\,{\rm g}^{-1}$.

2.4 The survival of the dark matter subcluster

For the observed subcluster mass and velocity, the most likely result of a particle collision is the loss of a particle by the subcluster. We can put an upper limit on the accumulated mass loss and thereby on the collision cross-section. C03 have derived a B-band mass-to-light ratio for the subcluster within its truncation radius, $M/L \simeq 260 \pm 90$, which is in good agreement with the universal cluster value from the lensing analyses (e.g., Mellier 1999), and a factor of 1.1 ± 0.3 from the main cluster value derived from the same data. If the subcluster had been continuously losing DM particles, we would expect an anomalously low M/L value for the subcluster — and in particular, a value lower the main cluster's. From this agreement, we infer that the subcluster could not have lost more than $\mu \approx 0.2 - 0.3$ of its initial mass within its present tidal truncation radius. This fraction is $\mu = (1 - 2\,v_{\rm esc}^2/v^2)\,\Sigma_{\rm m}\ \sigma/m$, where $v_{\rm esc}$ is the subcluster's escape velocity (M03). This gives our most sensitive limit, $\sigma/m < 1\ {\rm cm}^2\,{\rm g}^{-1}$.

This limit is comparable to other, less direct astrophysical constraints (e.g., Hennawi & Ostriker 2002 and references therein), and excludes most of the $0.5 - 5\ {\rm cm}^2\,{\rm g}^{-1}$ interval proposed to explain the galaxy mass profiles (e.g., Dave et al. 2001). It is only an order-of-magnitude estimate which involves a number of conservative simplifying assumptions; a more accurate, and possibly stronger, limit may be derived through hydrodynamic simulations of this unique merger.

References

Clowe, D., et al. 2003, in preparation (C03)
Dave, R., Spergel, D.N., Steinhardt, P.J., & Wandelt, B.D. 2001, ApJ, 547, 574
Gnedin, O.Y., & Ostriker, J.P. 2001, ApJ, 561, 61
Hennawi, J.F., & Ostriker, J.P. 2002, ApJ, 572, 41
Markevitch, M., et al. 2002, ApJ, 567, L27
Markevitch, M., et al. 2003, ApJ, submitted; astro-ph/0309303 (M03)
Mellier, Y. 1999, Ann. Rev. Astron. Astrophys., 37, 127
Spergel, D.N., & Steinhardt, P.J. 2000, Phys. Rev. Lett., 84, 3760
Yoshida, N., Springel, V., White, S.D.M., & Tormen, G. 2000, ApJ, 544, L87

SCALING LAWS IN X-RAY GALAXY CLUSTERS AT REDSHIFT > 0.4

S. Ettori,[1] P. Tozzi,[2] S. Borgani,[3] and P. Rosati[1]

[1]*ESO, Karl-Schwarzschild-Str. 2, D-85748 Garching, Germany,* [2]*INAF, Osservatorio Astronomico di Trieste, via G.B. Tiepolo 11, I-34131 Trieste, Italy,* [3]*Dip. di Astronomia, Universita di Trieste, via G.B. Tiepolo 11, I-34131 Trieste, Italy*

Abstract We present a study of the integrated physical properties of a sample of 26 X-ray galaxy clusters observed with *Chandra*at redshift between 0.4 and 1.3. We focus particularly on the properties and evolution of the X-ray scaling laws. We fit a $\beta-$model, which provides a good representation of the observed surface brightness profiles, and measure an emission-weighted temperature (in the range 3–11 keV) to recover gas luminosity, mass and total gravitating mass out to R_{500}. We observe scaling relations steeper than expected from self-similar model by a significant ($> 3\sigma$) amount in the $L-T$ and $M_{\rm gas}-T$ relations, by a marginal value in the $M_{\rm tot}-T$ and $L-M_{\rm tot}$ relations. The degree of evolution of the $M_{\rm tot}-T$ relation is found to be consistent with the expectation based on the hydrostatic equilibrium for gas within virialized dark matter halos. We detect hints of *negative* evolution in the $L-T$ and $M_{\rm gas}-T$ relations, thus suggesting that systems at higher redshift have lower X-ray luminosity and gas mass for fixed temperature. In this sample of hot clusters, we also find a significant evidence of *positive* evolution, such as $(1+z)^{0.3}$, in the entropy value estimated at $0.1\ R_{200}$ at fixed temperature. Such results point toward a scenario in which a relatively lower gas density is present in high–redshift objects, thus implying a suppressed X–ray emission, a smaller amount of gas mass and a higher entropy level. This represents a non–trivial constraint for models aiming at explaining the thermal history of the intra–cluster medium out to the highest redshift reached so far.

Keywords: galaxies: cluster: general – galaxies: fundamental parameters – intergalactic medium – X-ray: galaxies – cosmology: observations – dark matter.

1. X-ray scaling laws at high redshift

The physics of the intracluster medium (ICM) is mainly driven by the infall of the cosmic baryons trapped in the deep gravitational potential of the cluster dark matter halo. Through a hierarchical formation, that starts from the primordial density fluctuations and generates the largest virialized structures via gravitational collapse and merging, galaxy clusters maintain similar properties

M. Plionis (ed.), Multiwavelength Cosmology, 267-270.

when they are rescaled with respect to gravitational mass and epoch of formation. The shock–heated X-ray emitting intra–cluster medium (ICM) accounts for most of the baryons collapsed in the cluster potential. Under the assumptions that the smooth and spherically symmetric ICM is heated only by gravitational processes (adiabatic compression during DM collapse and shock heating from supersonic accretion), obeys hydrostatic equilibrium within the underlying dark matter potential and emits mainly by bremsstrahlung, one derives the following scaling relations between the observed (bolometric luminosity $L_{\rm bol}$, and temperature $T_{\rm gas}$) and derived (gas entropy S, gas mass $M_{\rm gas}$, and total gravitating mass $M_{\rm tot}$) quantities: $E_z^{4/3}\ S \propto \Delta_z^{-2/3}\ T_{\rm gas}$; $E_z^{-1}\ L_{\rm bol} \propto \Delta_z^{1/2}\ T_{\rm gas}^2$; $E_z\ M_{\rm tot} \propto \Delta_z^{-1/2}\ T_{\rm gas}^{3/2}$; $E_z^{-1}\ L_{\rm bol} \propto \Delta_z^{7/6}\ (E_z M_{\rm tot})^{4/3}$; $E_z\ M_{\rm gas} \propto \Delta_z^{-1/2}\ T_{\rm gas}^{3/2}$.[1]

We present here the main results of our work (Ettori et al. 2003, A&A, in press) that focuses on the general behaviour of the scaling laws for X-ray clusters at redshift in the range 0.4–1.3 and the evolution of these relations with respect to what is measured in nearby samples. To do this, we have assembled a set of 26 clusters observed at arcsec angular resolution with *Chandra*. This allows us to resolve on scales of few tens of kpc the surface brightness (i.e. the gas density) of these systems, and to determine a single emission weighted temperature, that is measured in the range 3–11 keV (median value of 6.0 keV). The data reduction and analysis are described in Tozzi et al. (2003) and Ettori et al. (2003).

2. Results

The surface brightness is well fitted, in general, by a single $\beta-$model. A double $\beta-$model is statistically not required, suggesting that these distant clusters do not show any significant central brightness excess. By collecting all the photons within $\sim$0.5 $\times R_{500}$, we have measured a single emission-weighted temperature (median value of 6.0 keV). The combination of the spatial and spectral analysis allows to recover the gas luminosity (observed in the range $1.0 - 54.6 \times 10^{44} h_{70}^{-2}$ erg s^{-1}), gas mass ($0.6 - 17.1 \times 10^{13} h_{70}^{-5/2} M_{\odot}$) and total gravitating mass ($1.2 - 21.4 \times 10^{14} h_{70}^{-1} M_{\odot}$) out to R_{500}. These quantities are used to quantify the behaviour and evolution of the X–ray scaling relations in hot ($T_{\rm gas} > 3$ keV) clusters.

[1]The factors that indicate the dependence on the evolution of the Hubble constant at redshift z, $E_z = H_z/H_0 = \left[\Omega_{\rm m}(1+z)^3 + 1 - \Omega_{\rm m}\right]^{1/2}$ (for a flat cosmology with matter density $\Omega_{\rm m}$), and on the overdensity Δ_z account for the fact that all the quantities are estimated at a given overdensity Δ_z with respect to the critical density estimated at redshift z, $\rho_{c,z} = 3H_z^2/(8\pi G)$. The cosmological parameters $H_0 = 70 h_{70}^{-1}$ km s^{-1} Mpc^{-1} and $\Omega_{\rm m} = 1 - \Omega_\Lambda = 0.3$ are assumed hereafter.

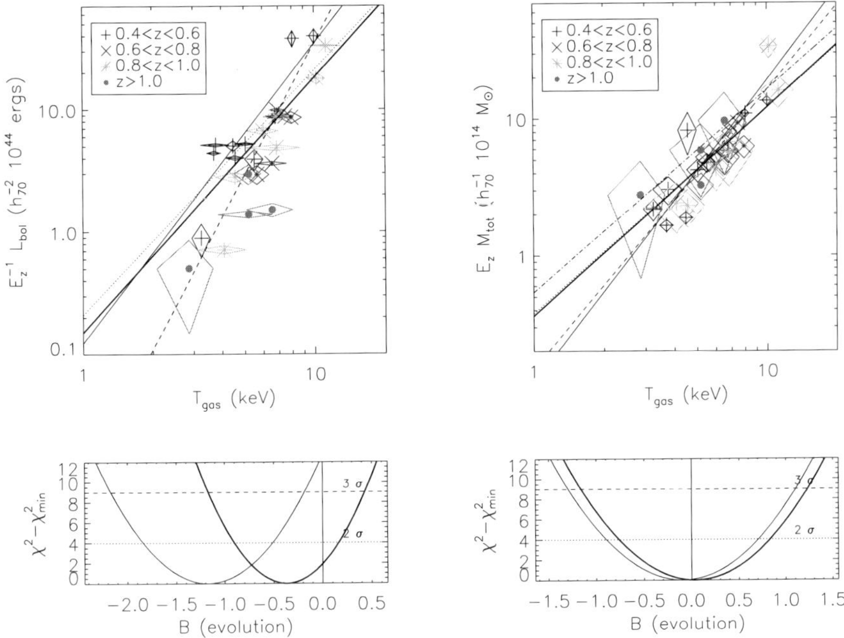

Figure 1. Dotted line: slope fixed to the predicted value. Dashed line: slope free. (Left) $L - T$ and (right) $M_{tot} - T$ relation. Solid lines indicate best-fit relations for nearby samples (see Ettori et al. 2003 for discussion). The evolution is parametrized through B in the relation $\log Y = \overline{\alpha} + \overline{A} \log X + B \log(1 + z)$, where $(\overline{\alpha}, \overline{A})$ are the best-fit values measured at lower redshift.

The slopes of all the investigated correlations tend to be higher than what predicted by self-similar models, with larger deviations observed in the $L - T$ (Fig. 1) and $M_{gas} - T$ relations. We notice that the values of the slope propagate through the relations in a self-consistent way: for example, the observed dependence $L \sim M_{tot}^{1.77 \pm 0.43}$ and $M_{tot} \sim T^{2.04 \pm 0.35}$ should imply $L \sim T^{3.61 \pm 1.07}$ that is well in accordance with the measured slope of 3.58 ± 0.50. The observed slopes suggest that simple gravitational collapse is not the only process that governs the heating of baryons in the potential well of clusters. To explain these correlations, a raise in the central entropy (and a correspondent suppression of the X-ray emissivity) is required and can be achieved either by episodes of non-gravitational heating due to supernovae and AGN, or by selective removal of low-entropy gas through cooling, possibly regulated by some mechanism supplying energy feedback. When the slope of the $M_{tot} - T$ relation (Fig. 1) is fixed to 1.5, its high–z normalization is consistent with the lack of expected evolution, well in agreement with results from low–z samples which follows from the prediction of hydrostatic equilibrium within virialized halos. In turn, this is consistent with the expectation that the $M_{\rm tot}$–T relation is the least sensitive to the thermal status of the gas and, therefore, not significantly affected by non–gravitational processes.

We do not observe any evolution at redshift around 1 in the $M_{\rm tot}$–T relation (Fig. 1), whereas $L - T$ (Fig. 1) and $M_{\rm gas} - T$ relations show a mild negative evolution [$B \sim (-0.1, -1.2)$, with a deviation from the no–evolution case of 0.7–3.6 σ; the correction by E_z is included in the relations], with clusters at higher redshift having in general lower X-ray luminosities and gas masses than what predicted from the self-similar model. This trend appears more clearly when the entropy of these hot ($T_{\rm gas} > 3$ keV) systems, evaluated at $0.1 R_{200}$, is compared with local determinations. A mild *positive* evolution ($B \simeq 0.3$) is then detected in particular when objects at redshift larger than 0.6 are considered. The evolution of the $L_{\rm bol}$–T and $M_{\rm gas}$–T relations is a sensitive test for non-gravitational heating models (e.g. Tozzi & Norman 2001, Bialek et al. 2001) and carries information about the source of this heating and the typical redshift at which it took place.

As a further diagnostic to trace the evolution of the ICM properties, we analyze the relation between gas temperature and entropy, S, as a diagnostic of the physical processes that establish the thermal status of the diffuse baryons in clusters (e.g. Ponman et al. 2003). For the first time, we measure and trace the evolution of the ICM entropy at high redshift. We obtain an overall fit to the data of $S = (427 \pm 20) \times T_{\rm gas}^{0.96 \pm 0.21}$ keV cm^2, well in agreement with the predicted scaling. No significant evolution is observed when all our sample is compared with the *local* estimate of the "entropy ramp" by Ponman et al. (2003). If we consider only the objects at $z > 0.6, 0.8$ and 1, respectively, we observe consistently $B \approx 0.3$ with a deviation from no evolution at the $2 - 3\sigma$ confidence level. This result, together with the scenario emerging from the entropy properties of low–z systems, seems to require either pre–heating effects with differential entropy amplification or a suitable combination of radiative cooling and heating by feedback energy release (e.g., Tornatore et al. 2003 and references therein).

References

Bialek J.J., Evrard A.E., Mohr J.J. 2001, ApJ, 555, 597

Ettori S., Tozzi P., Borgani S., Rosati P., 2003, A&A, in press

Ponman T.J., Sanderson A.J.R., Finoguenov A., 2003, MNRAS in press (astro-ph/0304048)

Tornatore L. et al., 2003, MNRAS, in press (astro-ph/0302575)

Tozzi P., Norman C. 2001, ApJ, 546, 63

Tozzi P. et al. 2003, ApJ, in press

THE EVOLUTION OF CLUSTER SUBSTRUCTURE

Tesla E. Jeltema[1], Claude R. Canizares[1], Mark W. Bautz[1], and David A. Buote[2]
[1]*Massachusetts Institute of Technology,* [2]*University of California at Irvine*
tesla@space.mit.edu

Abstract

Using Chandra archival data, we quantify the evolution of cluster morphology with redshift. To quantify cluster morphology, we use the power ratio method developed by Buote and Tsai (1995). Power ratios are constructed from moments of the two-dimensional gravitational potential and are, therefore, related to a cluster's dynamical state. Our sample will include 40 clusters from the Chandra archive with redshifts between 0.11 and 0.89. These clusters were selected from two fairly complete flux-limited X-ray surveys (the ROSAT Bright Cluster Sample and the Einstein Medium Sensitivity Survey), and additional high-redshift clusters were selected from recent ROSAT flux-limited surveys. Here we present preliminary results from the first 28 clusters in this sample. Of these, 16 have redshifts below 0.5, and 12 have redshifts above 0.5.

1. Introduction and Sample Selection

Clusters form and grow through mergers with other clusters and groups. Substructure or a disturbed cluster morphology indicates that a cluster is dynamically young (i.e. it will take some time for it to reach a relaxed state), and the amount of substructure in clusters in the present epoch and how quickly it evolves with redshift depend on the underlying cosmology. In low density universes, clusters form earlier and will be on average more relaxed in the present epoch. Clusters at high redshift, closer to the epoch of cluster formation, should be on average dynamically younger and show more structure. In addition, the evolution of cluster morphology is important to the understanding of many cluster properties including mass, gas mass fraction, lensing properties, and galaxy morphology and evolution.

Several studies have been done to quantify substructure in clusters at low redshift (e.g., Jones & Forman 1992; Mohr et al. 1995; Buote & Tsai 1996). However, it is only with recent X-ray and optical surveys that we are beginning to find tens of clusters with $z > 0.8$, and it is becoming possible to study

M. Plionis (ed.), Multiwavelength Cosmology, 271-274.

the evolution of substructure. Using the power ratio method (Buote & Tsai 1995), we are studying structure in a sample of 40 clusters observed with the Chandra X-ray Observatory. As a first cut, our sample includes only clusters with a redshift above 0.1 so that a reasonable area of each cluster will fit on a Chandra CCD. In order to have a reasonably unbiased sample, clusters were selected from the BCS (Ebeling et al. 1998) and EMSS (Gioia & Luppino 1994) surveys. They were also required to have a luminosity greater than 5×10^{44} ergs s^{-1}, as listed in those catalogs. Additional high-redshift clusters were selected from recent ROSAT flux-limited surveys (Rosati et al. 1998; Perlman et al. 2002; Voges et al. 2001; Vikhlinin et al. 1998). This led to a sample of 40 clusters with redshifts between 0.11 and 0.89. Here we present the results from 28 of these clusters. Sixteen of these have redshifts below 0.5 with an average redshift of 0.26; the other twelve have redshifts above 0.5 and an average redshift of 0.72.

2. Power Ratios

Power ratios are constructed from moments of the two-dimensional gravitational potential. They are capable of distinguishing many cluster morphologies, and they have been shown to distinguish different cosmological models (Buote & Xu 1997; Valdarnini, Ghizzardi, & Bonometto 1999; Suwa et al. 2003). Essentially, this method involves calculating multipole moments of the X-ray surface brightness. The moments, a_m and b_m, are given below. Here Σ is the surface brightness. These are calculated in a circle of radius R centered on the centroid of emission.

$$\begin{aligned} a_m(R) &= \int_{R'\leq R} \Sigma(\vec{x}')\,(R')^m \cos m\phi' d^2x' \\ b_m(R) &= \int_{R'\leq R} \Sigma(\vec{x}')\,(R')^m \sin m\phi' d^2x' \end{aligned}$$

The moments are sensitive to asymmetries in the surface brightness distribution and are, therefore, sensitive to substructure. The powers (shown below) are the sum of the squares of the moments, and the power ratios are formed by dividing by P_0 to normalize out flux. The physical motivation for the power ratio method is that it is based on the multipole expansion of the two-dimensional gravitational potential (Ψ). With Σ as the surface mass density, the powers are the squares of the multipole moments of Ψ evaluated over a circle of a given radius. Below $\Psi_m^{\rm int}$ is the *m*th multipole of the 2D gravitational potential due to matter interior to the circle of radius R, and $\langle\cdots\rangle$ represents the azimuthal average around the circle. Ignoring factors of 2G, the powers are

$$P_0 = [a_0 \ln(R)]^2 = \langle(\Psi_0^{\rm int})^2\rangle, \tag{1}$$

$$P_m = \frac{1}{2m^2 R^{2m}} \left(a_m^2 + b_m^2\right) = \langle(\Psi_m^{\rm int})^2\rangle. \tag{2}$$

In the case of X-ray studies, X-ray surface brightness replaces surface mass density in the calculation of power ratios. X-ray surface brightness is proportional to gas density squared and generally shows the same qualitative structure as the projected mass density, allowing a similar quantitative classification of clusters.

3. Preliminary Results

For each cluster in our initial sample of 28, we calculated P_2/P_0, P_3/P_0, and P_4/P_0 centered on the cluster centroid (where P_1 vanishes). We use an aperture radius of 0.5 Mpc. At larger radii, the high-z clusters have low S/N, and the low-z clusters become too large to fit on a Chandra CCD. Figure 1 shows a plot of P_2/P_0 versus P_3/P_0. High-redshift clusters (z>0.5) are plotted with diamonds and have solid error bars. Low-redshift clusters are plotted with asterisks and have dashed error bars. The error bars were found using Monte Carlo simulations and represent 90% confidence. The different power ratios are sensitive to different types of structure and looking at correlations among them can help distinguish cluster morphologies. In these plots, the most disturbed clusters appear at the upper-right, and the most relaxed clusters at the lower-left.

The high-z and low-z cluster samples have similar P_2/P_0 ratios, but the high-z clusters tend to have higher P_3/P_0. One possible reason that P_3/P_0 is better at distinguishing the high-redshift clusters from the low-redshift ones is that it is not sensitive to ellipticity: a purely elliptical cluster will only have even multipoles. Large odd multipoles unambiguously indicate asymmetry (substructure) in a cluster. P_4/P_0, which is similar to P_2/P_0 but sensitive to smaller scale structure, also appears to distinguish the two samples, especially when correlated with P_3/P_0. Figure 1 also shows a plot of P_3/P_0 versus P_4/P_0. Here the high-z clusters tend to be in the upper corner (more structure) and the low-z clusters tend to be in the bottom corner (less structure). A Mann-Whitney rank-sum test shows that the P_3/P_0 and P_4/P_0 ratios for the high-redshift clusters are on average larger than those for the low-redshift clusters at 95% significance.

At this point, we have analyzed 28 out of 40 clusters in our sample, and it appears that the amount of structure in clusters increases with redshift. Specifically, P_3/P_0 and P_4/P_0 are higher for the high-redshift clusters. These clusters were all selected to have high luminosities, and we do not believe the difference in power ratios is due to a difference in luminosities between the two samples. Using a radius of a fixed over-density rather than a fixed physical size also does not account for the difference. These and other possible systematic effects will be addressed in more detail later. In addition to completing the analysis of our

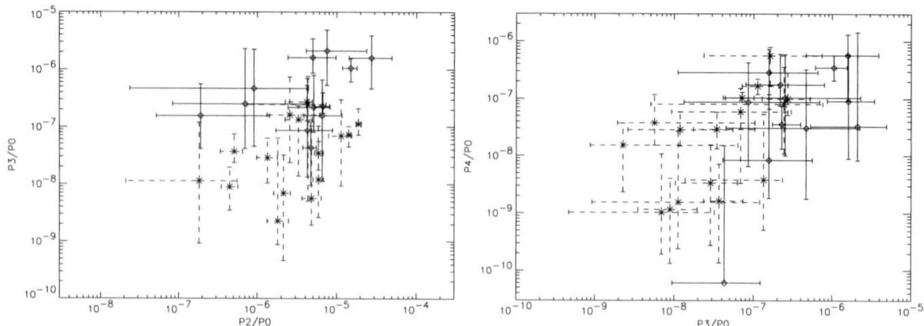

Figure 1. Power ratios computed in a 0.5 Mpc aperture for the first 28 clusters in our sample. High-redshift clusters (z>0.5) are plotted with diamonds and have solid error bars. Low-redshift clusters are plotted with asterisks and have dashed error bars. Left: P_2/P_0 versus P_3/P_0. Right: P_3/P_0 versus P_4/P_0.

Chandra sample, we also plan to compare the sample to numerically simulated clusters provided by Greg Bryan.

References

Buote, D. A., & Tsai, J. C. 1995, ApJ, 452, 522

Buote, D. A., & Tsai, J. C. 1996, ApJ, 458, 27

Buote, D. A., & Xu, G. 1997, MNRAS, 284, 439

Ebeling, H., Edge, A. C., Böhringer, H., Allen, S. W., Crawford, C. S., Fabian, A. C., Voges, W., & Huchra, J. P. 1998, MNRAS, 301, 881

Gioia, I. M., & Luppino, G. A. 1994, ApJS, 94, 583

Jones, C., & Forman, W. 1992, in Clusters and Superclusters of Galaxies (NATO ASI Vol. 366), ed. A. C. Fabian, (Dordrecht/Boston/London: Kluwer), 49

Mohr, J. J., Evrard, A. E., Fabricant, D. G., & Geller, M. J. 1995, ApJ, 447, 8

Perlman, E. S., Horner, D. J., Jones, L. R., Scharf, C. A., Ebeling, H., Wegner, G., & Malkan, M. 2002, ApJS, 140, 265

Rosati, P., Della Ceca, R., Burg, R., Norman, R., & Giacconi, R. 1998, ApJ, 492, L21

Suwa, T., Habe, A., Yoshikawa, K., & Okamoto, T. 2003, ApJ, 588, 7

Valdarnini, R., Ghizzardi, S., & Bonometto, S. 1999, New Astronomy, 4, 71

Vikhlinin, A., McNamara, B. R., Forman, W., Jones, C., Quintana, H., & Hornstrup, A. 1998, ApJ, 502, 558

Voges, W., Henry, J. P., Briel, U. G., Böhringer, H., Mullis, C. R., Gioia, I.M., & Huchra, J. P. 2001 553, L119

GALAXIES BEYOND THE DETECTION LIMITS OF DEEP X-RAY SURVEYS

Richard E. Griffiths, Takamitsu Miyaji, Adam Knudson
Carnegie Mellon University, Pittsburgh, PA15213, USA
griffith, miyaji & knudson@astro.phys.cmu.edu

Abstract

The great sensitivities of the Chandra X-ray Observatory and XMM-Newton are allowing us to explore the X-ray emission from galaxies at moderate to high redshift. By using the stacking method, we show that we can detect the ensemble emission from normal elliptical, spiral and irregular galaxies out to redshifts approaching one. The average X-ray luminosity can then be compared with the results of models of the evolution in the numbers of X-ray binaries and can possibly be used to constrain models of star formation.

Keywords: deep surveys, X-rays

1. Introduction

Deep surveys in X-ray astronomy had the initial goal of solving the problem of the origin of the extragalactic X-ray background, and these surveys have now shown that the XRB is largely comprised of the evolving populations of AGN, some heavily absorbed.
But the deep surveys with the Chandra X-ray Observatory (CXO) have shown that normal galaxies are also detected, and the 2Ms survey of the Hubble Deep Field (HDF) North demonstrated that about a third of the X-ray sources were identified with galaxies (Horneschmeier et al. 2002). We can thus begin to explore the evolution of galaxies in addition to the AGN.

2. Deep Surveys and Source Counts

The number counts in the HDF-N have been measured by Miyaji & Griffiths (2002), and extended to fluxes below 10^{-17} ergs cm^{-2} s^{-1} in the soft band (0.5 - 2 keV) and to 10^{-16} ergs cm^{-2} s^{-1} in the hard band (2 - 10 keV) by analysis of the fluctuations which remain after removal of the individual discrete source detections. Below this limit for the soft band, the fluctuation

M. Plionis (ed.), Multiwavelength Cosmology, 275-278.

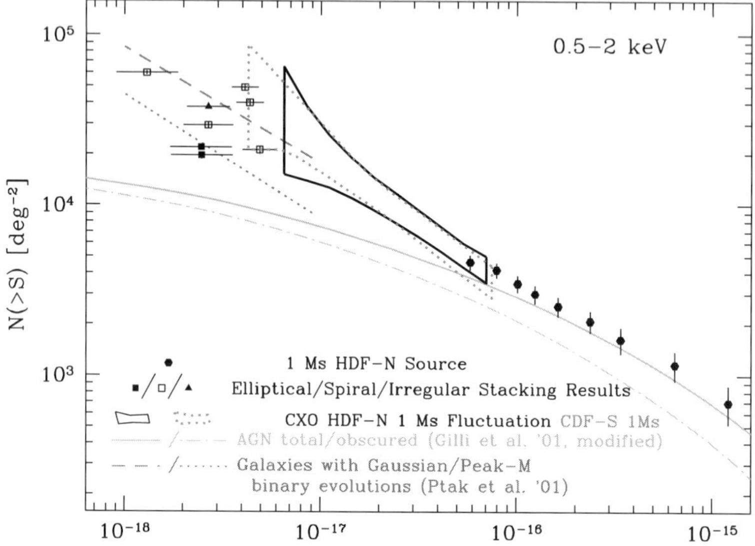

Figure 1. X-ray Number Counts from HDF-N

analysis shows that the number counts continue to rise, as shown in Fig. 1, with a slope consistent with that between 10^{-15} and 10^{-16} ergs cm^{-2} s^{-1}. At these latter X-ray fluxes, the optical identifications in the HDF-N (Horneschmeier et al. 2002) show that starburst and normal galaxies begin to dominate the number counts (0.5–8 keV). We infer that the number counts in the region explored with fluctuations (the boxed area in Fig. 1) are unlikely to be due to AGN. Furthermore, the current best models for the AGN number counts fall well below the fluctuations, which indicate counts of 20,000 – 40,000 per sq. deg. at X-ray fluxes of 10^{-17} ergs cm^{-2} s^{-1}, matching optical counts of galaxies at B = 24. We therefore explore the possibility of X-ray detection of faint galaxies in the HDF-N by using the stacking method.

3. Results of 'Stacking'

'Stacking' is the summation of X-ray sub-images centered on objects selected at another wavelength.

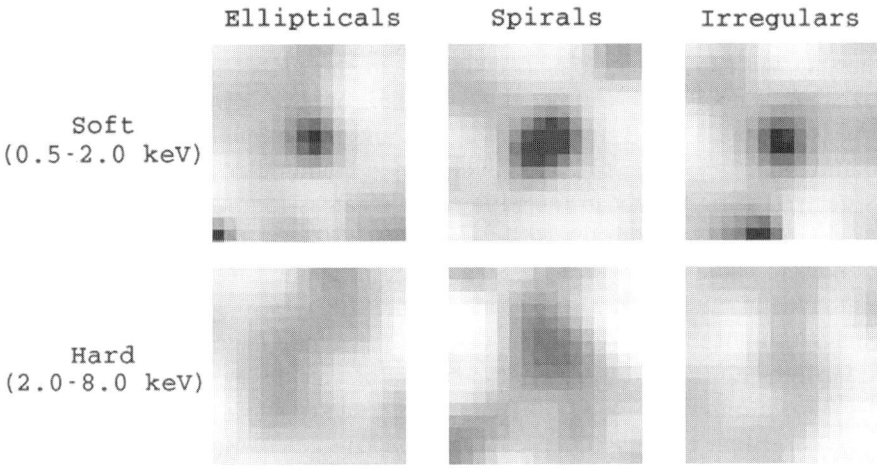

Figure 2. Stacked X-ray images of ensembles of galaxies

Brandt et al. (2001) and Horneschmeier et al. (2002) have used this method on early CXO data on the HDF-N. In the HDF, we have the advantage of being able to use the HST images themselves to select various types of galaxies for the stacking process, using the software developed as part of the HST Medium Deep Survey (Ratnatunga, Griffiths and Ostrander 1999), together with and redshift information from the spectroscopy of Cohen et al. (2000) and the photometric redshifts of Fernandez-Soto, Lanzetta and Yahil (1999). We have now done this for elliptical, spiral and irregular galaxies, and some of the results are shown in figure 2. The spiral and elliptical galaxies are detected at high confidence in both the soft and hard energy bands, but the irregular galaxies are detected in the soft band only. The median redshifts are 0.77 for the 29 ellipticals, 0.58 for the 53 spirals and 1.47 for the 50 irregulars in these stacked images. Monte Carlo simulations have been used to verify the statistical confidence in these results. In such simulations the stacking is done on 'null' candidates, i.e. positions where there are no optical sources but where the background level is similar to that around the actual optical sources. This Monte Carlo procedure is repeated 10^5 times in order to understand the distribution of false detections (Gaussian). The median X-ray luminosities are 2.5×10^{-18} ergs cm^{-2} s^{-1}for the ellipticals, 4.4×10^{-18} ergs cm^{-2} s^{-1}for the spirals and 2.7×10^{-18} ergs cm^{-2} s^{-1}for the irregulars, consistent with their B-band luminosities and the average values for L_X/L_B for the galaxy types.

4. X-ray Evolution of Galaxies

Starting with the cosmological evolution in the global star formation rate, Ghosh & White (2001) estimated the numbers of HMXRB and LMXRB, with a peak in the LMXRB numbers at $z \sim 1$ and in the numbers of HMXRB at slightly higher z closer to 2. The evolution in the number counts of LMXRBs is delayed with respect to the SFR peak, whereas the peak in the HMXRB numbers coincides with the SFR peak. The SFR itself was about ten times its present value at $z \sim 1$ and reached a peak about 10 – 100 times the current value at $z \sim 2 - 3$, beyond which the SFR was roughly flat or else declined to higher z. The estimates of Ghosh & White have been used by Ptak & Griffiths (2001) to estimate the numbers of galaxies which should be near the detection limit in the CXO HDF-N field and to show that the predictions were consistent with the observed numbers. This calculation was based on the observed B-band luminosities in the HDF and conversion to estimated X-ray fluxes. Grimm, Gilfanov and Sunyaev (2002) have shown that HMXB can be used effectively as an indicator of the SFR in distant galaxies.

5. Conclusions

Results from the stacking analysis of normal galaxy populations applied to the CXO deep survey of the HDF-N show that normal galaxy populations are observable in these stacks out to redshifts of ~ 1. The average X-ray fluxes observed in these stacks are consistent with the numbers and fluxes inferred from the fluctuation analysis of the CXO data. We conclude that the fluctuations are therefore caused primarily by normal galaxy populations and that such deep X-ray surveys will eventually allow us to constrain the evolution of the binary source populations within these galaxies, using the relationship between HMXB numbers and the SFR of nearby galaxies.

Acknowledgments

We acknowledge support from NASA grants NAG5-9902, NAG5-10875 and subcontract 2247-CMU-NASA-1128 from PSU (under NAS8-00128).

References

Brandt, W. N., et al. *Astrophys. J.*, 558: L5–8, 2001.

Ghosh, K. and White, N. *Astrophys. J.*, 559: L97–100, 2001.

Hornschemeier, A. E., et al. *Astrophys. J.*, 568: 82-87, 2002.

Hornschemeier, A. E. and the CDF-N team *X-rays at Sharp Focus: ASP Conf. Series, eds. S. Vrtilek, E. M. Schlegel, L. Kuhi* in press, 2002.

Miyaji T. and Griffiths, R. E. *Astrophys. J.*, 564: L5–8, 2002.

Ptak, A. and Griffiths, R. E. *Astrophys. J.*, 559, L91-94, 2001.

Ratnatunga, K., Griffiths, R. E. and Ostrander, E. J. *Astron. J.*, 118, 86–107, 1999.

THE X-RAY PROPERTIES OF 'NORMAL' GALAXIES

A. Georgakakis[1], I. Georgantopoulos[1], M. Plionis[1], S. Basilakos[1], V. Kolokotronis[1], G. C. Stewart[2], T. Shanks[3], B. J. Boyle[4]
[1] *National Observatory of Athens, 15236, Athens, Greece*
[2] *Department of Physics and Astronomy, University of Leicester Leicester LE1 7RH*
[3] *Physics Department, University of Durham, Science Labs, South Road, Durham, DH1 3LE*
[4] *Anglo-Australian Observatory, PO Box 296, Epping, NSW 2121, Australia*

Abstract We explore the X-ray properties of 'normal' (ie non AGN dominated) galaxies using a shallow $[f_X(0.5-8\,\mathrm{keV}) \approx 10^{-14}\,\mathrm{cgs}]$ wide area ($\approx 2.5\,\mathrm{deg}^2$) XMM-Newton survey. Two 'normal' galaxy candidates are *detected* in our survey. Despite the small number statistics the sky density of these systems at the flux limit above is low consistent with previous surveys. The number density estimated here is in good agreement with both the $\log N - \log S$ of 'normal' galaxies in deeper surveys and model predictions based on the X-ray luminosity function of local star-forming galaxies. Stacking analysis is employed to constrain the mean X-ray properties of *optically selected* galaxies at $z \approx 0.1$ that remain undetected to the limit of our survey. Comparison of our results with similar studies at higher redshifts suggest X-ray evolution of spiral galaxies out to $z \approx 1$.

Keywords: X-rays, Galaxies, Surveys

1. Introduction

X-ray selected 'normal' (ie non AGN dominated) galaxies have been reliably identified in ultra deep Chandra surveys $[f(0.5-2\,\mathrm{keV}) \approx 10^{-17}\,\mathrm{cgs}]$ either as direct detections (Hornschemeier et al. 2003) or indirectly using stacking analysis methods (Brandt et al. 2001; Hornschemeier et al. 2002).
The studies above however, primarily probe 'normal' galaxies at faint fluxes and/or high redshifts ($z \approx 1$). At relatively bright fluxes $[f(0.5-2\,\mathrm{keV}) > 10^{-15}\,\mathrm{cgs}]$ and/or low redshifts ($z < 0.3$) the X-ray properties of 'normal' galaxies remain ill constrained. This is due to the low surface density of these systems at bright fluxes requiring wide area surveys to identify 'normal' galaxies. Such surveys also provide sufficient volume coverage at low-z to explore the X-ray properties of the nearby 'normal' galaxy population. XMM-Newton with 4 times the Chandra field of view is clearly better suited for such studies.

M. Plionis (ed.), Multiwavelength Cosmology, 279-282.

In this paper we present results on the X-ray properties of 'normal' galaxies using a wide field ($2.5\,\text{deg}^2$) shallow [$f_X(0.5-8\,\text{keV}) \approx 10^{-14}$ cgs] XMM-Newton survey (the XMM/2dF survey) carried out in the North and South Galactic Pole regions. This survey has the advantage of multiwavelength data including radio (FIRST), high quality optical spectroscopy (2dFGRS, 2QZ, SDSS) and 5-band optical photometry (SDSS). These data are of key importance to identify 'normal' galaxies and to explore their properties. The X-ray sample comprises a total of 514 sources (5σ). Using the optical data above we have identified a total of 221 (42%) with an optical counterpart.

2. X-ray detected 'normal' galaxy candidates

We identify 'normal' galaxy candidates *detected* in our XMM/2dF survey by looking for low X-ray–to–optical flux ratio sources ($\log f_X/f_{opt} < -1$). This is demonstrated in Fig. 1a plotting B-band magnitude as a function of 0.5-8 keV X-ray flux. The diagonal lines represent constant X-ray–to–optical flux ratios of $\log f_X/f_{opt} \pm 1$ and delineate the region of the parameter space occupied by powerful AGNs. At lower X-ray–to–optical flux ratios $\log f_X/f_{opt} < -1$ (the shaded region in Fig. 1a) one starts picking LLAGNs and 'normal' galaxies.

As can be seen in Fig. 1a the majority of the $\log f_X/f_{opt} < -1$ sources are either Galactic stars or distant QSOs. However, we also find a small number of four X-ray sources with optical spectra showing either absorption or narrow emission lines. These four systems are 'normal' galaxy candidates. We further exploit the multiwavelength coverage of our XMM/2dF survey to identify obscured or weak AGN contaminating our 'normal' galaxy sample. As a result we end up with a sample of 2 bona fide 'normal' galaxy candidates with X-ray, optical and radio properties consistent with stellar processes rather than AGN activity. The candidate 'normal' galaxies in the present sample have $L(0.5-8\,\text{keV}) \approx 10^{41}$ cgs, $\log f_X/f_{opt} \approx -2$, narrow emission line optical spectra and radio luminosity consistent with the $L_X - L_{1.4}$ relation of local star-forming galaxies (Rannalli et al. 2003). We note that although the X-ray and optical properties of 'normal' galaxy candidates are consistent with stellar origin for the X-ray emission we cannot exclude the possibility of LLAGN.

In Fig. 1b the $\log N - \log S$ of the 'normal' galaxy candidates identified in the present study (2 galaxies) at fluxes $\approx 10^{-14}$ cgs is compared with the deeper sample of Hornchemeier et al. (2003). Despite the small number statistics our results are consistent with deeper samples (e.g. Hornschemeier et al. 2003) extrapolated to the flux limits probed here. Also shown are the model predictions using the X-ray luminosity function of H II galaxies (Georgantopoulos et al. 1999) under the no evolution scenario ($k = 0$) and pure luminosity evolution of the form $L_X \sim (1+z)^3$ ($k = 3$). The latter model is in better agreement with the data over a wider flux range suggesting X-ray evolution of starforming galaxies.

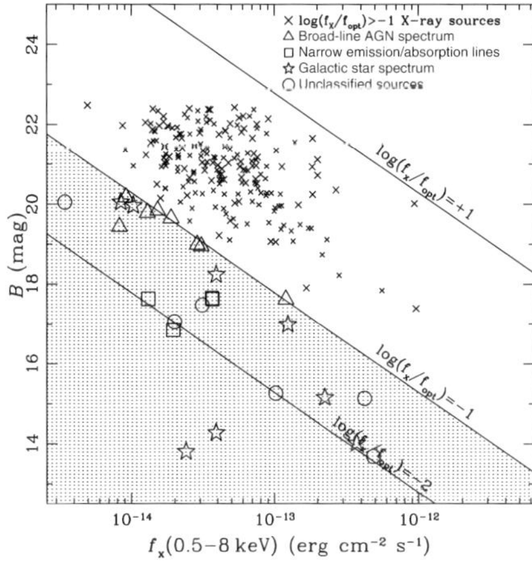

Figure 1a. B-band magnitude against 0.5-8 keV flux. The $\log f_X/f_{opt} < -1$ region studied here is shaded. The lines indicate constant X-ray–to–optical flux ratios of +1, –1 and –2. The lines $\log f_X/f_{opt} = \pm 1$ delineate the region of the parameter space occupied by powerful AGNs.

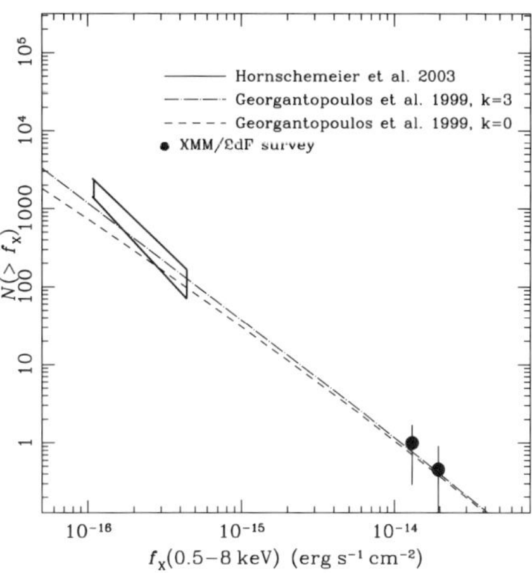

Figure 1b. Cumulative 'normal' galaxy counts. Filled circles are the 'normal' galaxy candidates in the present study. The solid lined rectangle marks the region occupied by the 'normal' galaxy counts of Hornchemeier et al. (2003).

3. Stacking results

The two X-ray *detected* 'normal' galaxy candidates found in our XMM/2dF survey are only the brightest analogues of the X-ray selected 'normal' galaxy population. Therefore, their properties are by no means representative of the whole population. Indeed, non-AGN galaxies are, on average, extremely faint at X-ray wavelengths ($< 10^{-16}$ cgs) and hence, are difficult to detect. However, for these X-ray faint sources we employ the stacking analysis method to explore their mean X-ray properties.

This technique is applied to *optically* selected galaxies from the 2dF Galaxy Redshift Survey (2dFGRS) overlapping with our XMM/2dF observations. 2dFGRS provides high quality spectra, redshifts and spectral classifications (e.g. E/S0, spirals) for galaxies brighter than $b_j = 19.4$ mag. The mean redshift of the 2dFGRS survey is $z \approx 0.1$. X-ray stacking analysis of 2dFGRS galaxies offers the opportunity to constrain the mean X-ray properties of 'normal' spirals and E/S0 at $z \approx 0.1$.

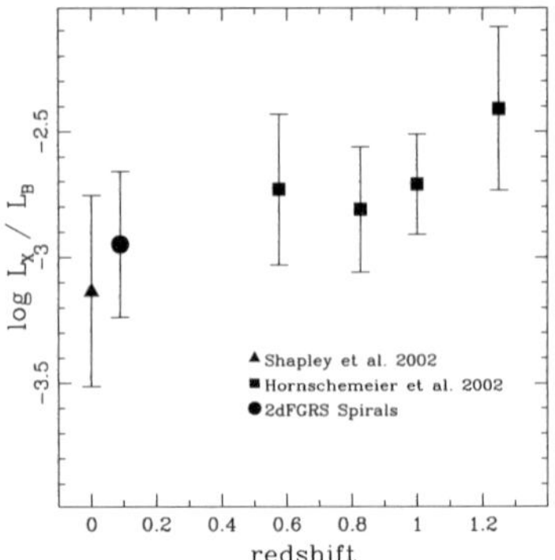

Figure 2. $\log L_X/L_B$ against z for spiral galaxies. The stacking analysis results for the 2dFGRS spirals are shown with the filled circle. This is compared with local spirals (triangle; Shapley et al. 2001) and high-z late type galaxies in the Chandra Deep Field North (squares; Hornschemeier 2002).

Stacking is independently performed on the sub-samples of E/S0 and spirals after excluding 2dFGRS sources associated or lying close to X-ray sources. A statistically significant signal is found for both sub-samples corresponding to: **(i)** $L_X = 5 \times 10^{39}\,\mathrm{erg\,s^{-1}}$ and $L_X/L_B \approx -3$ for spirals and **(ii)** $L_X = 3 \times 10^{40}\,\mathrm{erg\,s^{-1}}$ and $L_X/L_B \approx -2.4$ for E/S0. Note that *optically selected* spirals have mean X-ray properties 1 dex lower than *X-ray detected* sources (see section 1).

The mean X-ray properties of spirals can be further used to constrain their evolution. This is demonstrated in Fig. 2 plotting the mean L_X/L_B versus redshift. The 2dFGRS *optically selected* spirals at $z \approx 0.1$ are compared with both deeper higher-z spiral galaxy samples (Hornschemeier et al. 2002) and local spirals (Shapley et al. 2002). The spiral L_X/L_B remains roughly constant with z. Any small differences between high and low-z samples can be attributed to differences in the mean L_B of the galaxies in these sub-samples. This suggests that that the spiral galaxy L_X evolves with z in the same rate as L_B at least to $z \approx 1$. Since the spiral galaxy L_B evolves as $\approx (1+z)^3$ (Lilly et al. 1991), this is evidence for *X-ray evolution* of spirals.

References

Brandt W. N., et al., 2001a, AJ, 122, 1.
Hornschemeier A. E., et al, 2002, ApJ, 568, 82.
Hornschemeier A. E., 2003, AJ, 126, 575
Georgantopoulos I., Basilakos S., Plionis M., 1999, MNRAS, 305, L31.
Lilly S. J., Le Fevre O., Hammer F., Crampton D., 1996, ApJ, 460, L1.
Ranalli P., Comastri A., Setti G., 2003, A&A, 399, 39.
Shapley A., Fabbiano G., Eskridge P. B., 2001, ApJS, 137, 139.

THE HELLAS2XMM 1DF SURVEY: ON THE NATURE OF HIGH X-RAY/OPTICAL FLUX SOURCES

F. Fiore, M. Brusa, F. Cocchia, A. Baldi, N. Carangelo, P. Ciliegi, A. Comastri, F. La Franca, R. Maiolino, G. Matt, S. Molendi, M. Mignoli, G.C. Perola, P. Severgnini, C. Vignali

Abstract We present results from the photometric and spectroscopic identification of 122 X-ray sources recently discovered by XMM-Newton in the 2-10 keV band (the HELLAS2XMM 1dF sample). One of the most interesting results (which is found also in deeper surveys) is that $\sim$ 20% of the sources have an X-ray to optical flux ratio (X/O) ten times or more higher than that of optically selected AGN. Unlike the faint sources found in the ultra-deep Chandra and XMM-Newton surveys, which reach X-ray (and optical) fluxes $\gtrsim$ 10 times lower than in the HELLAS2XMM sample, many of the extreme X/O sources in our sample have R $\lesssim$ 25 and are therefore accessible to optical spectroscopy. We report the identification of 13 sources with extreme X/O values. While four of these sources are broad line QSO, eight of them are narrow line QSO, seemingly the extension to very high luminosity of the type 2 Seyfert galaxies.

1. Introduction

Hard X-ray surveys are the most direct probe of supermassive black hole (SMBH) accretion activity, which is recorded in the Cosmic X-ray Background (CXB), in wide ranges of SMBH masses, down to $\sim 10^6 - 10^7\ M_\odot$, and bolometric luminosities, down to $L \sim 10^{43}$ erg/s. X-ray surveys can therefore be used to: study the evolution of the accreting sources; measure the SMBH mass density; constrain models for the CXB (Setti & Woltjier 1989; Comastri et al. 1995), and models for the formation and evolution of the structure in the universe (Haehnelt, 2003; Menci et al. 2003). These studies have so far confirmed, at least qualitatively, the predictions of standard AGN synthesis models for the CXB: the 2-10 keV CXB is mostly made by the superposition of obscured and unobscured AGNs (Hasinger, 2003; Fiore 2003 and references therein). Quantitatively, though, rather surprising results are emerging: a rather narrow peak in the range z=0.7-1 is present in the redshift distributions from ultra-deep Chandra and XMM-Newton pencil-beam surveys, in contrast to the

M. Plionis (ed.), Multiwavelength Cosmology, 283-286.

broader maximum observed in previous shallower soft X-ray surveys made by ROSAT, and predicted by the above mentioned synthesis models. However, the optical identification of the faint sources in these ultra-deep surveys is rather incomplete, especially for the sources with very faint optical counterparts, i.e. sources with high X-ray to optical flux ratio (X/O). Indeed, the optical magnitude of $\approx 20\%$ of the sources, those having the higher X/O, is R$\gtrsim 26 - 27$, not amenable at present to optical spectroscopy. This limitation leads to a strong bias in ultra-deep Chandra and XMM-Newton surveys against AGN highly obscured in the optical, i.e. against type 2 QSOs, and in fact, only 10 type 2 QSOs have been identified in the CDFN and CDFS samples (Cowie et al. 2003; Hasinger, 2003). To help overcoming this problem, we are pursuing a large area, medium-deep surveys, the HELLAS2XMM serendipitous survey, which, using XMM-Newton archival observations (Baldi et al. 2003) has the goal to cover $\sim$ 4 deg^2 at a 2-10 keV flux limit of a few$\times 10^{-14}$ erg cm^{-2} s^{-1}. At this flux limit several sources with X/O$\gtrsim 10$ have optical magnitudes R=24-25, bright enough for reliable spectroscopic redshifts to be obtained with 10m class telescopes.

2. The HELLAS2XMM 1dF sample

We have obtained, so far, optical photometric and spectroscopic follow-up of 122 sources in five XMM-Newton fields, covering a total of $\sim$ 0.9 deg^2 (the HELLAS2XMM '1dF' sample), down to a flux limit of $F_{2-10keV} \sim 10^{-14}$ erg cm^{-2} s^{-1}. We found optical counterparts brighter than R$\sim$ 25 within $\sim 6''$ from the X-ray position in 116 cases and obtained optical spectroscopic redshifts and classification for 94 of these sources (Fiore et al. 2003). The source breakdown includes: 61 broad line QSO and Seyfert 1 galaxies, and 33 *optically obscured AGN*, i.e. AGN whose nuclear optical emission, is totally or strongly reduced by dust and gas in the nuclear region and/or in the host galaxy (thus including objects with optical spectra typical of type 2 AGNs, emission line galaxies and early type galaxies, but with X-ray luminosity $\gtrsim 10^{42}$ erg s^{-1}). We have combined the HELLAS2XMM 1dF sample with other deeper hard X-ray samples including the CDFN (Barger et al. 2002), Lockman Hole (Mainieri et al. 2002; Baldi et al. 2003), and SSA13 (Barger et al. 2001) samples, to collect a "combined" sample of 317 hard X-ray selected sources, 221 (70%) of them identified with an optical counterpart whose redshift is available. The flux of the sources in the combined sample spans in the range $10^{-15} - 4 \times 10^{-13}$ erg cm^{-2} s^{-1} and the source breakdown includes 113 broad line AGN and 108 optically obscured AGN.

Fig. 1 shows the X-ray (2-10 keV) to optical (R band) flux ratio (X/O) for the combined sample. About 20% of the sources have X/O$\gtrsim 10$, i.e ten times or more higher than the X/O typical of optically selected AGN. At the flux limit

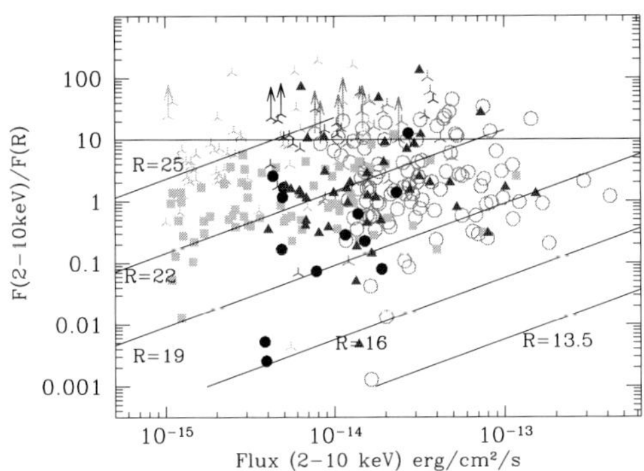

Figure 1 The X-ray (2-10 keV) to optical (R band) flux ratio X/O as a function of the X-ray flux for the combined sample (HELLAS2XMM = open circles; CDFN = filled squares; LH = filled triangles; SSA13 = filled circles, skeleton triangles are sources without a measured redshift). Solid lines mark loci of constant R band magnitude.

of the HELLAS2XMM 1dF sample several sources with X/O$\gtrsim$10 have optical magnitudes R=24-25, bright enough for reliable spectroscopic redshifts to be obtained with 8m class telescopes. Indeed, we were able to obtain spectroscopic redshifts and classification of 13 out of the 28 HELLAS2XMM 1dF sources with X/O$>$ 10; *8 of them are type 2 QSO at z=0.7-1.8*, to be compared with the total of 10 type 2 QSOs identified in the CDFN (Cowie et al. 2003) and CDFS (Hasinger, 2003). Fig. 2 show the X-ray to optical flux ratio as a function of the X-ray luminosity for broad line AGN (left panel) and non broad line AGN and galaxies (central panel). While the X/O of the broad line AGNs is not correlated with the luminosity, a striking correlation between log(X/O) and log($L_{2-10keV}$) is present for the obscured AGN: higher X-ray luminosity, optically obscured AGN tend to have higher X/O. A similar correlation is obtained computing the ratio between the X-ray and optical luminosities, instead of fluxes (because the differences in the K corrections for the X-ray and optical fluxes are small in comparison to the large spread in X/O). All objects plotted in the right panel of Fig. 2 do not show broad emission lines, i.e. the nuclear optical-UV light is completely blocked, or strongly reduced in these objects, unlike the X-ray light. Indeed, the optical R band light of these objects is dominated by the host galaxy and, therefore, *X/O is roughly a ratio between the nuclear X-ray flux and the host galaxy starlight flux*. The right panel of Figure 2 helps to understand the origin of the correlation between X/O and $L_{2-10keV}$. While the X-ray luminosity of the optically obscured AGNs spans about 4 decades, the host galaxy R band luminosity is distributed over less than one decade. The ratio between the two luminosities (and hence the ratio between the two fluxes, see above) results, therefore, strongly correlated with the X-ray luminosity.

3. Summary

We have obtained spectroscopic redshifts and classification of 13 out of the 28 HELLAS2XMM 1dF sources with X/O$\gtrsim$10: the majority of these sources (8) are type 2 QSOs at z=0.7-1.8, a fraction of type 2 QSOs much higher than

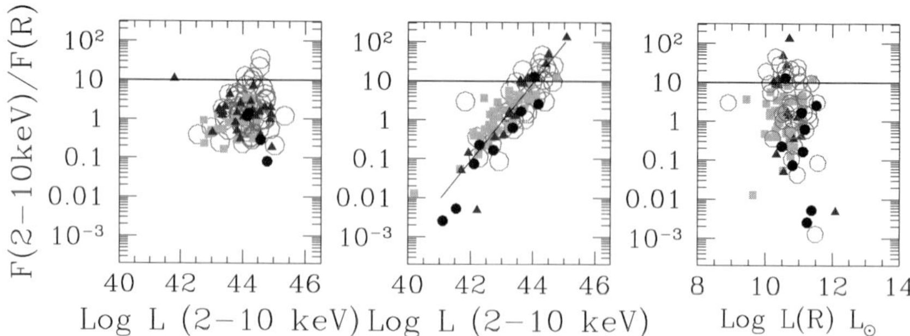

Figure 2. The X-ray to optical flux ratio X/O versus the X-ray luminosity for type 1 AGN (left panel), and non type 1 AGN and galaxies (central panel); X/O versus the optical luminosity for non type 1 AGN and galaxies (right panel). Symbols as in Fig. 1. The horizontal lines mark the level of X/O=10, $\sim$ 20% of the sources in the combined sample have X/O higher than this value. The diagonal line in the right panel is the best log(X/O)–log($L_{2-10keV}$) linear regression.

at lower X/O values. We find a strong correlation between X/O and the X-ray luminosity of optically obscured AGN, X/O=10 corresponding to an (average) 2-10 keV luminosity of 10^{44} erg s^{-1}. Sources of this luminosity and flux $\approx 10^{-15}$ erg cm^{-2} s^{-1} , reachable in Chandra and XMM-Newton ultra-deep surveys, would be at z$\sim$ 3. Although only 20% of the X-ray sources have such high X/O, they may carry the largest fraction of accretion power from that shell of Universe. Intriguingly, Mignoli et al. (2003 in preparation) find a strong correlation between the R-K color and the X/O ratio, all the 10 X/O>10 HELLAS2XMM 1dF sources studied having R-K$\gtrsim$5, i.e. they are all Extremely Red Objects.

References

Setti, G., & Woltjer, L. 1989, A&A, 224, L21

Comastri, A., Setti, G., Zamorani, G., & Hasinger, G. 1995, A&A, 296, 1

Haehnelt, M. Carnegie Observatories Astrophysics Series, Vol. 1: Coevolution of Black Holes and Galaxies, ed. L. C. Ho (Cambridge Univ. Press), 2003, astro-ph/0307378

Menci, N. et al. 2003, ApJ, 587, L63

Hasinger, G. 2003, proceedings of the Conference: The Emergence of Cosmic Structure, Maryland, Stephen S. Holt and Chris Reynolds (eds), astro-ph/0302574

Fiore, F. 2003, proceedings of the symposium "The Restless High-Energy Universe", E.P.J. van den Heuvel, J.J.M. in 't Zand, and R.A.M.J. Wijers Eds, astro-ph/0309355

Cowie L., Barger A., Bautz, M.W., Brandt, W.N., & Garnire, G.P. 2003, ApJ, 584, L57

Fiore, F. Brusa, M, Cocchia, F. et al. 2003, A&A in press, astro-ph/0306556

Baldi, A., Molendi, S., Comastri, A., Fiore, F., Matt, G., & Vignali, C. 2002, ApJ, 564, 190

Barger A., et al. 2002, AJ, 124, 1839

Barger, A., Cowie, L., Mushotzky, R.F., & Richards, E.A. 2001, AJ, 121, 662

Mainieri, V. et al. 2002, A&A, 393, 425

REDSHIFT SPIKES IN THE CHANDRA DEEP FIELD SOUTH

Roberto Gilli
Istituto Nazionale di Astrofisica - INAF
Osservatorio Astrofisico di Arcetri
Largo E. Fermi, 5 - 50125 Firenze, Italy
gilli@arcetri.astro.it

Abstract The redshift distribution of the X-ray sources in the 1Msec *Chandra*Deep Field South (CDFS) shows two prominent spikes at z=0.67 and z=0.73, mainly populated by Active Galactic Nuclei (AGN). Other significant spikes are detected at z=1.04, 1.62 and 2.57. Part of the CDFS has been covered by the K20 near infrared survey, where similar structures of galaxies (mostly early-type) are observed at z=0.67 and z=0.73. There is also a very good correlation between less prominent X-ray and K20 peaks suggesting that AGN and early-type galaxies are tracing the same underlying structures.

Keywords: Large-scale structure of universe, X-rays:general

1. Introduction

While the large scale structure of the Universe at $z < 1$ is usually mapped through galaxy surveys, AGN surveys are a powerful tool to study the clustering of high redshift objects. AGN clustering has been extensively studied and detected at optical wavelengths (Shanks et al. 1987; La Franca et al. 1998; Croom et al. 2001), where objects are mainly selected by means of their strong UV excess and include almost exclusively unobscured-type 1 AGN. An advantage of the X-ray selection, especially in the hard X-rays, resides in the capability of detecting also obscured AGN.

While angular clustering of X-ray selected AGN was detected by several authors (Akylas et al. 2000; Vikhlinin et al. 1995) in the past years, only recently a measurement of their spatial clustering became available (Mullis et al. 2002). The *Chandra*Msec surveys in the Deep Field South (CDFS, Rosati et al. 2002) and North (CDFN, Brandt et al. 2001) will allow a step forward in measuring the clustering of X-ray selected AGN, which constitute about 80-90% of the sources detected in these fields (Szokoly et al. 2003; Barger et al. 2003).

M. Plionis (ed.), Multiwavelength Cosmology, 287-290.

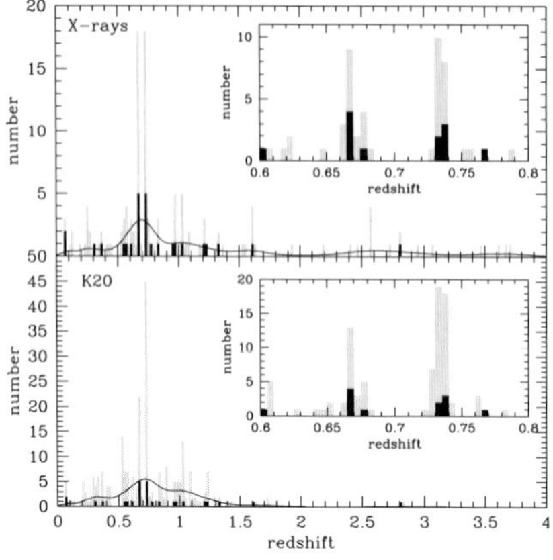

Figure 1. Redshift distribution for X-ray sources (Upper Panel) and K20 sources (Lower Panel) with high quality optical spectra. The insets show a zoom on the two main redshift spikes at z=0.67 and z=0.73. The black histograms represent the matches between the X-ray and the K20 catalogs.

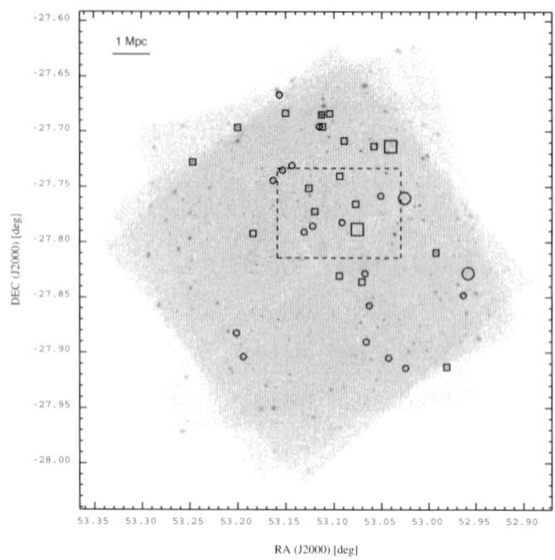

Figure 2. *Chandra* ACIS-I image of the CDFS with sources at z=0.67 and z=0.73 marked with circles and squares, respectively. Extended sources in the spikes are represented as big symbols. The dashed box indicates the 6.7×4.8 arcmin region covered by the K20 survey.

Throughout this paper we will use a cosmology with $H_0 = 70$ km s^{-1} Mpc^{-1}, $\Omega_m = 0.3$, $\Omega_\Lambda = 0.7$.

2. X-ray and K-band sources

The 1Msec observation of the CDFS with *Chandra*ACIS-I represents one of the deepest X-ray surveys to date (Rosati et al. 2002; Giacconi et al. 2002). The optical follow-up was primarily performed using the FORS1 camera at the VLT (Szokoly et al. 2003). As shown in Fig. 1, the redshift distribution of the 131 sources with unambiguous redshift measurement is dominated by two large concentrations of sources at $z \sim 0.67$ and $z \sim 0.73$. As shown in Fig. 2, these two structures extend to a minimum scale of ~ 17 arcmin, corresponding to a linear physical size of 7.3 Mpc at $z \sim 0.7$. Two extended sources, identified as galaxy groups/clusters, are present in each of the structures (see Fig. 2 and 3).

Smaller peaks are also recognized in the X-ray source redshift distribution of Fig. 1. A detection procedure to assess the significance of the redshift peaks was worked out by (Gilli et al. 2003). Besides the already mentioned structures at z=0.67 and z=0.73, other 5 peaks were detected, the most remarkable at z=1.04, 1.62 and 2.57.

About 1/10 of the X-ray field has been covered by the K20 near-infrared survey (see Cimatti et al. 2002 and references therein). Spectroscopic identification of the K20 sources has been performed with FORS1 and FORS2 at the VLT. The redshift distribution of the 258 galaxies with unambiguous redshift determination is also shown in Fig. 1. In strict analogy with the X-ray results, the redshift distribution of the K20 galaxies has two prominent peaks at z=0.67 and z=0.73.
While in the X-rays the two structures appear equally populated, in the smaller K20 field sources at z=0.73 (47 objects) are a factor of $\sim$ 2 more numerous than sources at z=0.67 (24 objects). Also, the structure at z=0.73 appears to be dominated by a relaxed galaxy cluster, showing a central cD galaxy and significant spectral segregation, with massive early type objects concentrated in the center (see Fig. 3). On the contrary, sources at $z = 0.67$ constitute a loose structure with early and late type galaxies uniformly distributed across the K20 field.
The procedure described in Gilli et al. 2003, allowed the detection of other 6 K20 peaks besides the two already mentioned. The full list of the X-ray and K20 detected peaks is given in Table 1 of Gilli et al. 2003. Overall, 5 out of 8 peaks in the K20 source redshift distribution have an X-ray peak counterpart, and all the 4 most significant K20 peaks have an X-ray counterpart. Then, in general, the structures seen in the K20 survey are traced on wider scales by the X-ray sources. It is worth noting that high redshift peaks are not detected in the K20 observations, whose sensitivity drops dramatically above $z \sim 1.3$.

3. The X-ray source fraction in the redshift structures

We have studied in more detail the correlation between the redshift distribution of the X-ray and K20 sources, since enhanced nuclear or star forming activity (both marked by X-ray emission) are likely to be produced by galaxy interactions in large scale structures.
The cross correlation between X-ray and K20 sources produced 30 matches. In the low redshift spike the fraction of K20 sources with an X-ray counterpart is a factor of $\sim$ 2.6 higher than in the high redshift spike and in the neighboring redshift bins. Because of the limited statistics, this overdensity is significant only at the $\sim 2\sigma$ level.
Significant improvements in the understanding of the large scale structures in the CDFS are expected from the on going and planned multiwavelength observations of the field. The XMM deep pointing (370 ksec in a 30 arcmin diameter region; Hasinger et al. 2002) is actually expanding the X-ray sky coverage by a factor of $\sim$ 2. Accurate photometric redshifts for the optically faint X-ray sources are being obtained from FORS optical images and ISAAC near-IR data

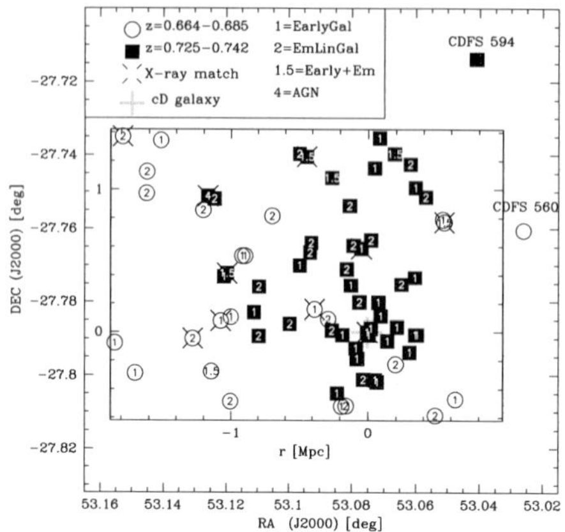

Figure 3 Spatial distribution of the K20 sources at z=0.67 (open circles) and at z=0.73 (filled squares). Their optical classification is indicated by the numbers inside the symbols. Crosses are over-plotted on sources with X-ray counterparts. One galaxy group at z=0.67 (CDFS 560) and one at z=0.73 (CDFS 594) detected in the X-rays just outside the K20 field are also indicated.

and will be refined further using data from the Advanced Camera for Survey (ACS) on HST.

Acknowledgments

I gratefully acknowledge all the members of the CDFS and K20 collaborations involved in this project.

References

Akylas, A., Georgantopoulos, I., Plionis, M. 2000, MNRAS, 318, 1036

Barger, A.J., et al. 2003, AJ, 126, 632

Brandt, W.N., et al. 2001, AJ, 122, 2810

Cimatti, A., et al. 2002, A&A, 392, 395

Croom, S.M., et al. 2001, MNRAS, 325, 483

Giacconi, R., et al. 2002, ApJS, 139, 369

Gilli, R., et al. 2003, ApJ, 592, 721

Hasinger, G., et al. 2002, ESO Msngr, 108, 11

La Franca, F., Andreani, P., & Cristiani, S., 1998, ApJ, 497, 529

Mullis, C.R., 2002, PASP, 114, 688

Shanks, T., et al. 1987, MNRAS, 277, 739

Szokoly, G., et al. 2003, ApJS, submitted

Rosati, P., et al. 2002, ApJ, 566, 667

Vikhlinin, A., Forman, W. 1995, ApJ, 455, L109

THE 2 MS CHANDRA DEEP FIELD-NORTH

Moderate-luminosity AGNs and dusty starburst galaxies

D. M. Alexander,[1] F. E. Bauer,[2] W. N. Brandt,[2] and A. E. Hornschemeier[3]

[1] *Institute of Astronomy, Madingley Road, Cambridge, CB3 0HA, UK*

[2] *Department of Astronomy & Astrophysics, The Pennsylvania State University, PA 16802, USA*

[3] *Department of Physics and Astronomy, Johns Hopkins University, MD 21218, USA*

Abstract The 2 Ms Chandra Deep Field-North survey provides the deepest view of the Universe in the 0.5–8.0 keV X-ray band. In this brief review we investigate the diversity of X-ray selected sources and focus on the constraints placed on AGNs (including binary AGNs) in high-redshift submm galaxies.

Keywords: surveys — cosmology — X-rays: active galaxies — X-rays: galaxies

1. Introduction

The 2 Ms Chandra Deep Field-North (CDF-N) survey provides the deepest view of the Universe in the 0.5–8.0 keV band. It is ≈ 2 times deeper than the 1 Ms *Chandra* surveys (Brandt et al. 2001; Giacconi et al. 2002) and ≈ 2 orders of magnitude more sensitive than pre-*Chandra* surveys. Five hundred and three (503) highly significant sources are detected over the 448 arcmin^2 area of the CDF-N, including 20 sources in the central 5.3 arcmin^2 Hubble Deep Field-North region (Alexander et al. 2003a; see Fig 1). The on-axis flux limits of $\approx 2.3 \times 10^{-17}$ erg cm^{-2} s^{-1} (0.5–2.0 keV) and $\approx 1.4 \times 10^{-16}$ erg cm^{-2} s^{-1} (2–8 keV) are sensitive enough to detect moderate-luminosity starburst galaxies out to $z \approx 1$ and moderate-luminosity AGNs out to $z \approx 10$.

In addition to deep X-ray observations, the CDF-N region also has deep multi-wavelength imaging (radio, submm, infrared, and optical) and deep optical spectroscopy (e.g., Barger et al. 2003a). Most recently, the CDF-N has been observed with the ACS camera on *HST* and will be observed with the IRAC and MIPS cameras on *SIRTF* as part of the GOODS project (Dickinson & Giavalisco 2003). The *HST* data, in particular, are providing key morphological and environmental constraints on the X-ray detected sources.

M. Plionis (ed.), Multiwavelength Cosmology, 291-294.

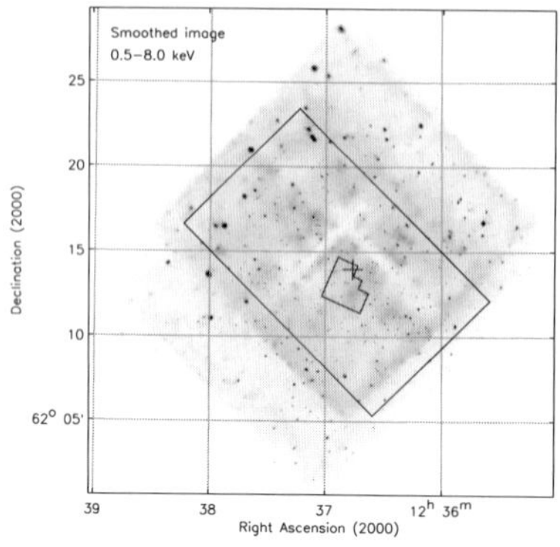

Figure 1. Adaptively smoothed (2.5σ) 0.5–8.0 keV image of the CDF-N (see Fig 3 of Alexander et al. 2003a).

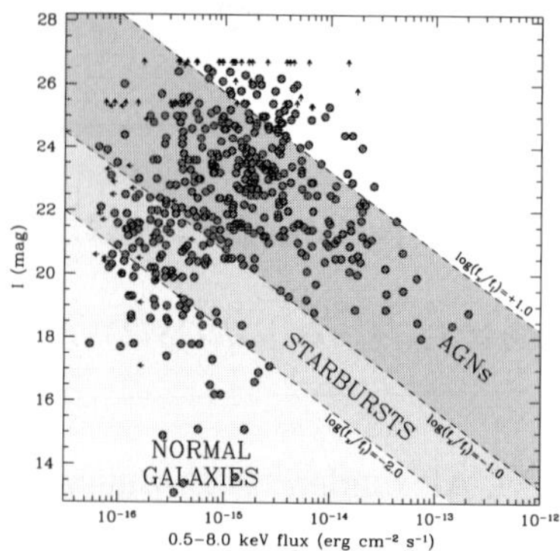

Figure 2. I-band magnitude versus X-ray flux. The shaded regions show approximate flux ratios for different source types.

2. The diversity of X-ray selected sources

The X-ray-to-optical flux ratios of the faintest X-ray sources span up to five orders of magnitude, indicating a broad variety of source types (including AGNs and starburst galaxies; see Fig 2). Many of the AGNs show the optical characteristics of AGN activity (e.g., broad or highly ionised emission lines). However, a large fraction (perhaps $> 50\%$) are either too faint for optical spectroscopic identification or do not show typical AGN optical features (e.g., Alexander et al. 2001; Hornschemeier et al. 2001; Comastri et al. 2003). It is the X-ray properties of these sources that identify them as AGNs [e.g., luminous, and often variable and/or hard (i.e., $\Gamma < 1$) X-ray emission]. The corresponding AGN source density (≈ 6000 deg^{-2}) is ≈ 10 times higher than that found by the deepest optical surveys (e.g., Wolf et al. 2003).

X-ray spectral analyses of the X-ray brightest AGNs indicate that both obscured and unobscured sources are found (Vignali et al. 2002; Bauer et al. 2003). However, few Compton-thick sources have been identified, and current analyses suggest that they are rare even at faint X-ray fluxes (e.g., Alexander et al. 2003a). Whilst AGNs are identified out to $z = 5.189$ fewer high-redshift moderate-luminosity AGNs are found than many models predict (e.g., Barger et al. 2003b). Indeed, current analyses suggest that moderate-luminosity AGN activity peaked at comparatively low redshifts (e.g., Cowie et al. 2003).

A large number of apparently normal galaxies are detected at faint X-ray fluxes (e.g., Hornschemeier et al. 2001). The properties of these sources at infrared, radio, X-ray, and optical wavelengths are consistent with those expected from starburst and normal galaxies (e.g., Alexander et al. 2002; Bauer et al. 2002; Hornschemeier et al. 2003). Furthermore, their X-ray and radio luminosi-

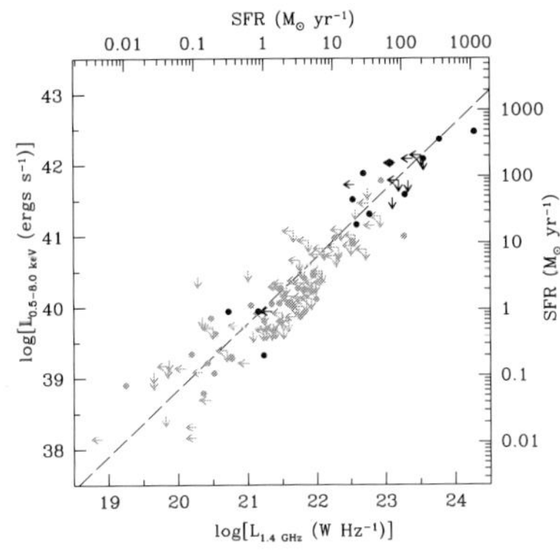

Figure 3. X-ray-radio luminosity comparison for CDF-N (black) and local (grey) sources (see Fig 4 in Bauer et al. 2002).

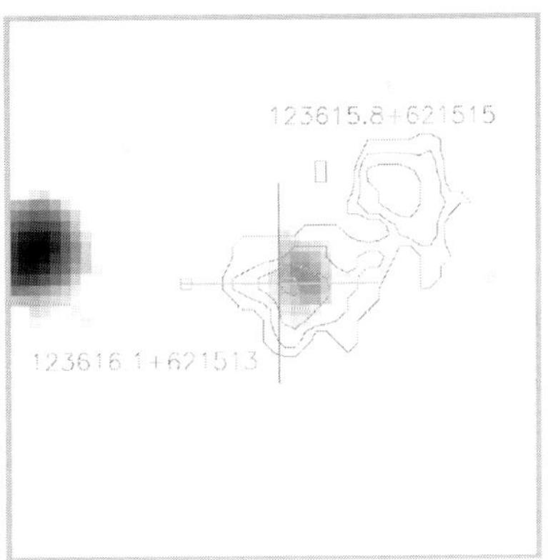

Figure 4. Submm source (I-band image; X-ray contours) with likely binary AGNs (see Fig 2 in Alexander et al. 2003b).

ties are correlated in the same manner as for local starburst galaxies, suggesting that the X-ray emission can be used directly as a star-formation indicator (Bauer et al. 2002; Ranalli et al. 2003; see Fig 3). While the X-ray emission from the low-redshift, low-luminosity sources could be produced by a single ultra-luminous X-ray source (e.g., Hornschemeier et al. 2003), the majority of these sources have X-ray luminosities between those of M 82 and NGC 3256, implying moderate-to-luminous star-formation activity.

3. AGNs in high-redshift submm galaxies

Deep submm surveys have uncovered a population of dust-enshrouded, luminous galaxies at high-redshift ($z \approx 1$–4; e.g., Smail et al. 1997; Hughes et al. 1998). Both AGN and starburst activity can theoretically account for the large luminosities of these sources; however, since few ($< 10\%$) submm sources have X-ray counterparts in moderately deep X-ray surveys, AGNs can only be bolometrically important if they are Compton thick (e.g., Fabian et al. 2000; Hornschemeier et al. 2000). The CDF-N is sensitive enough to place direct constraints on the presence and properties of AGNs in submm galaxies.

Seven bright ($f_{850\mu m} \geq 5$ mJy; S/N$\geq$ 4) submm sources have X-ray counterparts in a 70.3 arcmin^2 area centred on the CDF-N (Alexander et al. 2003b); using the most recent submm catalog of Borys et al. (2003), this corresponds to 54% of the bright submm sources! The X-ray emission from five of these sources is clearly AGN dominated, while the X-ray emission from the other two sources may be star-formation dominated (Alexander et al. 2003b). X-ray spectral analyses of the five AGNs indicate that all are heavily obscured; however, with 1–2 possible exceptions, the absorption appears to be Compton thin and the AGNs are of moderate luminosity (Alexander et al. 2003b). Consequently, the AGNs make a negligible contribution to the bolometric luminosity. This may imply that the central massive black holes are in their growth phase. Interestingly, two ($\approx$ 30%) of the seven submm sources are individually associated with X-ray pairs (Alexander et al. 2003b; see Fig 4). The small angular separations of these pairs ($\approx$ 2–3″) correspond to just $\approx$ 20 kpc at $z = 2$ (approximately one galactic diameter); the probability of a chance association is $<$1%. We may be witnessing the interaction or merging of AGNs in these sources (a low-redshift example of this binary AGN behaviour is NGC 6240; Komossa et al. 2003). Since only five ($\approx$ 3%) of the 193 X-ray sources in this region are close X-ray pairs ($<$3″ separation), binary AGN behaviour appears to be closely associated with submm galaxies (see also Smail et al. 2003).

Acknowledgments

Support came from NSF CAREER award AST-9983783, CXC grant G02-3187A, Chandra fellowship grant PF2-30021, and the Royal Society. We thank D. Schneider and C. Vignali for their considerable help in the CDF-N project.

References

Alexander, D. M., et al.: 2001, AJ, 122, 2156
Alexander, D. M., et al.: 2002, ApJ, 568, L85
Alexander, D. M., et al.: 2003a, AJ, 126, 539
Alexander, D. M., et al.: 2003b, AJ, 125, 383
Barger, A. J., et al.: 2003a, AJ, 126, 632
Barger, A. J., et al.: 2003b, ApJ, 584, L61
Bauer, F. E., et al.: 2002, AJ, 124, 2351
Bauer, F. E., et al.: 2003, AN, 324, 175
Borys, C., Chapman, S. C., Halpern, M., & Scott, D.: 2003, MNRAS, accepted (astro-ph/0305444)
Brandt, W. N., et al.: 2001, AJ, 122, 2810
Comastri, A., et al.: 2003, AN, 324, 28
Cowie, L. L., Barger, A. J., Bautz, M. W., Brandt, W. N., & Garmire, G. P.: 2003, ApJ, 584, L57
Dickinson, M., & Giavalisco, M.: 2003, The Mass of Galaxies at Low and High Redshift. Proceedings of the ESO Workshop held in Venice, Italy, 24-26 October 2001, 324
Fabian, A. C., et al.: 2000, MNRAS, 315, L8
Giacconi, R. et al.: 2002, ApJS, 139, 369
Hornschemeier, A. E., et al.: 2000, ApJ, 541, 49
Hornschemeier, A. E., et al.: 2001, ApJ, 554, 742
Hornschemeier, A. E., et al.: 2003, AJ, 126, 575
Hughes, D. H., et al.: 1998, Nature, 394, 241
Komossa, S., et al.: 2003, ApJ, 582, L15
Ranalli, P., Comastri, A., & Setti, G.: 2003, A&A, 399, 39
Smail, I., Ivison, R. J., & Blain, A. W.: 1997, ApJ, 490, L5
Smail, I., et al.: 2003, ApJ, in press (astro-ph/0307560)
Vignali, C., et al.: 2002, ApJ, 580, L105
Wolf, C., et al.: 2003, A&A, in press (astro-ph/0304072)

INFERRING THE STAR-FORMATION HISTORY FROM X-RAY OBSERVATIONS OF CLUSTERS

Alexis Finoguenov
Max-Planck-Institut für extraterrestrische Physik
Giessenbachstraie, 85748 Garching, Germany
alexis@xray.mpe.mpg.de

Abstract We discuss the direct implications of the observed state of the gas accreting on to the clusters of galaxies on the galaxy transformation. We explain why there is a truncation of the thin disk in spirals long before the galaxy enters the dense cluster environment. XMM observations thus provide a direct support for the revision of the picture of environmental effects in clusters of galaxies.

Thermodynamics of groups. Analysis of combined ROSAT and ASCA data on the entropy profiles for groups revealed very high entropy level at outskirts, 400 keV cm^2, supporting a need of energy injection into the gas before its accretion onto potentials of the groups (Finoguenov et al. 2002). After comparison with a larger sample of groups and clusters, we conclude that there is a variation in the entropy ($kT_e/n_e^{2/3}$) between $\sim$ 100 keV cm^2 and $\sim$ 400 keV cm^2 within every system due to non-gravitational heating. Using models of cluster formation as a reference frame, we established that the accreted gas reaches an entropy level of 400 keV cm^2 by redshift $2.0 - 2.5$, while such high entropies where not present at redshifts higher than $2.8 - 3.5$, favoring nearly instantaneous preheating.

To check our suggestions, we carry out a comparison between observations and hydrodynamic simulations of entropy profiles of groups and clusters of galaxies (Finoguenov et al. 2003a). We use the Tree+SPH GADGET code (Springel, Yoshida & White 2001) to simulate four halos of sizes in the $M_{500} = 1 - 16 \times 10^{13} h^{-1} M_\odot$ range, corresponding to poor groups up to Virgo-like clusters. We adopt a flat ΛCDM cosmology with $\Omega_m = 0.3$, $h = 0.7$, $\Omega_b = 0.02h^{-2}$, and $\sigma_8 = 0.8$. With the heating energy budget of ~ 0.7 keV/particle injected at $z_h = 3$, we are able to reproduce the entropy profiles of groups. We obtain the flat entropy cores in a combined effect of preheating and cooling, while we achieve the high entropy at outskirts by preheating. Figure 1 summarizes the comparison between the entropy properties of the ICM for simulated and observed galaxy systems and also compares the evolution

M. Plionis (ed.), Multiwavelength Cosmology, 295-298.

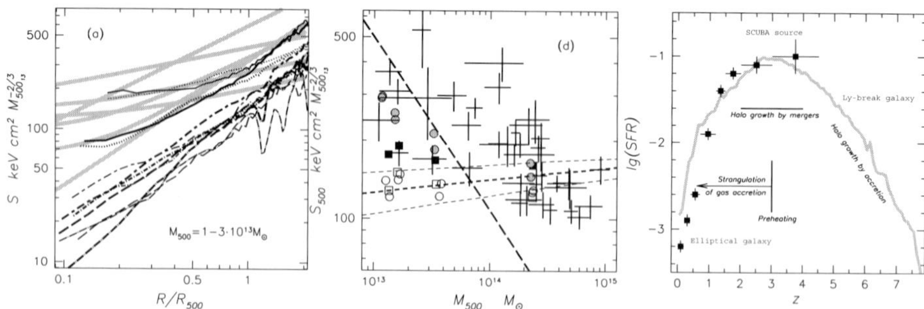

Figure 1. Two left panels: entropy profiles scaled by the total mass in units of $10^{13} M_\odot$, measured at overdensity Δ_{500} and plotted against a scaling radius measured as a fraction of r_{500}. Gray lines indicate the data. Black lines denote various simulation runs: dot-long-dash for the gravitational heating runs, short-long-dash for including the effect of cooling runs, long dash and dot-short-dash for preheating at z=9, dotted and solid for preheating at z=3. Panel (d) shows a comparison at r_{500}, where data points are shown as crosses, circles indicate simulations with preheating, with filled circles representing the $z_h = 3$. Open squares show the purely gravitational run and filed squares indicate the effect of cooling. Dashed lines represent the prediction for the shock heating. The long-dashed line indicates the effect of a preheating value of 500 keV cm^2. Right panel: comparison of the evolution of the star-formation rate density observed for early-type galaxies (squares with errorbars; Franceschini et al. 1998) and predicted in our model at z=3 for the group, normalized to the $\Omega_{groups+clusters}/\Omega_m$ (grey line).

of the star-formation rate density observed for early-type galaxies (Franceschini et al. 1998) and predicted in our model with preheating at z=3 for the group, normalized to the $\Omega_{groups+clusters}/\Omega_m$. Initial halo growth dominating the star-formation density at $z > 5$, coresponds to the so called quiescent mode of the star-formation, which is locally seen in spiral galaxies and at high-redshift is attributed to the Ly-break galaxies. The Ly-break stage of elliptical is followed by a dominance of the merger growth, characterized by bursts of star-formation (Sommervile et al. 2001), and thought to be observed in hyperluminous high-redshift infrared galaxies, such as SCUBA sources. Effect of intoduction of the preheating consists in *strangulation of the gas accretion*, leading to a complete squelching of the quiescent star-formation mode and reducing the effectiveness of the late-time burst mode at later epochs, reproducing therefore the observed decline by two orders of magnitude in the star-formation rate density of early-type galaxies.

Smoking gun evidence. XMM-Newton observations of the outskirts of the Coma cluster of galaxies (Finoguenov et al. 2003b) confirm the existence of a soft X-ray excess claimed previously and show it comes from warm thermal emission. Our data provide a robust estimate of its temperature ($\sim$ 0.2 keV) and oxygen abundance ($\sim$ 0.1 solar). Using a combination of XMM-Newton and ROSAT All-Sky Survey data, we rule out a Galactic origin of the soft X-ray emission. Associating this emission with a 20 Mpc region in front of Coma, seen in the skewness of its galaxy velocity distribution (Fig.2), yields an estimate of the density of the warm gas of $\sim 50 f_{\rm baryon}\rho_{\rm c}$. Our measurement of the

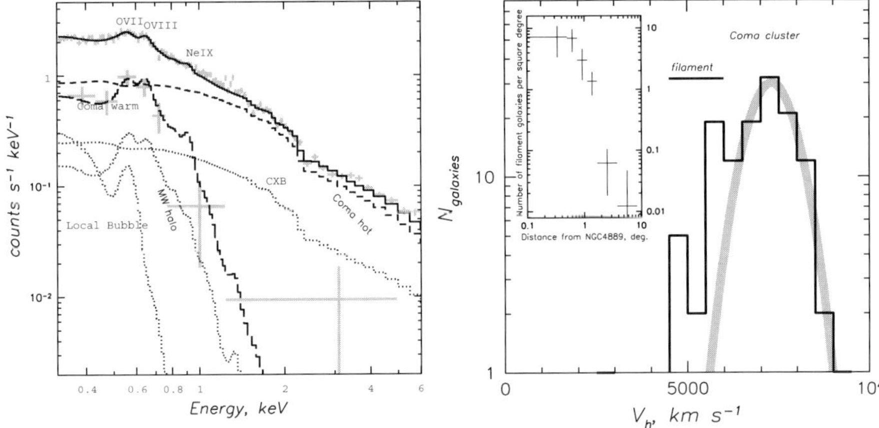

Figure 2. Left panel: detailed decomposition of the pn spectrum (small grey crosses) in the Coma-11 field into foreground and background components, obtained from the analysis of the blank field, plus hot and warm emission from the Coma cluster. Large crosses are the result of the in-field subtraction of Coma emission and detector background, possible due to differences in the spatial distributions between the Coma warm, Coma hot and detector background components. MW halo means Milky Way halo and CXB means cosmic X-ray background. Right panel: galaxy distribution in the direction to the Coma cluster from CfA2 survey. Central 2 degrees in radius field centered on NGC4889 are shown, excluding the 60 degrees cone in the South-West direction from the center, to avoid the effect of the NGC4839 subcluster. An excess of galaxies over the Gaussian approximation of the velocity dispersion of the Coma cluster is seen in the $4500 - 6000$ km s^{-1} velocity range. The insert shows the spatial distribution of the excess galaxies in the $4500 - 6000$ km s^{-1} velocity bin over the galaxies in the $8500 - 10000$ km s^{-1} velocity bin, this time out to 10 degrees.

gas mass associated with the warm emission strongly support its nonvirialized nature, suggesting that we are observing the warm-hot intergalactic medium (WHIM).

The resulting star-formation will proceed by consumption of the previously accumulated gas, in either quiescent mode as in the disk, or merger-induced bursts leading to formation of the spheroid. Galaxy mergers are frequent inside galaxy groups, but in a filament the infall of the gas will primarily be recorded in the star-formation history of the disk (Kennicutt et al. 1994). A relevant observation could therefore shed light on the feedback epoch, which is crucial in understanding the relation between the X-ray filaments and OVI absorbers. The recent Sloan Digital Sky Survey (SDSS) discovery of passive spirals, in the same filament in front of Coma (Goto et al. 2003), is exactly what is expected from this strangulation process. The existence of these passive spirals lends further support to the association of the soft X-ray excess with the Coma filament. As passive spirals are starting to be found in the outskirts of many clusters, the presence of the soft X-ray excess might be a widespread phenomenon among the massive clusters, as indicated by first results on XMM-Newton REFLEX-DXL survey (Zhang et al. 2003).

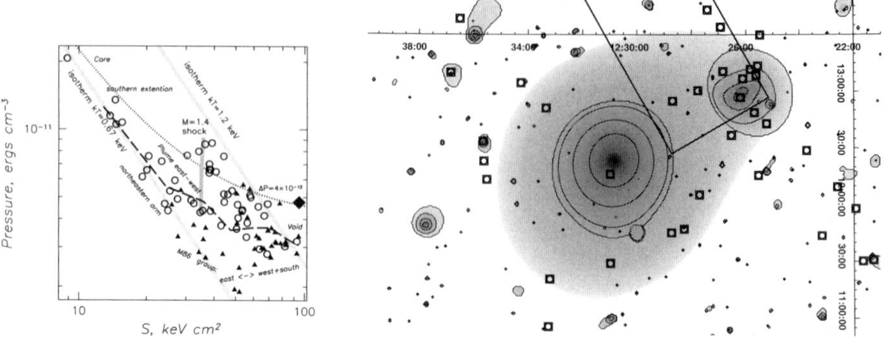

Figure 3. Left panel: pressure vs entropy for all the regions in M86. Only the points with significance greater than 4 sigma are shown. Open circles and filled triangles denote the points with high and low Fe abundance, respectively. Light gray lines show isotherms and a dark-gray line shows the shock adiabat of Mach number of 1.4. Dotted line shows the prediction on imposing an additional pressure of 4×10^{-12} ergs cm^{-3} to an initially isothermal gas with $kT = 0.67$ keV. Right panel: large-scale structure around M86. The wavelet-reconstructed image reproduces a part of ROSAT All-Sky Survey in the 0.5–2 keV energy band, close to M86. Black squares indicate the galaxies falling onto Virgo from behind ($V_{M87} - V_{galaxy} > 600$ km/s). Large rectangular region denotes the position of the excess X-ray emission identified in Böhringer et al. (1994).

Examples of other recently proposed gas removal mechanisms. An in-depth understanding of the processes in the hot interstellar medium of M86, provided by XMM-Newton observations of M86 has suggested a number of past and on-going interactions to take place. Comparison with the position of M86 in the Virgo cluster, promotes the environment outside the virial radius of M87 cloud to bear the prime responsibility for the apparent morphological transformation of the X-ray appearance of M86. Some of the past interactions in M86 are characterized by 10–50 kpc scale, which is only suitable for galaxy-galaxy interactions on one hand, setting limits on a degree of interaction, on the other, thus providing us with understanding of the processes leading to establishment of the morphology-density relations in clusters of galaxies.

References

Böhringer, H., et al. 1994, Nature, 368, 828

Finoguenov, A., Jones, C., Böhringer, H., Ponman, T.J. 2002, ApJ, 578, 74

Finoguenov, A., Borgani, S., Tornatore, L., Böhringer, H. 2003a, A&A, 398, L35

Finoguenov, A., Briel, U., Henry, J.P. 2003b, astro-ph/0309019

Franceschini, A., et al. 1998, ApJ, 506, 600

Goto, T., et al. 2003, astro-ph/0301303

Kennicutt, R., Tamblyn, P., & Congdon, C. 1994, ApJ, 435, 22

Somerville, R.S., Primack, J.R., Faber, S.M. 2001, MNRAS, 320, 504

Springel, V., Yoshida, N., White, S.D.M. 2001, NewA, 6, 79

Zhang, Y.-Y., Finoguenov, A., Böhringer, H., et al. 2003, astro-ph/0306115

THE XMM-NEWTON HARD BAND WIDE ANGLE SURVEY

Carangelo N., Molendi S.and the Hellas2XMM collaboration
Istituto di Astrofisica Spaziale e Fisica Cosmica (IASF)
Sez. di Milano - CNR, Via Bassini, 15, I-20133 Milano, Italy

1. Introduction

The origin of the hard XRB as a superposition of unabsorbed and absorbed AGNs is now widely accepted as the standard model. It has been recognized that a self consistent AGN model for the XRB (e.g. Comastri et al. 1995; Gilli et al. 2001) requires a combined fit of several observational constraints in addition to the XRB spectral intensity, such as the number counts, the redshift and the average spectra. Furthermore, a key ingredient is the absorption distribution of the hydrogen column density N_H: this is a very critical parameter in the model, the only one not constrained by any direct measurement and treated as a free parameter in the fitting procedure. To date, synthesis models have been based on the local N_H distribution of Piccinotti et al. (1982) sample or more recently on the Risaliti et al. (1999) distribution for Seyferts 2, even if several works have tackled the issue of N_H distribution (Maiolino et al. 1998, Bassani et al. 1999, Bauer et al. 2003). However all of these kinds of study are still affected by selections effects or incompleteness problems or by biases related to the absorption.

We designed an XMM-Newton serendipitous survey in the hard band [7-11] keV in order to improve the Piccinotti et al. (1982) distribution and to obtain a more reliable N_H distribution according to criteria independent of absorption effects. The selection of the hard [7-11] keV band can allow us to be unbiased against absorbed objects as much as possible. Moreover, the high throughput of XMM combined with the selection of bright objects, can allow us to perform an accurate spectral analysis and characterization of the sources.

The goal was to build a local reference sample the least affected than any other by N_H bias in order to study its properties and compare them to those of other deeper and non local samples; to derive, from spectral modeling, a solid determination of N_H followed by the N_H distribution of our sample (the N_H distribution has been usually derived from hardness ratio measurements in the previous surveys performed in the hard band); to help constrain the X-ray Lu-

M. Plionis (ed.), Multiwavelength Cosmology, 299-302.

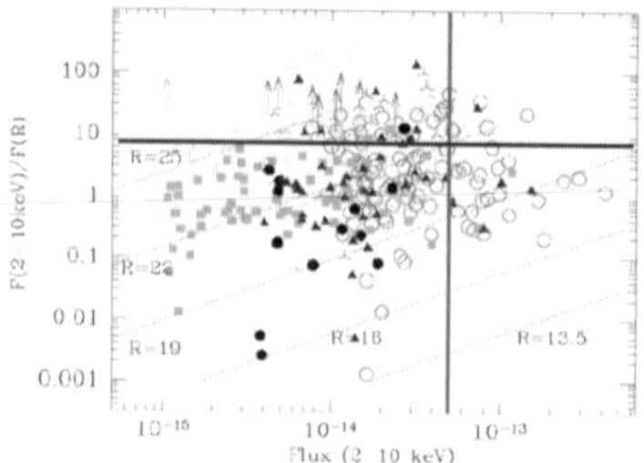

Figure 1. The X-ray ([2-10] keV) to optical (R band) flux ratio as a function of the X-ray flux combined sample:open circles= HELLAS2XMM, filled squares=CDFN, filled triangles=LH, filled circles=SSA13 (see Fiore et al. 2003 for details).Solid lines mark loci of constant R band magnitude. The vertical solid line mark the limiting [2-10] keV band flux of our survey.

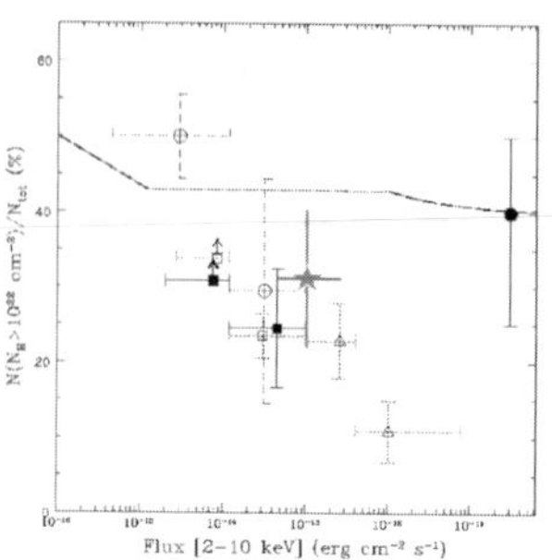

Figure 2. The fraction of sources with $N_H > 10^{22}$ (at z=0) as a function of 2-10 keV flux (see Baldi et al. 2003 for details). Filled squares=HELLAS2XMM, void triangles= ASCA MSS, void circles =CDFS, filled circle= value of Piccinotti et al. (1982), filled star=our preliminary average. The thick dash-dotted line represents the prediction of the XRB synthesis model (Comastri et al. 2001)

minosity Function (XRLF) of absorbed objects and consequently the AGN unification schemes and the synthesis models of the XRB; finally to address other important issues such as the relative importance of the reflection component and the equivalent width (EW) of the iron lines.

2. The sample

We analyzed all of the XMM-EPIC fields publicly available before March 1st 2003, selected at high galactic latitude $|b| > 27$ (excluding the Magellanic Clouds), with $t_{exp} > 10$ ksec and with a full frame observation in PN+MOS1+MOS2. Our survey covers about 30 square degrees for a total of 112 fields observed and 45 sources detected in the [7-11] keV band down to a limiting flux of about 3×10^{-14} erg cm^{-2} s^{-1}. We cross-correlated the X-ray positions of the 45 sources with the NASA Extragalactic Database (NED), finding the optical counterpart for 10 objects with a measured redshift. For 16 sources we found an optical counterpart visible in the R plates (which have a limiting magnitude of 21) of the Digitized Sky Survey (DSS) and as far as the other sources with no DSS counterpart are concerned, we estimated the limiting optical magnitude from the Fiore et al. (2003) diagram (see figure 1). Given the high limiting flux of our survey ($F_{lim} \sim 5\times10^{-14}$ in the [2-10] keV band), we expected the remainder of our sources to have an optical magnitude of R< 22-23 and therefore, to be able to perform an optical follow-up with a 4m class telescope.

3. Data reduction and spectral characterization

The data were processed using the XMM-Science Analysis Software (XMM-SAS v5.4) and the Baldi et al. (2002) pipeline implemented for the [7-11] keV band. The excellent astrometry between the three cameras allowed us to merge together the MOS and PN images in order to increase the S/N of the sources. We constructed an automatic spectral pipeline to extract, from the cleaned event files, the source spectrum, within a radius r_{sou}=30-45$''$ and the corresponding background spectrum within a radius $r_{bkg}=3\times r_{sou}$. We obtained spectra for the PN, for the two MOS cameras and for the combined MOS1 and MOS2 cameras. With the final products we performed spectral characterization in the 0.4-10 keV energy range.
We carried out a preliminary spectral characterization in the [2-10] keV band, using a simple absorbed power law model in order to achieve a preliminary N_H distribution: about 30-35% of our sources have a value of $N_H > 10^{22}$. We report the our preliminary average value in the figure of Baldi et al. (2003) (see figure 2): a clear inconsistency is present between the fraction derived from both the HR and spectral analysis and the prediction of the XRB synthesis model. We expected that our survey can allow us to put a more reliable constraint on this fraction by a direct spectral characterization in the flux range 5×10^{-14}-5×10^{-13} erg cm^{-2} s^{-1} in the 2-10 keV band and to test whether this inconsistency is true or related to bias effect.

4. A peculiar source

The very wide angle covered by our XMM-Survey allows us to search for very rare objects, like type 2 Quasars (QSO2): only a handful of QSO2 are known to date. Here we present the example of a very peculiar and interesting object found during our serendipitous survey. The complex spectrum of this source has been fitted in the 0.3-10 keV range with a *Model* described with the following formula:

$$Model = A_G \times (M_C + ZA_T \times (PL + R_C + GL))$$

where A_G is the absorption associated with the Galactic column, M_C is the thermal component,which describes the soft excess (MEKAL in XSPEC); ZA_T is the absorption acting on the nuclear emission; PL is the power-law modeling the nuclear component with Γ=1.8; R_C is the cold reflection component (PEXRAV in XSPEC); GL is a Gaussian line modeling the iron line at 6.4 keV. In figure 3 we report the PN spectrum and the best fitting model and residuals. It is clear that only the primary component is relevant. The iron line, detected at 97% significativity allowed us to measure the redshift of the source and the best fit value is z=0.31$\pm$0.03. We obtained a value of $N_H \simeq 10^{23}$ and at this redshift a luminosity in the 2-10 keV band of $L_{[2-10]}=1.4\times10^{44}$ erg

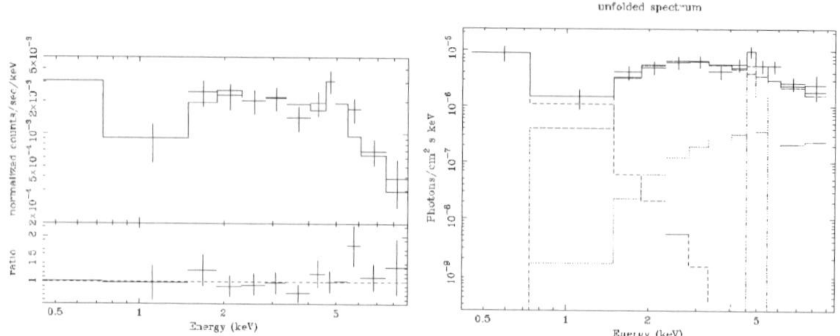

Figure 3. **Left Pannel**: The best fitting model and residual of QSO2. **Right Pannel**: The unfolded spectrum of QSO2. The dashed line is the thermal component, the dot-dashed line is the power law component, the dotted line is the reflection component and the dot-dot-dashed line is the iron line at 6.4 keV.

cm^{-2} s^{-1}. We conclude that this object is a QSO2. In particular, this is the first QSO2 discovered and completely spectral characterized in the X-ray (typically QSO2 were discovered with combined X-ray and optical observations).

5. Conclusion

We would like to stress that our XMM-Survey performed in the hard [7-11] keV band allows us to perform an accurate spectral analysis and characterization for the first time of a local reference sample, the least affected than any other by N$_H$ bias in order to derive a N$_H$ distribution and constrain the XRLF of absorbed objects. Considering the knowledge and scientific equipment available today, this is the best result we could achieve. Future study will cover the optical follow up and we plan to extend our survey to about 40 square degrees before October 1st 2003.

References

Baldi A., Molendi S., Comastri A., Fiore F., Matt G., Vignali C., 2002, ApJ,564, 190

Baldi A., et al. 2003, in preparation

Bassani L., Dadina M., Maiolino R., et al., 1999, ApJSS,121,473

Bauer et al. 2003, submitted

Comastri A., Setti G., Zamorani G., Hasinger G., 1995, A&A, 296,1

Comastri A., Fiore F., Vignali C., Matt G., Perola G.C., La Franca F., 2001, MNRAS, 327, 781

Fiore F., et al. 2003, astro-ph/0306556, A&A in press

Gilli R., Salvati M., Hasinger G., 2001, A&A,366,407

Maiolino R., Salvati M., Bassani L., et al., 1998, A&A 338, 781

Piccinotti G., Mushotzky R.F., Boldt E.A. et al. 1982, ApJ, 253, 485

Risaliti G., Maiolino R., Salvati M., 1999, ApJ, 522, 157

FUTURE MISSIONS

COSMOLOGY WITH ESA'S FUTURE HIGH-ENERGY ASTRONOMY PROGRAMME (*INVITED*)

Arvind N. Parmar
Research and Scientific Support Department of ESA
ESTEC, 2000 AG Noordwijk
The Netherlands

Abstract The European Space Agency, ESA, is currently studying 4 high-energy astrophysics missions. These are Lobster-ISS, an all-sky imaging X-ray monitor, the Extreme Universe Space Observatory (EUSO) which will study the highest energy cosmic rays by using the Earth's atmosphere as a giant detector, Rosita (ROentgen Survey with an Imaging Telescope Array), and XEUS - the X-ray Evolving Universe Spectroscopy Mission, a potential successor to ESA's XMM-Newton X-ray observatory. These missions will study many types of objects including black holes and clusters of galaxies. The origin of the high-energy cosmic rays to be observed by EUSO is unknown and may have cosmological implications. These first 3 missions will be attached to external platforms on the International Space Station (ISS), while XEUS will visit the ISS to attach additional X-ray mirrors to enlarge the original 4.5 m diameter mirrors to the 10 m diameter required for deep cosmological studies of the early Universe.

Keywords: Cosmology, X-ray astronomy, Space Station, Cosmic rays, Instrumentation

1. Introduction

The European Space Agency (ESA) has ambitious plans to utilize facilities offered by the International Space Station (ISS) for high-energy astronomy. Currently four such high-energy astronomy missions are being jointly studied by ESA's Manned Space flight and Science Directorates. These are Lobster-ISS, an all-sky imaging X-ray monitor, the Extreme Universe Space Observatory (EUSO) which will study the highest energy cosmic rays by using the Earth's atmosphere as a giant detector, Rosita - a medium energy imaging X-ray survey mission and XEUS - the X-ray Evolving Universe Spectroscopy Mission, a potential successor to ESA's XMM Newton X-ray observatory.

A mission is studied before final approval to allow its overall design to be elaborated, the scientific and technical feasibility demonstrated and most importantly the costs evaluated and commitments obtained for all the necessary

M. Plionis (ed.), Multiwavelength Cosmology, 305-312.

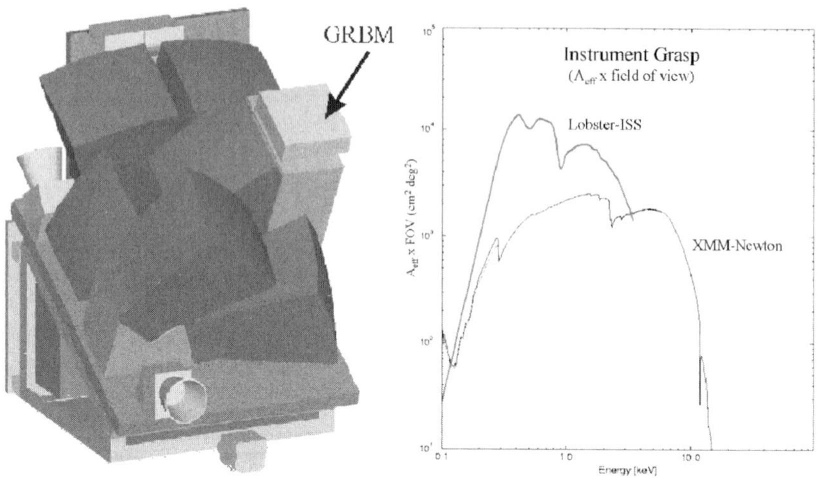

Figure 1. The baseline configuration of the approximately 1 m^3 Lobster-ISS (left panel). The 6 telescope modules can be seen each pointing in slightly different directions. The small box at the front is a particle monitor and the radiator can be seen at the rear. The right panel shows the impressive grasp (Effective area x FOV) of Lobster-ISS <3 keV compared to XMM-Newton.

elements. These activities are normally part of a so-called "Phase A Study" following a successful outcome of which, a project can hopefully move forward into detailed design and build phases as an approved mission. The scientific goals of the X-ray missions include the study of AGN and their cosmological evolution, groups and clusters of galaxies, the X-ray background, the warm-hot intergalactic medium and gamma-ray burst afterglows. All these topics can play an important role in understanding the nature and evolution of the Universe.

2. Lobster-ISS

Lobster-ISS in an imaging all-sky monitor which will utilize a novel form of micro-channel plate X-ray optics and thin window microwell anode gas proportional counters to provide a 0.5–3.5 keV sensitivity of ~0.15 mCrab in a one day observation. This is more than an order of magnitude better than previous all-sky monitors. Even though the effective area for any particular position on the sky is small, the enormous FOV of Lobster-ISS means that its "grasp" (effective area x FOV) is more than that of XMM-Newton below 3 keV (Fig. 1), right panel). The angular resolution of 4′ FWHM will allow source location to 1′ to allow the rapid identification of new transient sources. Lobster-ISS will produce a catalog of 200,000 X-ray sources every 2 months which will be made available to the astronomical community via the WWW. As well as

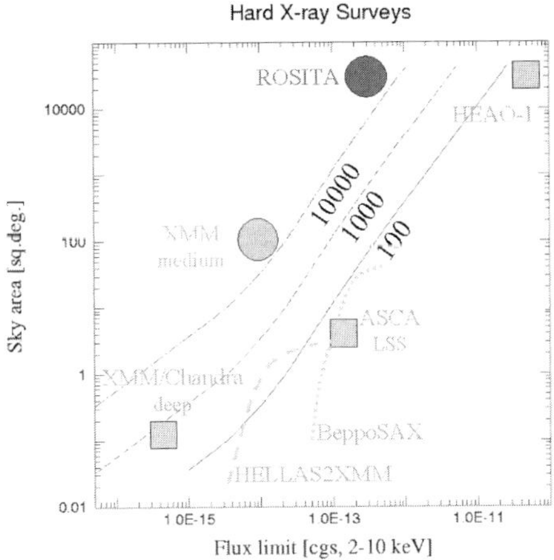

Figure 2. Existing and planned (circles) medium-energy X-ray surveys. The lines give the number of expected sources in the total survey.

providing an alert facility, Lobster-ISS will allow the study of long-term AGN variability, and stellar activity, new X-ray transients, X-ray bursts and super-bursts, black holes in AGN, Type II supernovae shock break out, flares from non-active galaxies, X-ray background mapping and observations of the X-ray afterglows of gamma-ray bursts.

The instantaneous FOV of Lobster-ISS is 162 x 22.5° and is synthesized from six identical offset telescope modules each with a 27 x 22.5° FOV. A gamma-ray burst monitor (GRBM) is being considered for Lobster-ISS in order to provide simultaneous high-energy spectral and temporal measurements of all the gamma-ray bursts that occur within this large FOV. The proposed configuration of the instrument is shown in the left panel of Fig. 1. It is envisaged that Lobster-ISS will be located on the zenith pointing external platform on ESA's Columbus module. Unlike a conventional satellite which orbits the Earth pointing in the same direction, unless commanded otherwise, the ISS orbits rather like an airplane, keeping its main axis parallel to the local horizon. This is a great advantage for an all-sky monitor since it means that the FOV will automatically scan most of the sky during every 90 minute ISS orbit. The Lobster-ISS web site is to be found at http://www.src.le.ac.uk/lobster.

3. Rosita

Rosita is an all-sky survey instrument proposed as an external scientific payload on the ISS X-ray all-sky surveys were first performed in the late seventies using non-imaging detector systems. The highest sensitivity survey in the classical 2–10 keV band is still the HEAO-1 A2 survey, with an angular resolution of about 2.5 degrees. Rosita will conduct the first imaging all-sky survey in the 0.3–10 keV medium energy X-ray range with an unprecedented angular resolution of $\sim 40''$. The main scientific goals are (1) to detect systematically all obscured accreting Black Holes in nearby galaxies and many (>170000) new, AGN, (2) detect the hot intergalactic medium in rare, very massive clusters of galaxies at high redshifts as well as in several ten thousand nearer clusters and potentially hot filaments in between clusters and (3) study in detail the physics of galactic X-ray source populations, such as stellar coronae, supernova remnants and X-ray binaries. Above 2 keV the Rosita survey will have a hundred times more sensitivity and have a hundred times better angular resolution than the HEAO-1 A2 survey, which was performed about 30 years ago. In the 0.5–2 keV energy range the Rosita survey will be more sensitive and have a substantially better energy and angular resolution than the ROSAT all-sky survey (Fig. 2).

The telescope will consist of a copy of the seven 27-fold nested Wolter-1 mirror system already flown on the ABRIXAS mission, which failed soon after launch due to a battery conditioning fault. The telescope will provide a 7 x $42'$ FOV and give an average exposure on the sky of 5800 s for a 3 year mission duration. The CCD detectors are currently being developed at MPE Garching on the basis of the successful XMM-Newton pn-CCD technology, and are also a potential prototype for the XEUS wide field imager. The Rosita web site is: http://wave.xray.mpe.mpg.de/rosita.

4. EUSO

EUSO is an ultra-high energy cosmic ray observatory also proposed to fly on the Columbus External Payload Facility on the ISS. EUSO will detect cosmic rays with energies $>4 \times 10^{19}$ erg s^{-1} by observing the flash of fluorescence light and the reflected Cerenkov light produced when the particles interact with the Earth's atmosphere (Fig. 3). Direct imaging of the light track and its intensity variations will allow the sky position of the event as well as the overall energy to be reconstructed. The origin, nature, method of propagation, and acceleration mechanism of such cosmic rays are all highly uncertain, and either involves extreme objects such as massive black holes or gamma-ray bursts as accelerators or, even more interestingly, implies the presence of exotic physics such as the collapse of massive relics, or topological defects, left over from the Big Bang. If EUSO could make the connection between high-energy cosmic rays and the collapse of topological defects, it would provide a great step forward in understanding the early Universe.

By looking down from the ISS with a 60° FOV EUSO will detect around 1000 events per year - a factor of around 7 more than the most sensitive planned ground based facility - Auger, now under construction. Since neutrinos prop-

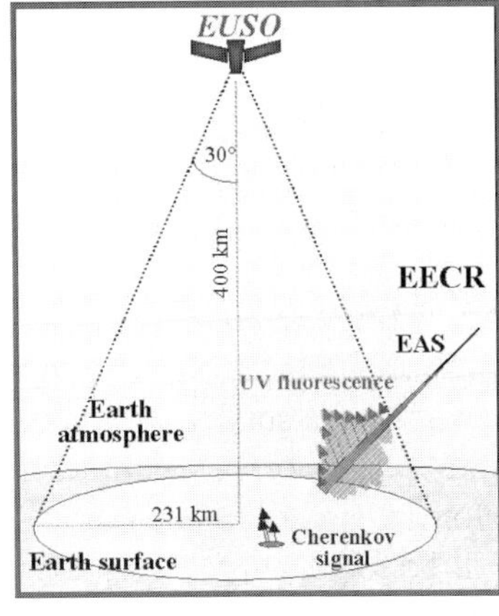

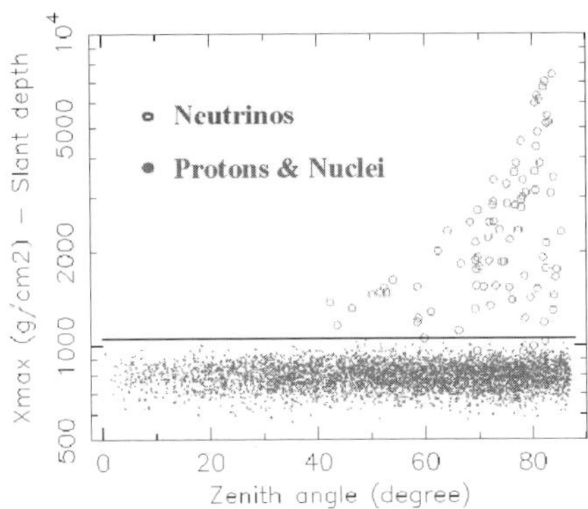

Figure 3. Left: The method of operation of EUSO. EUSO will look down from the ISS with a 60° FOV and detect the fluorescent and reflected Cerenkov radiation produced when an Extreme Energy Cosmic Ray (EECR) interacts with the Earth's atmosphere. Right: Cosmic ray shower depth simulations showing how particle and neutrino induced events can be distinguished.

agate, on average, much deeper into the atmosphere than protons before interacting, EUSO will be able to distinguish between the two types of events by selecting on interaction depth (Fig. 3) opening up the new field of high-energy neutrino astronomy. Unlike Lobster-ISS which fits neatly into the ISS standard external accommodation package which basically provides for about a cubic meter of volume, EUSO is substantially bigger with a 2.5 m diameter cylinder necessary to enclose the double Fresnel lens optics and photo-multiplier tube focal plane. A view of what EUSO might look like attached to ESA's Columbus module is shown in Fig. 4. More information on EUSO is available on http://www.euso-mission.org.

5. XEUS

The fourth high-energy mission under study by ESA that utilizes the ISS is the X-ray Evolving Universe Spectroscopy mission, or XEUS. This mission is under study as part of ESA's long-term Horizon 2000+ science programme. The aims is to place a long lived X-ray observatory in space with a sensitivity comparable to the next generation of ground and space based observatories such as ALMA and JWST (Fig. 5). By making full use of the facilities available at the ISS and by ensuring in the design a significant growth and evolution potential, the overall mission lifetime of XEUS could be as long as 25 years.

A key goal of this mission is to study of the hot baryons and dark matter at high red-shift through spectroscopic investigations of some of the first massive black holes. These may form at red-shifts of 10–20 and have X-ray luminosities of $\sim 10^{44}$ erg s^{-1}. In order to have sufficient sensitivity to derive their

Figure 4. An idea of the size of EUSO may be obtained from this cut-away image of EUSO shown attached to ESA's Columbus module and viewing towards the ISS nadir.

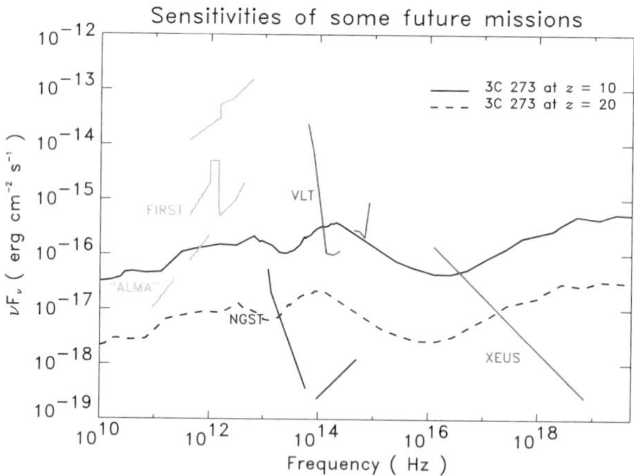

Figure 5. Comparison of the sensitivities of future missions in different wave-bands. A horizontal line corresponds to equal power output per decade of frequency. For ALMA an 8 hr integration was assumed, for FIRST (Herschel) a 5σ detection in 1 hr, for NGST (JWST) a 5σ detection in 10 ksec, and for XEUS a 100 ksec exposure

Figure 6. XEUS in its operational configuration. The detector spacecraft (foreground) maintains its position at the focus of the of the X-ray mirrors 50 m away to within ± 1 mm.

masses, spins and red-shifts through studies of intensity variability and Fe-K lines broadened by strong gravity effects, XEUS will need a 10 m diameter X-ray mirror with a spatial resolution of 2 to 5$''$ half-energy width. This will provide a limiting 0.1–2.5 keV sensitivity of around 4×10^{-18} erg cm^{-2} s^{-1}. Other key science goals include (1) the study of the formation of the first gravitationally bound, dark matter dominated, systems ie. small groups of galaxies and tracing of their evolution into today's massive clusters. (2) The study of the evolution of metal synthesis down to the present epoch, using in particular, observations of the hot intra-cluster gas and (3) The characterization of the mass, temperature, and density of the intergalactic medium, much of which may be in hot filamentary structures, using absorption line spectroscopy. High red-shift luminous quasars and the X-ray afterglows of gamma-ray bursts may be used as background sources.

XEUS will be a long-term X-ray observatory consisting of separate formation flying detector and mirror spacecraft separated by the 50 m focal length of the optics (Fig. 6). XEUS will be launched by an Ariane V and have an initial mirror diameter of 4.5 m, limited by the launch shroud of the rocket. The mirror will be divided into segments, or "petals". Each petal will be individually calibrated and aligned in orbit. Narrow and Wide field imagers will provide FOVs of 1$'$ and 5$'$, and energy resolutions of 1–2 eV and 50 eV at 1 keV. It is likely that the narrow field imager will be a cryogenic detector such as an array of bolometers or super-conducting tunneling junctions and the wide field device will be based on more conventional semi-conductor technology. An additional ring of CCD detectors is being considered to extend the FOV to 15$'$. so increasing significantly the amount of serendipitous science. By super-coating the X-ray mirrors the high-energy response will be extended to $\sim$50 keV. The

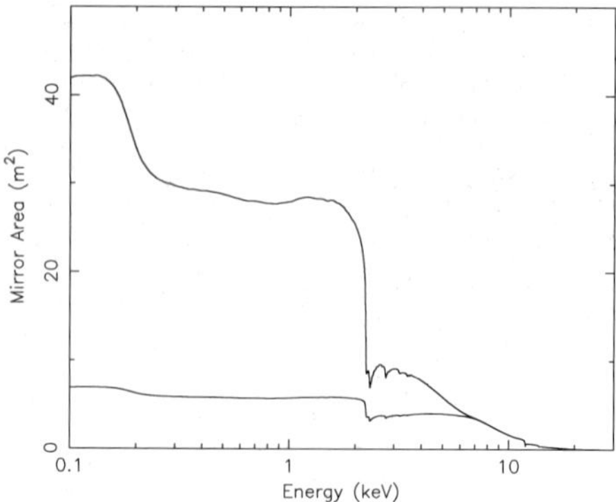

Figure 7. The importance of mirror growth at the ISS. The lower curve shows the initial XEUS mirror area. The upper curve shows the factor 5 increase in area <2 keV after growth.

high-energy X-rays will be imaged by a separate detector, located below the wide field imager. Finally, the large mirror throughput opens up other interesting possibilities such as high time resolution studies and X-ray polarimetry.

The detector spacecraft will have a sophisticated attitude and orbit control system, maneuvering itself to remain at the focus of the optics. After using most of its fuel the detector spacecraft will dock with the mirror spacecraft and the mated pair will transfer to the same orbit as the ISS. The mirror spacecraft then docks with the ISS and additional mirror segments that have been previously transported to the ISS, are attached around the outside of the spacecraft. This increases the mirror diameter to 10 m and the effective area at 1 keV from 6 m^2 to 30 m^2 (Fig. 7). The huge increase in sensitivity that is associated with this expansion at the ISS means that once the mirror spacecraft has left the ISS to be joined by a new detector spacecraft, complete with the latest generation of detectors, the study of the early X-ray Universe can begin in earnest.

Following an initial feasibility study, the many new and challenging technologies that are needed for XEUS are being studied within Europe and Japan. The XEUS web site is at http://astro.esa.int/XEUS. There is great excitement at the prospect of a mission with a sensitivity some 200 times better than that of XMM-Newton, ESA's current X-ray observatory.

Acknowledgments

I thank the EUSO, Lobster and Rosita Principal Investigators, L. Scarsi, G. Fraser, and G. Hasinger for their contributions, together with O. Catalano, N. Bannister and P. Predehl. The XEUS Science Advisory Group (M. Arnaud, X. Barcons, J. Bleeker, G. Hasinger, H. Inoue, G. Palumbo and M. Turner) and my ESA colleagues M. Bavdaz, D. Lumb, C. Erd, G. Gianfiglio and J. Schiemann are also thanked.

PROBING IGM REIONIZATION THROUGH THE 21 CM RADIATION

Benedetta Ciardi
Max-Planck-Institut für Astrophysik
Karl-Schwarzschild-Str. 1, 85741 Garching
Germany
ciardi@mpa-garching.mpg.de

Abstract We will present simulations of the IGM reionization process and discuss its observability through the 21 cm line. We find that the primordial stellar sources considered in this study give a value of the reionization epoch and of the electron optical depth consistent with the observations by *WMAP*, without invoking the presence of additional sources of ionization. Depending on the redshift of reionization, broad-beam observations at frequencies $< 100 - 150$ MHz with the next generation of radio telescopes should reveal angular fluctuations in the sky brightness temperature in the range 5–20 mK on scales < 5 arcmin.

1. Introduction

Despite much recent theoretical and observational progress in our understanding of the formation of early cosmic structures and the high redshift universe, many fundamental questions remain only partially answered. When did the first luminous objects form, and what was their impact on the surrounding intergalactic gas? While the excess HI absorption measured in the spectra of $z \sim 6$ quasars in the SDSS has been interpreted as the signature of the trailing edge of the cosmic reionization epoch (e.g. Fan et al. 2002), the recent analysis of the first year data from the *WMAP* satellite infers a mean optical depth to Thomson scattering $\tau_e \sim 0.17$, suggesting that the universe was reionized at higher redshift (e.g. Kogut et al. 2003).

The study of IGM reionization by primeval stellar sources has been tackled by several authors, both via semi-analytic and numerical approaches. Two main ingredients are required for a proper treatment of the reionization process: *i)* a model of galaxy formation and emission properties and *ii)* a reliable treatment of the radiative transfer of ionizing photons. Once a reionization history has been produced, different observational strategies can be proposed to test the model. It has long been known that neutral hydrogen in the IGM and gravita-

M. Plionis (ed.), Multiwavelength Cosmology, 313-316.

tionally collapsed systems may be directly detectable in emission or absorption against the CMB at the frequency corresponding to the redshifted 21 cm line (see for example Tozzi et al. 2000; Carilli, Gnedin & Owen 2002; Furlanetto & Loeb 2002; Iliev et al. 2003). Here, we discuss the observability of the reionization process through the 21 cm line emitted by neutral IGM.

2. Numerical Simulations of IGM Reionization

We have studied the galaxy formation and evolution process with a combination of high resolution N-body simulations and semi-analytic techniques (Ciardi, Ferrara & White 2003; Ciardi, Stoehr & White 2003, CSW). The simulations, based on a ΛCDM "concordance" cosmology, are run on a box of $20h^{-1}$ Mpc comoving side (the largest ever used to study the reionization process) representing a field region of the universe. The smallest resolved halos have masses of $M \simeq 10^9$ $M_\odot$ and start forming at $z \sim 20$. As for the galaxies emission properties, we have assumed a time-dependent spectrum of a simple stellar population of metal-free stars, with either a Salpeter or a mildly top-heavy, Larson IMF. Of the emitted ionizing photons, only a fraction f_{esc} will actually be able to escape into the IGM. This quantity is poorly determined both theoretically and observationally and may well vary with, e.g., redshift, mass and structure of a galaxy, as well as with the ionizing photon production rate. Finally, the propagation into the given density field of the emitted ionizing photons is followed with the radiative transfer code `CRASH` (Ciardi et al. 2001; Maselli, Ferrara & Ciardi 2003).

The simulations of galaxy formation and reionization described above have been run for the three different parameter combinations in Fig. 1. The critical parameter differentiating these runs is the number of ionizing photons escaping a galaxy into the IGM for each solar mass of long-lived stars which it forms. This number is maximized in the L20 run and minimized in the S5 one. We quantify the agreement between our simulations and the *WMAP* data through the optical depth to electron scattering, τ_e.

The redshift evolution of the volume-averaged ionization fraction, x_v, essentially coincident with mass-averaged one (CSW), is reported for each run in the inset of Fig. 1. The three runs reach complete ionization ($x_v \approx 1$) at $z_r \approx 13$ (L20), 11 (S20) and 8 (S5). Finally, Fig. 1 shows the evolution of τ_e corresponding to the above reionization histories. The three runs yield the values $\tau_e = 0.104$ (S5), 0.132 (S20) and 0.161 (L20). A value $\tau_e = 0.16$ is also obtained if one assumes instantaneous reionization at $z_r \approx 16$ (dotted line).

To assess the impact of environment on the reionization process, an additional simulation has been run (with the same parameters of the S5 run) on a box of $10h^{-1}$ Mpc comoving side, centered on a proto-cluster (CSW). The results will be presented in the Summary.

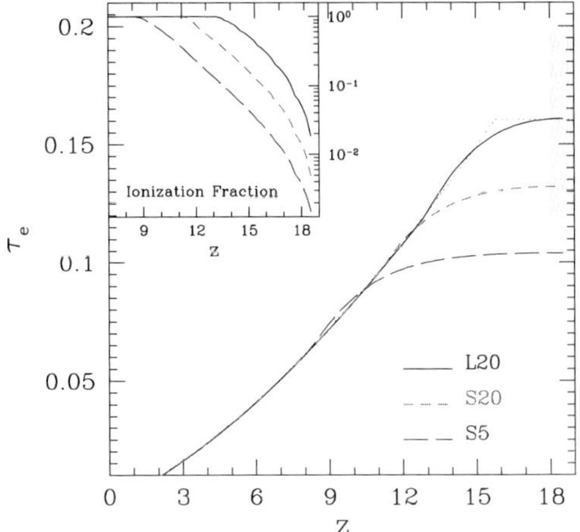

Figure 1. Redshift evolution of the electron optical depth, τ_e, for the S5 (Salpeter IMF, f_{esc}=5%; long-dashed line), S20 (Salpeter IMF, f_{esc}=20%; short-dashed) and L20 (Larson IMF, f_{esc}=20%; solid) runs. The dotted line refers to sudden reionization at $z = 16$. The shaded region indicates the optical depth $\tau_e = 0.16 \pm 0.04$ (68% CL) implied by the Kogut et al. (2003) "model independent" analysis. In the inset the redshift evolution of the volume-averaged ionization fraction, x_v, is shown for the three runs.

3. 21 cm Radiation Emission

The emission of the 21 cm line is governed by the spin temperature, T_S. In the presence of the CMB radiation, T_S quickly reaches thermal equilibrium with $T_{\rm CMB}$ and a mechanism is required that decouples the two temperatures. While the spin-exchange collisions between hydrogen atoms proceed at a rate that is too small for realistic IGM densities, the Lyα pumping dominates, by mixing the hyperfine levels of neutral hydrogen in its ground state via intermediate transitions to the $2p$ state. At the epochs of interest, Lyα pumping will efficiently decouple T_S from $T_{\rm CMB}$ if $J_\alpha > 10^{-21}$ ergs cm^{-2} s^{-1} Hz^{-1} sr^{-1}. We find (Ciardi & Madau 2003) that the expected diffuse flux of Lyα photons should satisfy the above requirement during the 'grey age', from redshift ~ 20 to complete reionization. As the IGM can be easily preheated by primordial sources of radiation (e.g. Lyα and X-ray heating, Chen & Miralda-Escudè 2003), the universe will be observable in 21 cm emission at a level that is independent of the exact value of T_S. In Ciardi & Madau (2003) we have followed the emission related to the reionization histories described above. The main results are outlined in the following Section.

4. Summary

The main results of these studies can can be summarized as follows.

i) Our field region of the universe gets reionized earlier than the proto-cluster, although the latter produces a higher number of ionizing photons. This is due to the fact that high density gas, which is more common in the proto-cluster, is more difficult to ionize and recombines much faster.
ii) While for the field region the mass and volume averaged ionization fractions, x_m and x_v respectively, are always comparable, x_m substantially exceeds x_v for the proto-cluster, at high redshifts. This is because the high density regions surrounding the sources have to be ionized first, before the photons can break out into the low density IGM.
iii) The primordial stellar sources considered in this study give a value of the reionization epoch and optical depth consistent with observations, without invoking the presence of additional sources of ionization.
iv) At epochs when the IGM is still mainly neutral, the simulated early galaxy population provides enough Lyα photons to decouple T_S from $T_{\rm CMB}$. As in the same redshift range the IGM is expected to be 'warm', the 21 cm line would be seen in emission.
v) The rms temperature fluctuations relative to the mean, increase with decreasing angular scale, as variance is larger on smaller scales. The signal peaks at an epoch when several high density neutral regions are still present, but HII occupies roughly half of the volume.
vi) Depending on the redshift of reionization breakthrough, broad-beam observations at frequencies $< 100 - 150$ MHz with the next generation of radio telescopes should reveal angular fluctuations in the sky brightness temperature in the range 5–20 mK (1σ) on scales < 5 arcmin.

Acknowledgments

I would like to thank my collaborators A. Ferrara, P. Madau, F. Stoehr and S. White.

References

Carilli, C., Gnedin, N. Y., and Owen, F. 2002, ApJ, 577, 22.
Chen, X. and Miralda-Escudè, J. 2003, astro-ph/0303395.
Ciardi, B., Ferrara, A., Marri, S. and Raimondo, G. 2001, MNRAS, 324, 381.
Ciardi, B., Ferrara, A., and White, S. D. M. 2003, MNRAS, 344, L7.
Ciardi, B., and Madau, P., ApJ, astro-ph/0303249.
Ciardi, B., Stoehr, F., and White, S. D. M. 2003, MNRAS, 343, 1101 (CSW).
Fan, X. et al. 2002, AJ, 123, 1247.
Furlanetto, S., and Loeb, A. 2002, ApJ, 579, 1.
Iliev, I. T., Scannapieco, E., Martel, H., and Shapiro, P. R. 2003, MNRAS, 341, 81.
Kogut, A. et al. 2003, ApJS, 148, 161.
Maselli, A., Ferrara, A. and Ciardi, B., MNRAS, astro-ph/0307117.
Tozzi, P., Madau, P., Meiksin, A., and Rees, M. J. 2000, ApJ, 528, 597.

SUMMARY

CONCLUDING REMARKS

P. J. E. Peebles
Joseph Henry Laboratories
Princeton University
Princeton NJ 08544 USA
pjep@princeton.edu

Abstract The mood at this conference is summarized in David Hughes' comment, "this decade will be amazing." We've just had a pretty good ten years of advances in cosmology and extragalactic astronomy; why should we expect a repeat, another decade of comparable or even greater progress? The obvious answer is that there still are many more questions than answers in cosmology, and a considerable number of the questions will be addressed by research programs planned and in progress: we certainly are going to learn new things. But beyond that is the fact that there is no practical limit to the hierarchy of interesting topics to explore in this subject. I organize my comments on the state of research, and the prospects for substantial new developments in the coming decade of multi-wavelength cosmology, around the concept of social constructions.

Keywords: Cosmology and extragalactic astronomy

Social scientists inform us that the alpha members of a community set the social standards and constructions, enforce them by the weight of their authority, and see to it that the young members of the community are taught the standards so they will be remembered and enforced by the next generation of alphas. You have experienced all this in your careers in physical science. There is one minor difference – we replace the phrase "social construction" with "working hypothesis" – and one big addition – the remarkable fact that some hypotheses become so thoroughly checked as to be convincing approximations to reality. These comments on the social constructions of cosmology include elements of their history, the present state of the promotions from constructions to established facts, and the prospects for continued additions to our understanding of the real world.

In their book, *The Classical Theory of Fields*, Landau and Lifshitz offer a very sensible caution about the assumption that the universe is close to homogeneous and isotropic on the scale of the Hubble length. When this book was published, in the 1940s, the evidence for homogeneity was sparse at best: this was a social construction. Now the observational tests are tight and believable.

M. Plionis (ed.), Multiwavelength Cosmology, 319-325.

Einstein was led to the picture of homogeneity by his reading of Mach's principle: he felt there had to be matter everywhere to fix inertial motion everywhere. This argument from a philosophical concept led Einstein to an aspect of reality. It is a mystery whether Einstein found the right picture for the right reason. Landau and Lifshitz assume without discussion that general relativity theory applies on the scales of cosmology, which is fair enough in a survey of theoretical physics. But at the time – the first revision of the Russian edition was published in 1948 – there was just one precision test of the theory, the precession of the perihelion of Mercury, and hints of two others, the gravitational redshift and deflection of light. It certainly made sense to consider the application of the theory to cosmology, but not to trust it.

The searching probes of gravity physics from the tests that commenced in the 1960s give convincing evidence that general relativity theory is a good approximation on length scales ranging from the laboratory to the size of the Solar System, let us say to 10^{13} cm. The Hubble length, $cH_o^{-1} \sim 10^{28}$ cm, is fifteen orders of magnitude larger. This spectacular extrapolation is a social construction, until checked, which is the purpose of the cosmological tests.

The results certainly look promising. An example is the broad concordance of evidence that the matter density parameter is in the range $0.15 \lesssim \Omega_m \lesssim 0.3$, from analyses of galaxy peculiar velocities, gravitational lensing measurements, the SNeIa redshift-magnitude relation, the cluster baryon mass fraction, the galaxy two-point correlation function, the cluster mass function and evolution, and the ratio $H_o t_o$ of stellar evolution and expansion time scales. A recent addition to the list comes from the wonderfully successful comparison of the theory and measurements of the anisotropy of the 3K thermal background radiation. This is a demanding test of the gravity theory, which has to describe the propagation of irregularities in the radiation distribution through strongly curved spacetime during the expansion factor $z_{\rm dec} \sim 1000$ since the last substantial interaction of matter and radiation, from initial conditions that have to agree with what grew into the structures observed in the distribution of gas at $z \sim 3$ – in the Lymanα forest – and in the present distribution of galaxies. This has given a new check on Ω_m, from the apparent detection of a contribution to the temperature anisotropy from the matter distribution at modest redshifts, an effect demanded by the theory if Ω_m is significantly below unity.

Each of these measures of the mean mass density is subject to the hazards of interpretation in astronomy. But it is hard to see how systematic errors could affect the many entries in this list all in the same way. Each measure depends on the assumed physics of gravity and the dark sector, which we are supposed to be testing. The test is the consistency: if we were using the wrong physics the broad concordance would be unlikely. The important thing from the point of view of the cosmological tests is not the value of Ω_m but rather the convergence of evidence that the estimates of this number are not seriously confused by systematic errors in the observations or by flaws in the underlying theory: we have a good approximation to one aspect of the real world. Physical science can't explain why reality is a meaningful concept, but we can produce examples of approximations to it. This now includes the measurement of Ω_m.

The evidence that the physics of the standard Friedmann-Lemaître CDM cosmology is on the right track is a considerable advance, but incomplete. An assignment for this decade is to put the tests of gravity physics applied to cosmology on a systematic basis, in analogy with the program of tests of general relativity on the scale of the Solar System and smaller, though one would of course have to replace the parametrized framework of that program with a framework – maybe parametrized – that is adapted to what is relevant to cosmology.

The cosmological principle and general relativity theory are examples of deep advances in physical science that grew out of concepts of philosophy and elegance, which is why we pay attention to such ideas. The lesson is slippery, of course, because our ideas of what is elegant are adaptable. If the cosmological tests had favored the Steady State cosmological model we would be celebrating the perceptive foresight of a different group of alphas. And recall the history of opinions of Einstein's cosmological constant, Λ. Einstein came to quite dislike it. Pauli agreed. And Landau and Lifshitz (in the 1951 English translation by M. Hammermesh) asserted that "it has finally become clear that there is no basis whatsoever" for the introduction of this term. Others at the time paid no attention to this impressive list of alphas, and they seem to have been on the right track: now there is serious evidence for the detection of Λ – or a term in the stress-energy tensor that acts like it.

Although most of us would agree that the universe could have done without Λ, the dark sector of the ΛCDM cosmology is strikingly simple: the dark energy density is close to constant and the dark matter collects in nearly smooth halos by the gravitational growth of small Gaussian departures from a homogeneous primeval dark matter distribution. This picture for the dark matter was introduced two decades ago, and for some years was one of a half dozen viable models for galaxy formation. We had useful analytic solutions for explosion models, but serious challenges in an analysis of the physics of a real cosmic explosion. A reliable analysis of the behavior of cosmic strings or monopoles or textures is even more difficult. The CDM model is easy: structure is dominated by particles that move on geodesics, which are readily simulated in numerical computations, and there is the added advantage that structure forms later than in isocurvature variants, so an interesting numerical simulation need not deal with a large expansion factor. Simplicity recommended the CDM model. Now we have substantial observational evidence that it is a useful approximation to what actually happened.

Is the CDM model complete? One line of thought is that since the dark matter consists of the particles that happen to interact too weakly to be readily observable the dark matter is of course well described as a gas of weakly interacting particles. Another is that the real world seldom is that simple, but that it makes sense to start with the simplest working hypothesis we can get away with, which we will plan to use as a basis for the search for a better approximation, which might in turn lead to a still more complete theory. This is how the physics of the visible sector was discovered.

The search for ideas about how the CDM model might be made more complete – if that is required – can be compared to what was happening in the 1930s when Fermi, Yukawa, and others were trying out ideas of how elementary particles interact. Ideas then and now may be represented by Lagrangian densities with forms like

$$L = \frac{1}{2}\phi_{,\nu}\phi^{,\nu} - V(\phi) + \sum_a \left[i\bar{\psi}_a \gamma \cdot \partial \psi_a - y_a(\phi - \phi_a)\bar{\psi}_a \psi_a \right].$$

Yukawa's scalar field (ϕ in this equation) is complex – charged – in order to couple neutrons and protons. Data were sparse in the 1930s, but Yukawa did know that a reasonable interaction length for nuclear physics – comparable to the size of an atomic nucleus – would follow from the potential $V(\phi) = \mu^2\phi^2/2$, where the meson mass μ is about 200 times that of an electron. The standard model for particle physics follows Yukawa, with considerable elaborations. The search for models for the dark sector is at an even simpler level than Yukawa. In the above Lagrangian the scalar field ϕ is real, so the Yukawa interaction $y_a(\phi - \phi_a)\bar{\psi}_a\psi_a$ just changes the momenta of dark matter particles (represented by the spin-1/2 field operator ψ_a for the $a^{\rm th}$ family, where y_a and ϕ_a are real constants). If the potential $V(\phi)$ in the dark sector is close to Yukawa's form, and μ is relatively large, this is a model for the self-interacting cold dark matter picture. If V varies only slowly with ϕ the $a^{\rm th}$ family behaves as particles with variable mass, $m_a = y_a(\phi - \phi_a)$, and the gradient of the mass is a fifth force – a long-range inverse square force of attraction of dark matter particles that adds to the gravitational attraction. This physics traces back to Nordstrøm's (1912) scalar field model for gravity in Minkowski spacetime, from there to the scalar-tensor gravity theories that were much discussed in the 1950s and 1960s, and from there to the present-day ideas of dilaton fields that would have observable effects – variable parameters – in the visible sector, and maybe a considerably stronger fifth force in the dark sector. A potential energy density $V(\phi)$ that varies slowly with ϕ also appears in a popular model for the dark energy or quintessence. The pedigree is impressive, and it suggests many scenarios for physics in the dark sector even without elaborations comparable to what happened to the model for particle physics after Fermi and Yukawa. To be seen is whether it might lead us to a model that can remedy apparent anomalies – some of which are mentioned below – in the standard ΛCDM cosmology.

If the present standard cosmology really differs from reality enough to matter it will appear in anomalies. But there is a problem, that we cannot in practice unambiguously connect given physics and initial conditions to the details of cosmic structure that are revealed in the observations. How do we decide whether apparent anomalies are only the result of the difficulty of modeling the physics, or whether real failures of the theory have been obscured by the modeling? We ned a new generation of tests of reliability of the hypotheses that are used to model the connection between the theory and observations. The situation is similar to condensed matter physics, where complexity also drives model building, but very different in the sense that we have excellent reason to trust the underlying physics of condensed matter. We will gain confidence in the physics of the dark sector by the accumulation of tests, including the examination of alternatives to CDM. This is another assignment for multi-wavelength cosmology.

Two apparent anomalies that fascinate me have to do with the cosmic web and the galaxy merger rate. The cosmic web is a striking visual feature of numerical simulations of the CDM model, and the web does predict the observed walls of galaxies. But in maps of halo distributions in the simulations I see chains of dwarfs running into the voids between the concentrations of the more massive halos, which I don't see in maps of the real galaxy distribution. Maybe this is a result of the complexity of modeling the connection between theory and observations, exploration of which is part of the research assignment. For now I'm counting the cosmic web of galaxies as a social construction.

I hasten to emphasize that I am deeply impressed by the successful account the cosmic web of gas offers for the statistical measures of the Lymanα forest. On the other hand, I wish I felt better about the apparent lack of disturbance by whatever added heavy elements to the hydrogen in the Lymanα forest clouds.

Another apparent anomaly is the rate of merging of closely bound galaxies. A pair of galaxies separated by a few half-light radii is routinely labeled a merging system, whether at high redshift or low, whether in a group or a rich cluster of galaxies. There is a good reason – simulations and analytic estimates predict the pair will merge in a few crossing times – but is it more than a social construction? The theoretical argument is sound, but only if we have the right physical model for the dark matter, as a nearly collisionless gas, which is not yet something we know. On the observational side, it is often said that the merger remnant of a compact group of spirals is an elliptical, but I also hear that the pattern of element abundances in the progenitors – typically late types – does not look like a promising match to the abundances in a typical early-type galaxy. Also puzzling is the effect of mergers on the shape of the low order galaxy correlation functions. The two-point function is a good approximation to a power law from 10 Mpc separation down to separations of a few half-light radii. Standard estimates of the cosmic merger rate assume close pairs merge in a few orbit times. If so, what preserves the power law form?

Again, I have to qualify these remarks. There is good evidence of mergers at low redshift: we see a clear example in the Centaurus group, where the big elliptical clearly has recently merged with a dusty galaxy, and there are several other classical examples of galaxies that surely are observed in the act of merging. But these spectacular systems do not seem to be all that common: the familiar examples are repeatedly cited. The assignment is to show whether the number of merging galaxies at low redshift really is consistent with the theoretical prediction that galaxies closer than a few half-light radii merge on time scales small compared to the Hubble time.

We might consider also that we have to live with quite a few coincidences within the standard ΛCDM Friedmann-Lema^tre cosmology. Heavily advertised nowadays is the coincidence that we flourish just as the universe is making the transition from matter-dominated expansion to an approximation to the de Sitter solution. Maybe related to this is the observation that we flourish just as the Milky Way is running out of gas for the formation of new generations of planetary systems like our own, along with the rapid collapse of the global star formation rate density since redshift $z = 1$. Less widely discussed these days is the possible relation to Dirac's large numbers coincidence: the ratio of the present Hubble length to the classical electron radius is close to another enormous number, the ratio of the electric and gravitational forces between a proton and electron. Another timing coincidence I suppose is unrelated, but also curious, is that in the standard cosmology optically selected galaxies have just now become good mass tracers: the theory seems to predict strong biasing at redshift $z = 1$. In the standard cosmology the mass of a large galaxy is dominated by dark matter in the outer parts, and by stars near the center. The conspiracy is that the distributions of these two components produce a net mass density run that shows no feature at the transition from high to low mass-to-light ratio. And finally, the ΛCDM cosmology predicts separation-dependent bias of light as a tracer of mass: the ratio of the mass autocorrelation function to the two-point correlation function of optically selected galaxies is a function of separation. But it is curious that the galaxies seem to give the better approximations to power law forms for the low order correlation functions deep in the nonlinear clustering limit, rather than the mass that is supposed to control the dynamics.

It is reasonable to expect that some of these curiosities are nothing more than accidents, and some will be seen not to be curious at all when we have a really good understanding of the theory and its relation to the observations. But it is sensible to be aware of the possibility that some are clues to improvements in the physics.

What might come from continued multi-wavelength research on such challenges to cosmology? I expect the paradigms will continue to rest on the relativistic Friedmann-Lema^tre model, or some good approximation to it, because general relativity theory has passed quite demanding checks on the scales of cosmology. But we owe it to our subject to turn these scattered checks into a systematic survey of the constraints on the physics of spacetime and gravity.
I do not expect a paradigm shift back to the Einstein-de Sitter model: the lines of evidence for low Ω_m are impressively well checked by many independent applications of the theory that depend on quite different elements of the astronomy. The evidence of detection of the cosmological constant is serious, too, but not as well checked as Ω_m. The Λ term has been debated for more than eight decades; we can wait a few more years before deciding whether it deserves a place in the list of convincingly established results.
In the next ten years multi-wavelength observations, including (in the words of a participant) "millimeter, submillimeter and FIR observations with the imaging fidelity currently enjoyed by X-ray, optical, IR and radio astronomers" will produce an enormous increase in our knowledge of cosmic structure, and that is going to drive the development of exceedingly detailed models to relate the theory to the observations. The theory of choice will continue to be ΛCDM, unless or until the observations drive us to something better. While waiting to see whether that happens a assignment for model builders is to develop a convincing case for how far they have gone beyond curve fitting.
After a major advance in a physical science, such as we have seen in the past decade in cosmology, there is the tendency to ask whether the subject has now reached completion, requiring only the "addition of decimal places." You don't hear this talk among astronomers, and I wouldn't expect it to be on astronomers' minds in the coming decades, because there is no practical limit to the layers of detail one may study about things like the populations of stars, planetary systems, and civilizations that are communicating by radio broadcasts in the Milky Way, in the Magellanic clouds, and on out. We have good reason to expect the first decade of the 21st century will be remembered as a golden time for cosmology, but we can be sure there will be room for productive applications of multi-wavelength cosmology for decades to come.
My overall conclusion is that you should pay attention to the alphas – their concepts of simplicity and elegance really have led to deep advances in our understanding of the material world – but then go make the measurements – the alphas have feet of clay like everyone else.

I am grateful to Manolis Plionis for inspiration, David Hogg and David Hughes for advice, and the USA National Science Foundation for financial support for this essay.

OPEN TALK

IONIAN PHILOSOPHERS AND EARLY GREEK COSMOLOGY (*INVITED*)

Nikolaos K. Spyrou
Astronomy Department, Aristoteleion University of Thessaloniki, 541 24 Thessaloniki, Macedonia, Hellas (Greece)

Abstract The contribution of the ancient Ionian Philosophers and particularly of Aristarchus of Samos to the astronomical-cosmological science is described with the purpose of restoring the historical truth concerning this contribution.

1. Introduction

It is a pleasure to express, from this position also, my sincere thanks to the Institute of Astronomy and Astrophysics of the National Observatory of Athens, in the name of its Director Professor C.Goudis, and mainly to Dr Plionis, the heart of this meeting, for their kind invitation to talk during it and for their really warm hospitality. I accepted the invitation, because it is my belief that, in an international cosmological meeting like this one, reference should be made to the founders of the astronomical-cosmological knowledge and science, so that their essential contribution to them be officially recognized and subsequently forwarded both in the interior of our country and abroad. I shall kindly ask you to permit to start this lecture with an introduction of rather personal nature. Let me tell you from the very beginning that, from a scientific point of view, I consider myself as belonging to the broad area of relativity, cosmology and astrophysics; by no means I consider myself a historian of science. The reason of my interest to the ancient Greek philosophers must be attributed to the work and sensitive teaching of the wise, I could say, teachers, I was lucky to have during my elementary and high-school studies. In fact they formulated my interest in this specific subject and to names like e.g. Aristotle, Pythagoras, Aristarchus, Eratosthenes et al. With this interest at hand, later in 1973, namely, thirty years ago, while attending, as a fresh Ph.D. holder of the University of Thessaloniki, the works of the Extraordinary General Assembly of the International Astronomical Union in Warsaw, Poland, I, from my point of view, was really surprised, realizing that the meeting was devoted to the otherwise great Polish astronomer *Nicolaus Copernicus* as the founder

M. Plionis (ed.), Multiwavelength Cosmology, 329-345.

of Astronomy, with, however, no reference at all (as far as I can recall after so many years) being made to Aristarchus of Samos, something that a young Greek astronomer like me would then naturally expect. That was the first time to realize that abroad the contribution of the ancient Greek philosophers to at least the astronomical science was neither accepted unquestionably nor even acknowledged. Thus, a few years later, begun more systematically my interest to the contribution of the ancient Greek philosophers, rather amateur, however, and, in any case, parallel to my main scientific research interests (astrophysics, relativity, cosmology). By the beginning of the 80s, the Vice-Rector of the University of Thessaloniki, who happened to be also the President of the Greek Mathematical Society, asked me to write an essay on Aristarchus of Samos on the occasion of the 2.3 millennia since his birth, which was published in the official journal of the Greek Mathematical Society for the high-school pupils. The first reaction of the pupils and their teachers to the content of the essay revealed generally ignorance of but also interest in the specific subject. In this way, I realized that, also in the interior of my country and especially to the young people, the contribution of the ancient Greek philosophers to at least the astronomical science was not known. The above two facts, namely, the 1973 IAU Extraordinary General Assembly and the ignorance and interest of the high-school pupils and teachers were decisive to me. Since then, the University lecturing revealed, among others, the interest of the students also in the contribution of the ancient Greek philosophers, the necessity for the as widely as possible informing on it of both the scientist and the layman , and, finally, of course, the need for collecting further details on this contribution. Now, after so many years, armoured with all this experience, I can say, quite objectively, that he organizers of this meeting correctly decided to include in the program also an open lecture of this kind. The structure of the lecture is as follows: In the next Section 2 I refer briefly to the astronomical knowledge of the ancients, and in Section 3 I refer briefly to the importance of their contribution to the astronomical-cosmological science. In Section 4 the contribution of especially Aristarchus of Samos is described, and in the final Section I make an effort to restore the historical truth on the contribution of Aristarchus of Samos. For those interested in the subject some indicative international literature is given at the end.

2. Astronomical Knowledge of the Ancients

It is true that we seldom see up-wards in the night sky. Some of us, being so "busy" with every-day life and all its problems or due to the light pollution, perhaps will never look at the skies with the intention of a scientific-philosophical meditation and the need of an analogous questioning. In the pre-industrial period, however, the night sky was something very important, in the sense that, for example, the shepherd, by simply looking at the position of some constellations, could know when the sunrise happens. A few thousands years ago, people did not have the basic knowledge that e.g. the Earth is not flat. They did not know that the clouds in the sky were formed by the evaporation of the water on the surface of the Earth. They did not know the various chemical and biological processes occurring in the plants and the animals. They knew nothing about the transistors, the microchips, or the nuclear energy. Of course, beyond the forests, the mountains, the oceans, the flowers, the animals, they watched the Sun, the Moon in the night sky, but they didn't know that the bright points in the night sky were stars like the Sun. Therefore, it was quite natural to them to respect and feel fear for the unknown. It was natural to feel fear for e.g. the fire, because the fire was hot and caused pain and, occasionally, catastrophes, and also because it moved up-wards from the interior of the Earth, while everything else was moving down-wards. Some of the ancient Greeks believed that the flat Earth was covered by a semispherical sky. Over that semispherical shield, there were hundreds of small holes, through which an external fire could be seen. This was their explanation for the stars. People, therefore, quite naturally, adored the fire and the stars as gods. In most of the human civilizations, the powerful beings of the sky were promoted to gods, and to each one of them there were given names, and relatives, and special responsibilities for the cosmic services they were expected to perform. For every human concern there was a god or a goddess. Nothing could happen without the direct intervention of gods, only through which Nature could function. If the gods were happy, there was plenty of food, and humans were happy. But, if something displeased gods, and this could happen quite easily, the consequences were awesome, like floods, droughts, storms, wars, earthquakes, volcanoes, epidemics. So the gods had to be propitiated, to be made less angry, and this was the work of the numerous priests and oracles. Since the gods were capricious, nobody could be sure about their intentions, about what they would do. *Nature was a mystery, and so the understanding of the world was very hard*. In the origins of every civilization there are many metaphors concerning what the ancients talked about the heavens, how they identified everything they observed in it, and how they watched it from night to night and from year to year. I shall give you three typical examples about the very elegant ideas and legends for the stars and the Milky Way, which, in one form or another, seem to be common in various human civilizations. The Kung Bushmen of the Kalahari Desert in Botsuana have an explanation for the Milky Way, which in their latitude is often overhead. They call it the *backbone of night*, as if the sky were some great beast inside which we live. Their explanation makes the Milky Way useful as well understandable. The Kung believe that the Milky Way holds up the night; that, if it were not for the Milky Way, fragments of darkness would come crashing down at our feet. Isn't it a really elegant idea?. On the island of Samos in the Aegean Sea, very close to Mykonos, there are the remains of the Heraion, one of the wonders of the ancient world. This is a great temple dedicated to Hera, the goddess of the skies. Hera was the patron deity of Samos, playing the same role in Samos, as Athena did in Athens. She married Zeus, the chief of the Olympian gods, and they honeymooned on Samos. The ancient Greek religion explained the diffuse band of light in the night sky as the milk of Hera, squirted from her breast across the heavens. This legend is the origin of the phrase Westerners still use, the *Milky* Way. Perhaps, it originally represented the important insight that the sky nurtures the Earth; if so, that meaning seems to have been forgotten millennia ago. Finally, it is extremely interesting to examine the position and effect of the star called Sirius on the scientific, agricultural, and religious life of the ancient Egyptians. Around 2000 B.C., the Egyptians worshipped Sirius: In the eastern horizon, Sirius rises up just before the Sunrise, a little before the beginning of the period of floods caused by the river Nile. Therefore, this specific rise of Sirius was marking the beginning the whole of the agriculture cycle of Egypt, with all its positive and negative consequences; on the one hand, enrichment of the fields with the river's mud during the floods, and, hence, rich produce and prosperity, and on the other hand, deaths and catastrophes caused by the river's overflows. So, quite naturally, the purely astronomical phenomenon of the rising up of the star Sirius, but also the star Sirius itself, were gradually acquired, or, preferably, given divine properties and capabilities. It is interesting that, in this way, the religious origin of Astronomy can be explained. It is necessary, however, to emphatically recall that the dead, useless and scientifically unfounded relic of that first form of Astronomy is Astrology; and as such Astrology must be treated and confronted.

3. Ancient Ionian Philosophers, Astronomy, and Heliocentric Theory

The simple watching, by the primitive observer on Earth, of the skies and all their impressive phenomena started its first evolutionary steps towards science through the observation of two fundamental phenomena. These are the diurnal rotation of the celestial sphere, due to the axial rotation of the Earth in twenty four hours, and the orbital motion of the Earth around the Sun in one year. For centuries, man struggled to understand and explain how these two phenomena were taking place. More precisely, whether they are due to the rotation of all the stars around the standing Earth, or they are due to the axial rotation of the Earth about its axis, as the Earth moves around the standing center of the cosmos, the Sun. As a result, two conflicting schools of thought arose, namely, the geocentric view of the world and the heliocentric view of the world, the first of which was the generally accepted and, hence, the dominant one. The faith to the correctness and validity of the geocentric model of the world was powerful, it had a purely religious origin, and it was based on two unshaken beliefs. The first belief was that the Earth was the home of the gods, and so it is the immovable center of the world, so that the tranquility and stillness of the gods never be disturbed. The second belief was that the celestial objects move around us (the Earth) in absolutely circular orbits, namely, perfect orbits. The conclusion about the geocentric view of the cosmos (not valid, not acceptable today, of course) was aesthetically extremely simple (namely, only circular orbits around us), and, for this reason, it was also scientifically easily acceptable (namely, simple physical laws). However, the proponents of both the geocentric and the heliocentric models did not have the necessary proofs for supporting the truth of their cosmic view. In fact, in both cases, the same known phenomena were used, and this contributed to the continuation of the conflict. The need of finding firm proofs intensified and reinforced the necessity of observing the celestial phenomena. This continuous observation of the sky had some very useful consequences of interest in the every-day life. These are the invention of many *instruments* (like e.g. the *astrolab*, the *gnomon*, the *sundial*, the *celestial sphere*, the *Antikythera mechanism*). Also the development of many new professions (e.g. the clock-maker, the engraver), and also of many methods for the measurement of time, for use in agriculture, in the determination of basic chronometric units, in the use of the constellations in navigation, in the invention of various systems of spherical coordinates et c. So the ancient Greeks considered that the celestial Universe existed in order to serve them. Today, all the above mechanisms are considered as the forerunner of modern watches and computational machines. It is true that for the construction of such complex instruments, a very advanced geometrical sense of ancient Greeks is required. Therefore, beyond the knowledge of practical geometry (which characterizes also other near-river civilizations), some very delicate notions and ideas are necessary, which, in the framework of Euclidean geometry, were given the form of existence theorems, and led to high technology, characterized by high mentality, abstract scientific thought, and mathematical skill. It must be stressed that, in contrast to the above, the study of the skies by the eastern, prehistoric people has been limited to simply the chronographic accumulation of data and knowledge related to occasionally important celestial phenomena. But, since the appearance of the Greek philosophers in ancient Ionia, *here*, about 2.5 millennia ago, the search of the heavens changes form, it acquires an explicit scientific character, and the scientific revolution replaces the mythological explanation of the celestial phenomena. What exactly was this revolution? Creation of the World out of Chaos. A Universe emerging out of chaos was in complete agreement with the faith of the ancient Greeks to a non-predictable Nature governed by capricious and perverse gods and goddesses. But, in the 6th century B.C., in Ionia the new concept developed, according to which the Universe is comprehensible, because it has internal order, because in Nature there are regularities permitting its secrets and functioning to be uncovered. It is exactly this order and miraculous character and nature of the Universe that were given by the ancients the name *Cosmos*, namely, beauty (ornament). It is remarkable that the revolution occurred in Ionia and not in one of the large cities of India, Babylonia, China or Mesoamerica. This, actually, is not a question of simply academic interest, because we all know the various important contributions of the ancient civilizations. For example, the contribution of the ancient Egyptians to Astronomy and Mathematics, of the ancient Babylonians to the compilation and forming of catalogues, or of the Arabs to Mathematics (as, of course, of the ancient Greeks to physics, geometry, medicine, zoology, and botany). Also the Chinese, which had an astronomical tradition of millennia, contributed to the understanding of magnetism and seismology, and invented the paper and printing technique, the rockets, the clocks, the porcelain, and the open-sea navy. However, according to some historians, China was a strictly traditional society, reluctant and unprepared to accept novelties. But also the very rich and mathematically charismatic civilization of India was captured by the idea of an infinite, ancient universe, condemned to an eternal cycle of deaths and recreations, and in which (universe) nothing new can happen. Analogous arguments hold for the Aztec and Mayas societies, which could not make mechanical inventions, since they could not invent even the wheel. In contradiction to the above, Ionia had many advantages. In the first place, this geographic region consisted of many islands. Isolation, even if imperfect, results in diversity. Due to its large number of islands, Ionia was characterized by a multitude of political systems. There was no force capable of imposing social and spiritual uniformity to all these islands. As a consequence, the free search and quest of truth was possible, so that the acceptance and promotion of prejudices could not be considered as political necessity. Unlikely other people, Ionians were located at the crossroads of civilizations, not at the center of one civilization. The Phoenician alphabet was adopted by the Greeks for the first time in Ionia, and in this way the wide spreading of education and culture became possible. So, the writing was not a privilege of only the priests and scribers. Thoughts and ideas originated in many different places were available for discussion, commenting, debating, and dispute (exactly as it happens in a meeting like this one). On the other hand, the political force was in the hands of merchants, who effectively promoted the technology necessary for the success of their plans and purposes. It was exactly in Eastern Mediterranean, where the great civilizations of Egypt and Mesopotamia, but also of Africa, Asia and Egypt met and influenced each other in the form of the intense and direct confrontation of prejudices, languages, ideas and gods. It was in this way that the great idea arose, (namely, the realization of the fact that the knowledge of the cosmos is possible without the a priori acceptance of the existence of gods), and that there must exist principles, forces, physical laws that can be understood without e.g. the necessity of the direct intervention of Zeus for explaining the flight of a bird. Ionia, therefore, was the place, where science was born, and where between 600 B.C and 400 B.C. the great revolution to the thinking occurred. The Ionians rejected prejudice and so did miracles. Generally, it can be argued that the key to the revolution was the hand, namely, the handicraft work, the experiment, the observation. Some of the brilliant Ionian thinkers were children of sailors, farmers, weavers used to do handicraft work, in contrast to the priests and scribers of other nations, who were grown up in luxury and were reluctant to dirty their hands.

4. The Contribution of the Ionian Philosophers

It is not possible to enumerate all the Greek philosophers and their contribution. However, at least the names should be mentioned of the great scientists and philosophers from *Thales* to *Democritus* (in chronological order, *Anaximander, Pythagoras, Anaxagoras, Empedocles, Hippocrates*; all of them during the period 650-350 B.C.) and those after *Aristotles* (namely, *Euclides, Aristarchus, Eratosthenes, Hipparchus, Ptolemy, Hypatia*; all of them during the period 300 B.C.-450 A.D.). From all of them I shall refer only to three, whose contribution appears to be closely related to the objectives of the present meeting. Thus the philosopher *Anaxagoras* (450 B.C.) from the city of Klazomenai, was called *Nous* (Mind), and was a rich, experimentalist scientist, indifferent to riches, but with a passion for science. When he was asked about the purpose of his life, he answered: The search of the Sun, the Moon, and the Heavens, exactly as a real astronomer would answer. Anaxagoras believed (this was a purely Ionian idea) that man is intellectually superior compared to the rest of the animals, because man has hands. He was the first one to declare explicitly that the Moon shines due to the reflection on it of the light of the Sun, and, as a consequence, he proposed a theory about the phases of the Moon. Also he considered that the Sun and the Moon were not deities, but, instead, fireballs. The heating from the stars is not perceived, because of their large distances from the Earth. Also the Moon has mountains, and the Sun is huge, probably much larger than Peloponnese's (!!!). Because of his ideas he was accused for impiety as introducing new evil spirits, and he died in exile (428 B.C.). *Eratosthenes*, on the other hand, from Cyreneia, belonged to the School of Alexandria, and contributed substantially to realizing in the 3rd century B.C. that the flat Earth is in reality a small spherical world. Eratosthenes has been a director of the great Library of Alexandria. There he read on a papyrus that on the noon of the 21st of June (summer solstice) at the southern limits of the city Syene (contemporary Assuan), close to the first waterfall of the river Nile, vertical columns do not cast shadow at all, and the Sun is reflected exactly from the bottom of a well (namely, it coincides with the Zenith cardinal point upwards). As a scientist he wondered, whether the same happens also in another city like e.g. Alexandria. He realized that this does not really happen, and that analogous vertical columns in Alexandria do cast shadow at that moment. But, if the surface of the Earth was flat, then the vertical columns in the two cities would be parallel to each other, and they should cast shadows and the lengths of their shadows should be equal. But since none of these is true, what can happen? Eratosthenes gave the answer himself, arguing that the surface of the Earth is not flat but spherical. This is a conclusion of really fundamental importance that, additionally, enabled Eratosthenes to calculate the radius and the length of the equatorial perimeter of the Earth. Actually, from the length of the shadow the difference of the geographical latitudes of the two places can be evaluated (approximately seven degrees). The distance between the two cities (being on the same meridian circle), about 800 km, was known to Eratosthenes from stories told by pacers (it is said that Eratosthenes hired pacers in order to measure the distance). So the length of the Earth's perimeter was calculated to be about 40.000 km. This is the correct answer, and Eratosthenes gave this answer using as tools only vertical columns, eyes, hands, and feet, and a mind characterized by simplicity of thought, ability of invention, and a feeling of experiment. The error in the calculation of Eratosthenes was only 2%, a really remarkable achievement 2.5 millennia ago. Therefore, Eratosthenes was the first man to measure the dimensions of the planet Earth, and, for this reason, he is considered as the inventor of mathematical geography. In essence, the method of Eratosthenes is used by the current geodets, who recognize Eratosthenes as the father of geodesy. Finally, *Aristarchus*, an astronomer, mathematician, and geometer of the School of Alexandria, was born in Samos, very close, to the east of Myconos, and lived in the beginning of the 3rd century B.C.(310-230 B.C.). Aristarchus was the introducer, crier, and supporter of the radical in his time heliocentric theory. In describing the differences between this theory and the generally accepted geocentric theory, we must take into account that the observed (from the Earth) motions of the planets are the consequence of the combination of the motions of the planets and of the Earth with respect to the Sun. Therefore, with respect to an observer on Earth, the observed (apparent) motion of a planet is periodically direct (namely, the planet is seen to move in the direction of the Earth's motion) and then for a while retrograde. For explaining the apparent motions of the planets, as well as those of the Sun and the Moon, the ancients invented the geocentric system of the world (or Ptolemaic system, after the name of *Claudius Ptolemy*). According to the geocentric system, the planets move around the Earth following the so- called epicycles. More precisely, the planet moves with constant speed around a small circle (the epicycle), whose center moves with constant speed around a larger circle (the deferent) centered on the stationary Earth. In this way, the ancients represented the apparent motions of a planet (both direct and retrograde) as the result of the simultaneous, uniform motion of the planet on at least two circles. This complex and marvellous system constituted the absolutely accepted and dominant system of the world's description. In reality, however the planets move (around the Sun) on elliptical orbits, so that the hypothesis of the uniform circular motions was not realistic. The observed differences, known to the ancients, remained unexplained. These differences are naturally explained as a physical reality, if, according the heliocentric system of Aristarchus, the Sun was put where the ancients believed the Earth was. Aristarchus composed many scientific works, most of which are lost. Especially his treatise entitled: *On the dimensions and distances of the Sun and the Moon* was based on the theoretical structure founded by Euclides and contained *eighteen geometrical "propositions" and six "hypotheses"*. Copies of some parts of the treatise of Aristarchus survived to date. This is due to the fact that, since antiquity, *Pappus* (3rd century A.D.) had included this treatise in a collection of works by *Appolonius, Archimedes, Autolycus, Euclides* et.al. in a volume with the title *Brief Astronomy*, to be distinguished from *The Maximal Astronomy* (or *Large Mathematical Syntaxis*, known as *Almagest*) of Claudius Ptolemy. Five copies of this manuscript are found today in the Apostolic Library in Vatican (with Codex Vaticanus Graecus No 204), and eight more copies are found in Paris (Paris, Gr.2348). The first typed version of the treatise of Aristarchus in Greek was edited in 1688 in Oxford by J.Wallis under the title *On the Dimensions and Distances of the Sun and the Moon*:

ΑΡΙΣΤΑΡΧΟΥ ΣΑΜΙΟΥ Περί Μεγεθών και Αποστημάτων Ηλίου και Σελήνης ΒΙΒΛΙΟΝ ΠΑΠΠΟΥ ΑΛΕΞΑΝΔΡΕΩΣ Του της Συναγωγής ΒΙΒΛΙΟΥ Β' Απόσπασμα

On the other hand, the treatise was published in 1913 A.D. as a book entitled *Aristarchus of Samos-The Ancient Copernicus*, with comments by Sir Thomas L. Heath. In the manuscript of the treatise, the Sun is not referred to as the center of the Solar System. The theory of Aristarchus about the Sun as the center of the Solar System was published in another work, which also is lost. However, the relative information is witnessed in an indisputable way, by other ancient authors. So, Archimedes, in his mathematical treatment *Psammites* writes:

Αρίσταρχος ο Σάμιος υποτίθεται γαρ τα μεν απλανέα των άστρων και τον Ἅλιον μένειν ακίνητον, ταν δε Γαν περιφέρεσθαι περί τον Ἅλιον κατά κύκλου περιφέρειαν, ος έστιν εν μέσω τω δρόμω κείμενος.

Also, Stovaeos, in his work *On Physics*, writes:

Αρίσταρχος τον Ήλιον ίστηση μετά των απλανών.

Finally, Plutarch, in his work *Peri Areskonton tis Filosofois*, writes:

Αρίσταρχος τον Ήλιον ίστησι μετά των απλανών, την δε Γην κινεί περί τον ηλιακόν κύκλον, εξελίττεσθαι δε κατά λοξού κύκλου την Γην, άμα δε και περί τον αυτής άξονα δινουμένην και κατά τας ταύτης εγκλίσεις σκιάζεσθαι τον δίσκον.

The general meaning of the above citations is that Aristarchus made the Sun standing among the stars and the Earth moving around the Sun in an elliptical orbit causing eclipses of the Sun. These really impressive citations do not cast doubt that the paternity of the heliocentric theory belongs to Aristarchus of Samos. But Aristarchus is known not only as the introducer of the heliocentric theory. Essentially, he is the "father" and founder of Astronomy, based on the logical reasoning, not on religious beliefs. He is the inventor of skafion, a spherical sundial of special form. With the aid of this sundial, he managed to determine the moment of the true noon in a place on Earth (and more generally to measure the time during a sunny day); also the geographic latitude of a place, the true value of the obliquity of the ecliptic, the daily declination of the Sun, and the exact dates of equinoxes and solstices of a specific year (281 B.C.). Along with Heraclides the Pontius, he is considered among the first, who explained the daily apparent axial rotation of the celestial sphere, considered as the result of the daily rotation of the Earth around an axis perpendicular to the Earth's equatorial plane. Also he explained the succession of the seasons, as the result of the inclination of the axis of rotation of the Earth with respect to the axis (perpendicular to the plane) of the ecliptic. He is the first Greek astronomer, who gave the most accurate value of the apparent diameters of the Sun and the Moon. It seems that he had a real sense of the really large distances of the stars, one of which, he believed, was the Sun itself. Finally, around 288 B.C., Aristarchus succeeded *Theophrastus* as the leader of the Peripatetic School, a post that he retained for eight years. From the book by Pappus we learn that Aristarchus invented a very remarkable method for determining the relative distances of the Sun, Moon, and Earth, as well as their relative dimensions. The way of the thinking of Aristarchus was based on the exact determination of the moments of the first and the third quarters of the phases of the Moon. From the difference between these two moments he determined the distance of Sun-Earth in units of the radius of the Moon's orbit. The result was twenty times smaller than the exact one, but since then it has been used for centuries. The above method of determination constitutes a really important contribution to Astronomy, and proves that Aristarchus had the ability of a geometrical viewing of the celestial phenomena. After the determination of the relative distances of the Sun, Moon, and Earth, Aristarchus invented an equally important method for determining the relative dimensions of these three bodies. The method was based on the determination, during an eclipse of the Moon, of the relative curvature of the shadow of the Earth on the Moon's surface and of the surface of the Moon. Using this result and the known (and approximately equal to each other) apparent diameters of the Sun and the Moon, Aristarchus determined the diameters of the Sun and the Moon, in units of the Earth's diameter. Most probably, the discovery that the real diameter of the Sun was twenty times the diameter of the Moon, in conjunction with the fact that the distance of the Sun from the Earth was twenty times larger than the distance of the Moon from the Earth, led Aristarchus to the conclusion that the Sun, not the Earth, is the center of the world. The proposition of the heliocentric theory reveals that Aristarchus could judge with clearness, and also interpret and explain correctly the observed celestial phenomena, without being affected by accepted for centuries, although incorrect conceptions and doctrines and beliefs of his contemporary scientists. On the other hand, the invention and use of skafion demonstrates that Aristarchus, could not only give successfully theoretical solutions to astronomical problems, but also invent and use the appropriate astronomical instruments. In other words, Aristarchus was also a skillful observer of the sky.

5. Restoration of the Historical Truth

After all the above, the conclusion is that Aristarchus was the first to introduce the correct and accepted today heliocentric theory, and that he founded Astronomy on the logical reasoning. This must be emphasized particularly, because part of the international community, astronomical or not astronomical, either justified due to ignorance, or even unjustified, does not share absolutely this point of view. Unfortunately for the heliocentric theory, strong supporters of the geocentric theory with proponent Pythagoras, also from Samos, were scientists of the authority of Aristotle, Hipparchus, Ptolemy and others. As a consequence, the revolutionary idea of Aristarchus could not be accepted. It fell in oblivion, but it was not forgotten, until the times of Renaissance, when two millennia later, in 1543 A.D., it was justified by the famous Polish astronomer *Nicolaus Copernicus*. Although Copernicus simply drew up the heliocentric theory from oblivion, repeating in this way the ideas of Aristarchus, he, namely, Copernicus is recognized as the introducer of the heliocentric theory, and the accepted heliocentric system is still named internationally "Copernican", not "Aristarchian", as it should be (the same is true for the so-called Copernican Principle of Cosmology). It must be emphasized that the survival of the heliocentric theory against the reaction by its opponents is to attributed much less to Copernicus, and mainly to the convincing arguments for its correctness given by Galileo, Kepler, Newton and others. Therefore the question arises, whether the work of Copernicus is original and what is its value. In order to answer this question responsibly, we must take into account the difficulties of the times of Copernicus, when the doctrines of Aristotle prevailed, and any disagreement with them was not allowed. In this sense, the contribution of Copernicus to the revival of the heliocentric theory is significant; but it does not suffice for attributing to Copernicus the paternity also of this theory. It is true that Copernicus was aware of the views of Aristarchus. This is verified by an extract of the manuscript of the treatise of Copernicus entitled *De Revolutionibus Orbium Coelestium*, which is still kept in the library of the University of Warsaw. In that extract, one can see delineated a paragraph referring to the treatise of Aristarchus; but this paragraph, quite strangely, has not been included in the printed version of the treatise of Copernicus edited in 1543A.D. In contemporary Greek, in translation by Professor S.Svolopoulos, the content of this paragraph is:

Αν και αναγνωρίζομεν ότι η πορεία του Ηλίου και της Σελήνης θα ήτο επίσης δυνατόν να εξηγηθή με την προϋπόθεσιν ότι η Γη είναι ακίνητος, τούτο είναι ολιγώτερον δυνατόν δια τους άλλους πλανήτας. Είναι πιθανόν ότι δι' αυτούς, ως και δι' άλλους λόγους, ο Φιλόλαος συνέλαβεν την ιδέαν της κινήσεως της Γης, η οποία, όπως μερικοί λέγουν, ήτο επίσης γνώμη του Αριστάρχου του Σαμίου και όχι δια τους λόγους, τους οποίοπυς αναφέρει ο Αριστοτέλης και τους απορρίπτει. Αλλά, αφού αυτά τα ζητήματα είναι τοιούτου είδους, δεν είναι δυνατόν να κατανοηθούν παρά μόνον από οξείς εγκεφάλους και κατόπιν μακράς προσπαθείας και κατ' εκείνους τους χρόνους παρέμενον μεταξύ των φιλοσόφων και δεν ήσαν ολίγοι εκείνοι, οι οποίοι κατενόησαν τον λόγον της κινήσεως των αστέρων, όπως μας πληροφορεί ο Πλάτων.Αλλ' εάν ο λόγος ήτο γνωστός εις τον Φιλόλαον ή εις κάποιον των Πυθαγορείων, είναι πιθανόν να μην ανεφέρθη εις τους νεωτέρους, δεδομένου ότι οι Πυθαγόρειοι δεν συνήθιζον να καταγράφουν τα ζητήματα που τους απησχόλουν.

The fact that the paragraph of the manuscript has not been included in the printed version could be characterized by some as plagiarism, while others consider that not referring especially to Aristarchus is lack of courage or fearfulness. It is, however, fair to be emphasized that it is not absolutely verified that the omission of the paragraph above must be attributed to Copernicus himself or to the editor of the book, because the book was published after (soon, however) the death of Copernicus. It is also remarkable that Copernicus for more than ten years did not give his consent for the publication of his treatise, because he was afraid his condemnation by the Roman Catholic Church. Finally, in 1540 A.D., Rheticus, an admirer of Copernicus and Professor of

Mathematics at the University of Wittenberg, managed in taking from Copernicus a copy of his manuscript, and with his consent, he published in Danzig a preliminary report on the ideas of Copernicus, under the title Narratio Prima. In 1542 A.D. Copernicus sent the preface of his compete manuscript to Rheticus, dedicating the book to Pope Paul the 3rd, in which he wrote: *I understand that as soon as some will be informed that in my book I attribute some motions to the Earth, they will cry that I and my theory must be rejected.* Also he explained that he had agreed on the edition of his book being encouraged by others, and that the reason for publishing the book was the uncertainty in the mathematical methods used for the determination of the motions of the celestial objects. Additionally, he appealed to the Pope for protecting him against the accusations of his libellers. Finally, in 1543 A.D., Rheticus, who had close relations with an editorial company in Nuremberg, managed in publishing a copy of the manuscript of Copernicus under the title *De Revolutionibus Orbium Coelestium.* It is real, but this monumental work was condemned in 1616 A.D. by the Roman-Catholic Church of Rome as heretic. The fact that Copernicus was very much afraid of the Roman Catholic Church becomes also transparent from the published book's preface written by Andreas Osiander, a well-meaning friend of Copernicus. He essentially wrote (I'm paraphrasing), *Dear reader, when you look at this book, it may appear that the author is saying that the Earth is not at the center of the universe. He doesn't really believe that. You see, this book is for mathematicians. If you wish to know where Jupiter will be two years from next Wednesday, you get an accurate answer by assuming that the Sun is at the center. But this is a mere mathematical fiction. It does not challenge our holy faith. Please, have no anxiety in reading this book.* This peculiar split-brain compromise between conventional wisdom and new ideas actually lasted for almost two centuries.

6. Conclusion and Perspectives

According to all the above, Copernicus is not the introducer but only the renovator of the heliocentric theory. The paternity of this theory, exclusively and originally, belongs to Aristarchus. One could only argue that the personal contribution of Copernicus is mainly that he introduced the geometric mechanism of Ptolemy's geocentric system to the heliocentric system of Aristarchus. But it is obvious that the whole effort was in a wrong way, because the real difficulty, namely, the faith tat the planets move uniformly on circular orbits, could not be overcome. For the restoration and forwarding of this truth, it is necessary to inform people, in the broader possible way, on the work of Aristarchus and more generally of the ancient Greek astronomers-mathematicians-philosophers. This is the purpose of the present lecture. At this point I wish to repeat what the late Zdenek Kopal, Professor at the University of Manchester, well-known admirer of the ancient Greek civilization and warm friend of Greece, wrote: **The first and essential steps that led to the definite formulation of a correct model of the solar system were made by Aristarchus of Samos in the 3rd century B.C. He declared the correct path, twenty centuries before this result of research become a permanent spiritual achievement of humanity. He lighted the first sparks of the divine fire that revealed our real position in space**.

I would like to finish this talk in a slightly different and unexpected way. After what has been said, it is commonplace today that the heliocentric system is the correct and generally accepted model of the solar system. However, it is true that the new scientific ideas and the progress in the understanding of the universe always had a large constant of time, namely, a large interval of time had to pass, before people realize and accept the reality in nature. Some simple examples are the following:

- *Two millennia ago, Eratosthenes proved and Aristotle accepted, that the Earth is spherical, but today we still have organized groups of supporters for the "Flat Earth".*
- *Aristarchus put the Sun at the center of the Solar System, but it took two and a half thousand years for humanity to accept it.*
- *Darwin explained the biological evolution about one hundred and fifty years ago, but most people today do not know even what he explained.*
- *Science today proves that the human being is a system of neurons, but it will take a long time until something like this will be accepted.*

Today we are all being taught that the Earth is not the center of the universe. However, I believe that there are many indications that we all are still concealed geocentricists under a heliocentric mantle. Actually we still say: The Sun rises or The Sun sets, as if, two and a half millennia after Aristarchus, our tongue pretends that the Earth does not move. But I am afraid that this is not simply pretence. Still more disappointing is currently the situation concerning our curious and subconscious, although unintended, persistence in anthropocentricism and geocentricism. It's a pity, but, unfortunately, a large percentage of the Earth's population does not know that the Earth revolves around the Sun in one year. Most of the inhabitants of this planet, deep in their hearts, consider that Earth is the center of the world. If you want a proof of this large ignorance of the society on astronomical matters, I can give you the following one: Once a Persian philosopher was asked: Which of the Sun and the Moon is more useful and important? His answer was: The Moon, of course, because during the night, when it is dark, the Moon sends us a little light, while during the daytime, when there is light, who needs the Sun? Is there any more evident proof of scientific ignorance? I can give you a few more, sad, in fact, examples of the really disappointing current level of the knowledge of even educated people, and of our persistence in the centrality of our position in the Universe: When about ten years ago the twenty four fragments of the comet Shoemaker-Levy (which had been assigned as a, b ,c,? etc.) were bombarding the planet Jupiter , the reader of a scientific journal wrote to the editor: *I can understand everything, everything, everything, except how is it possible that the fragments bombard Jupiter in alphabetical order.* When a student was asked who was Isaac Newton, he answered: *He was a famous British scientist, who verified the theories of relativity of Einstein.* A scientist was describing a total eclipse of the Sun and how it darkened during the day, and another one asked him: *But, why you scientists manage such ugly things happen ?* A group of people, charmed by the travels in space, where discussing that the time has come to visit the Sun as well. When one member of the group noticed that this is difficult because of the high temperature of the Sun, another member of the group, quite hurriedly and untormentedly, suggested that we go to the Sun during the night.

I really do not know whether with all these we must laugh or have tear drops in our eyes. I believe that a way out of this difficulty is that people must learn to appreciate and evaluate science as social reality. They should know the subject of science and how scientists reach their conclusions. The public accepts the useful products of science, and admires the impressive scientific discoveries and achievements, even if they do not understand how an airplane flies or why seasons occur. Therefore the public understands very little what science is, and mainly how science and technology contribute to our life, our ambitions, and our national targets. For this reason, the ignorance of the public loads with responsibility much more, perhaps, the scientists than the public itself. The public loves science. But, do scientists love the public? I think that we have to answer this question as a society. The humanity seems to walk on a thin crust of ice. In the beginning of the twenty first century the academic community faces difficulties. The confidence of the public to the university- level education continues to be shaken, the economic perspective for education is generally sad and dim, and, the worse, finally, the cultural level of the public is not improved. The twenty first century will be a decisive period that more or less will determine the future (happy or unhappy) of all the humanity. A realistic hope is based on education. If what we want is a peaceful and comfortable life for us and for future generations, then the public must learn to understand and evaluate the relative matters correctly. I think that there is no much time left for this. The president of the United States John F. Kennedy used to say the following story: A retired French general asked from his gardener to plant a tree immediately, and the gardener answered: *What's the hurry? This tree will grow very slowly, and for this it will take a hundred years*, whence the French general told him: *For exactly this reason, do not delay at all. Plant it right now.*

In conclusion, the necessity of education in the broader field of science looks like this story. We have to act immediately with an increased feeling of responsibility and with persistence, in order to impart the love of knowledge, informing, and teaching to everybody. If the present lecture indeed contributed a little towards the fulfilment of these objectives, this would constitute a real satisfaction to the speaker.

Acknowledgments

Dr Plionis, I thank you once more for your kind invitation and warm hospitality, and you all, dear colleagues, I thank you very much for your kind interest, attention, and patience. Thank you!!!

References

Abetti, G., 1954, The History of Astronomy, Sidgwick and Jackson, London.

Aristotle, 1960, The Physics, Vol. II, Chapter VIII, Loeb Classic Library, Harvard University Press.

Aristotle, 1986, On the Heavens, Book III, Chapter 6, 303-307 and 120, Loeb Classic Library, Harvard University Press.

Brecher, K., and Feirtag, M. (eds). 1980, Astronomy of the Ancients, The MIT Press, Cambridge, Massachusetts

Czartoryski, P., 1985, Nicholas Copernicus: Complete Works, Vol. III, 75-90, Polish Scientific Publishers, Warshaw-Cracow.

De Solla Price, D., 1997b, An Ancient Greek Computer, Scientific American (Special Issue: The Origins of Technology), 96-104.

Gingerich, O., 1985, Did Copernicus Owe a Debt to Aristarchus ?, Journal for the History of Astronomy, 16, 36-42.

Gingerich, O., 1992, From Aristarchus to Copernicus, in The Great Copernicus Chase and Other Adventures in Astronomical History, 63-68, Sky Publishing Corporation and Cambridge University Press.

Heath, Sir Thomas, 1912, The Works of Archimedes, Cambridge University Press, Cambridge, England, 1897, Reprinted by Dover, New York.

Heath, Sir Thomas, 1913, Aristarchus of Samos-The Ancient Copernicus, Constable, London, 1981, Reprinted by Dover Publcations, New York,

Heath, Sir Thomas, 1932, Greek Astronomy, Dent and Co., London, 1991, Dover Publications, Inc., New York.

King, H., 1957, The Background of Astronomy, Watts, London.

Kepler, J., 1952, Epitome of Copernican Astronomy, in Great Books of the Western World, University of Chicago Press, (Associate Editor: J. Adler, translated by C. G. Wallis).

Neugebauer, O., 1969, The Exact Sciences in Antiquity, Dover Publications, New York.

Neugebauer, O., 1975, A History of Ancient Mathematical Astronomy (Three Volumes), Springer Verlag, New York.

Pannekoek, A., 1961, A History of Astronomy, Ruskin House, London.

Paschos E. A., and Sotiroudis, P., 1998, The Schemata of the Stars: Byzantine Astronomy from A. D. 1350, World Scientific Publishing Company, Singapore. Pannekoek, A., 1961, A History of Astronomy, Ruskin House, London.

Rosen, E., 1973, Copernicus' Place in the History of Astronomy, Sky and Telescope, 45, Nr. 2, 72-75.

Rosen, E., 1959, Three Copernican Treatises, Dover Publications, New York.

Sagan, C, 1980, Cosmos, Wings Books, New York

Sagan, C., 1994, Pale Blue Dot: A Vision of Human Future in Space, Random House, New York

Sagan, C., 1997, The Age of Exploration, in Carl Sagan's Universe,Eds. Y.Terzian and E.Bilson, Cambridge University Press (In the same volume see also the contributions on the chapters Life in the Cosmos, Science Education, and Science, Environment and Public Policy)

Schwerdlow, N. M., and Neugebauer, 1984, O., Mathematical Astronomy in Copernicus' De Revolutionibus, Springer Verlag, New York.

Shipley, D. G. J., 1987, A History of Samos, 800-188 B. C., Clarendon Press, Oxford, (Reprinted 1998).

Spyrou, N., 1999, Aristarchus of Samos: Founder of Astronomy, Invited Talk, in Proceedings of the 4th Panhellenic Conference in Astronomy Pythagoreion, Samos, 16-18 September 1999) (Electronic Version) Ed. John Seimenis, Rhodes 2002

The Cambridge. Illustrated History of Ancient Greece, Cambridge University Press, 2003

Thurston, H., 1994, Early Astronomy, Springer Verlag, New York.

Toomer, G. J., 1984, Ptolemy's Almagest, Gerald Duckworth and Co. Ltd., London.

Van de Weygaert, R., 2002, Report to Anaximander: A Dialogue on the Origin of the Cosmos in the Cradle of Western Civilization, in Proceedings of "Modern Theoretical and Observational Cosmology", pp. 349-372, Eds. M. Plionis and S. Cotsakis, Kluwer Academic Publishers, The Netherlands.

LIST OF PARTICIPANTS

1	Akylas Thanassis (LOC)	NOA, Greece
2	Alexander David	Penn State University & Institute of Astronomy, USA
3	Allen Steve (*invited*)	IoA Cambridge, UK
4	Antonakis Manolis	UK
5	Arabadjis John	SMIT, USA
6	Aretxaga Itziar	INAOE, Mèxico
7	Arnouts Stephane	Lab. d'Astrophysique Obser. de Marseille, France
8	Babul Arif	University of Victoria, Canada
9	Baker Andrew	MPE Garching, Germany
10	Bamford Steven	University of Nottingham, UK
11	Bardelli Sandro	INAF - Oss. Astronomico di Bologna, Italy
12	Basilakos Spyros (LOC)	NOA, Greece
13	Belsole Elena	Department of Physics -University of Bristol, UK
14	Benoist Christophe	Nice Observatory, France
15	Blanchard Alain	LAOMP, France
16	Bludman Sidney	DESY-Theory Group/Univ. of Pennsylvania, USA
17	Bondi Marco	Istituto di Radioastronomia CNR, Italy
18	Bonometto Silvio	Universita' di Milano-Bicocca Italy
19	Borgani Stefano (*invited*)	Univ. of Trieste Italy
20	Bradac Marusa	IAEF Bonn, Germany
21	Braito Valentina	INAF-Osservatorio Astronomico di Brera, Italy
22	Branchesi Marica	IRA-CNR Italy
23	Branchini Enzo	Universita di Roma 3, Italy
24	Bridge Carrie	University of Toronto, Canada
25	Bunker Andrew	Institute of Astronomy, UK
26	Canavezes Alexandre	IoA, Cambridge, UK
27	Canizares Claude (SOC)	MIT, USA
28	Carangelo Nicoletta	IASF/CNR Sezione di Milano, Italy
29	Cassata Paolo	Dept. of Astronomy - Univ. of Padova, Italy
30	Castander Francisco Javier	IEEC/CSIC, Spain
31	Cattaneo Andrea	Institut d'Astrophysique de Paris, France
32	Chapin Edward	INAOE, Mèxico
33	Chartas George	Penn State University, USA
34	Chatzichristou Eleni	Yale University, USA
35	Ciardi Benedetta	Max-Planck-Institut für Astrophysik, Germany
36	Ciliegi Paolo	INAF - Osservatorio Astronomico di Bologna, Italy
37	Cimatti Andrea	INAF - Oss. di Arcetri, Italy
38	Cocchia Filomena	INAF-Osservatorio Astronomico Roma, Italy
39	Collister Adrian	IoA, Cambridge, UK
40	Connolly Andrew	University of Pittsburgh, USA
41	Conselice Christopher	Caltech, USA
42	Conway Edward	University of Nottingham, UK
43	Cox Thomas J.	UC Santa Cruz, USA

44	Da Angela Jose Antonio Cruz	Univ. of Durham, UK
45	De Grandi Sabrina	INAF-Oss Astronomico di Brera, Italy
46	De Lucia Gabriella	Max Planck Inst. für Astrophysik, Germany
47	D'Elia Valerio	Astronomical Observatory of Roma, Italy
48	Della Ceca Roberto	Osservatorio Astronomico di Brera, Italy
49	Dimopoulos Konstantinos	Lancaster University, UK
50	Dolag Klaus	Universitá di Padova, Italy
51	Domainko Wilfried	University of Innsbruck, Austria
52	Ebeling Harald	University of Hawaii, USA
53	Elbaz David	CEA Saclay, France
54	Ettori Stefano	ESO-Garching, Germany
55	Fabian Andy (SOC)	IoA Cambridge, UK
56	Fernández-Soto Alberto	Observatory, Univ. de Valencia, Spain
57	Finoguenov Alexis	MPE, Germany
58	Fiore Fabrizio	INAF - Osser. Astronomico di Roma, Italy
59	Floor Stephen	University of Kansas, USA
60	Foucaud Sebastien	IASF-Milano, Italy
61	Furusawa Hisanori	Subaru telescope, NAOJ, Japan
62	Gabasch Armin	Universitaetssternwarte Muenchen, Germany
63	Gaga Dora	NOA, Greece
64	Georgakakis Antonis (LOC)	NOA, Greece
65	Georgantopoulos Ioannis (LOC)	NOA, Greece
66	Giannakis Omiros (LOC)	NOA, Greece
67	Gilli Roberto	INAF - Osservatorio Astrofisico di Arcetri, Italy
68	Gioia Isabella Maria	IRA-Bologna, Italy
69	Girardi Marisa	Dept. of Astronomy - Univ. of Trieste, Italy
70	Gizani Nectaria	NOA, Greece
71	Gnedin Oleg	STScI, USA
72	Gorski Krzysztof (SOC)	JPL, USA
73	Goudis Christos	NOA & Univ. of Patras, Greece
74	Granato Gian Luigi	INAF - Oss. Astronomico di Padova, Italy
75	Griffiths Richard	Carnegie Mellon University, USA
76	Guetta Dafne	Osservatorio Astrofisico di Arcetri, Italy
77	Gutièrrez Carlos	IAC, Spain
78	Haardt Francesco	University of Como, Italy
79	Heinis Sebastien	Lab. d'Astrophysique de Marseille, France
80	Herve Bourdin	Observatoire de la Cote d'Azur, France
81	Hicks Amalia	University of Colorado - Boulder, USA
82	Hopkins Andrew	University of Pittsburgh, USA
83	Hoyos Carlos	Univ. Autónoma de Madrid, Spain
84	Hughes David (*invited*)	INAOE, Mèxico
85	Hughes John	Rutgers University, USA
86	Ishii Takako	Kwasan & Hida Obs., Kyoto, Japan
87	Ivezic Zeljko	Princeton Univ., USA
88	Jeltema Tesla	MIT, USA
89	Johansson Peter	IOA, University of Cambridge, UK
90	Kirpal Nandra	Imperial College London, UK
91	Kitsionas Spyros (LOC)	NOA, Greece

92	Knudsen Kirsten Kraiberg	Leiden Observatory, Holland
93	Koch Patrick	University of Zurich, Switherland
94	Kolokotronis Vaggelis (LOC)	NOA, Greece
95	Koopmans Leon	Caltech, USA
96	Kurk Jaron	Arcetri, Florence, Italy
97	La Franca Fabio	Universita‘ Roma Tre, Italy
98	Lamer George	Astrophysikalisches Institut Potsdam, Germany
99	Le Fevre Olivier (*invited*)	Lab. d'Astrophysique de Marseille, France
100	Lauger Sebastien	Laboratoire d'Astrophysique de Marseille, France
101	Lombardi Marco	IAEF, Bonn University, Germany
102	Lonsdale Carol (*invited*)	Caltech, USA
103	López-Cruz Omar	INAOE, Mèxico
104	Maccio' Andrea	Universita' di Milano-Bicocca, Italy
105	Magliocchetti Manuela	SISSA, Italy
106	Maier Christian	MP Institut für Astronomie Heidelberg, Germany
107	Maraschi Laura	Osservatorio di Brera, Italy
108	Markevitch Maxim	CfA, USA
109	Martin Crystal	University of California, S.Barbara, USA
110	Mason Brian	NRAO, USA
111	Mateos Silvia	Inst. de Física de Cantabria (CSIC-UC), Spain
112	Matthieu Tristram	LPSC, France ??
113	Menanteau Felipe	Jonhs Hopkins University, USA
114	Metcalf Ben	UC Santa Cruz, USA
115	Molendi Silvano	IASF/CNR Sezione di Milano, Italy
116	Molnar Sandor	Rutgers University, USA
117	Mueller Volker	Astrophysikalisches Institut Potsdam, Germany
118	Murante Giuseppe	Univ. of Torino, Italy
119	Ortiz Gil Amelia	Observatory, Univ. de València, Spain
120	Ouchi Masami	University of Tokyo, Japan
121	Outram Phil	University of Durham, UK
122	Panagia Nino	ESA - Space Telescope Science Institute, USA
123	Parmar Arvind (*invited*)	ESA/ESTEC
124	Peebles Jim (SOC)	Princeton Univ., USA
125	Pierpaoli Elena	Princeton Univ., USA
126	Plionis Manolis (LOC)	NOA, Greece
127	Pozzetti Lucia	INAF, Osservatorio Astronomico di Bologna, Italy
128	Prandoni Isabella	IRA-CNR, Italy
129	Pratt Gabriel	CEA/Saclay Service d'Astrophysique, France
130	Rauch Michael	Carnegie Observatories, USA
131	Reddy Naveen	Caltech, USA
132	Richard Johan	Observatoire Midi-Pyrenees, France
133	Ridgway Susan	Johns Hopkins University, USA
134	Rodighiero Giulia	Univ. of Padova, Italy
135	Romeo Alessio	Osservatorio Astrofisico Catania, Italy
136	Rosati Piero (SOC)	ESO - European Southern Observatory, Germany
137	Rowan-Robinson M. (SOC)	Imperial College, UK
138	Sadat Rachida	Observatoire de Midi-Pyrennees, France
139	Sanchez Sebastian	Astrophysikalisches Institut Potsdam, Germany

140	Santos Michael	Caltech, USA
141	Sawicki Marcin	Herzberg Institute of Astrophysics, Canada
142	Scannapieco Evan	Osservatorio Astrofisico di Arcetri, Italy
143	Schaerer Daniel	Geneva Observatory, Switherland
144	Schindler Sabine	University of Innsbruck, Austria
145	Scott Susie	IfA, University of Edinburgh, UK
146	Sekiguchi Kazuhiro	Subaru Telescope, NAOJ, Japan
147	Serjeant Stephen	University of Kent, UK
148	Shapley Alice	California Institute of Technology, USA
149	Silva Laura	INAF Italy
150	Smith Harding (Gene)	Univ. California San Diego, USA
151	Soechting Ilona K.	UCLan, UK
152	Sorrentino Gennaro	Dept. of Astronomy - Univ. of Bologna, Italy
153	Springel Volker (*invited*)	Max-Planck, Germany
154	Spyrou Nikolaos (*invited*)	Univ. of Thessaloniki, Greece
155	Sullivan Mark	University of Durham, UK
156	Sutherland Will	Royal Observatory, UK
157	Szalay Alexander	Johns Hopkins Univ., USA
158	Takamitsu Miyaji	Carnegie Mellon University, USA
159	Takata Tadafumi	Subaru Telescope, Japan
160	Takeuchi Tsutomu	T.National Astronomical Obs of Japan
161	Tasca Lidia	Max-Planck Inst.für Astrophysik, Germany
162	Thompson Rodger I.	Univ. of Arizona, USA
163	Tovmassian Hrant	INAOE, Mèxico
164	Tzanavaris Panayiotis	University of New South Wales, UK
165	Ueda Yoshihiro	Institute of Space and Astronautical Science, Japan
166	Vaccari Mattia	Univ. di Padova, Italy
167	Valdarnini Riccardo	SISSA, Italy
168	Van de Weygaert Rien	Kapteyn Institute, University of Groningen, Holland
169	Van Kampen Eelco	Inst. for Astrophysics, Innsbruck, Austria
170	Vanzella Eros	Dip. di Astronomia di Padova & ESO
171	Vauclair Sebastien	Osservatoire Midi-Pyrennees, France
172	Venturi Tiziana	Inst. di Radioastronomia, CNR, Italy
173	Vernet Joel	Osservatorio Astrofisico di Arcetri, Italy
174	Vikhlinin Alexey	CfA, USA
175	Volonteri Marta	UCSC, USA
176	Webb Tracy	Leiden Observtory, Holland
177	Wild Vivienne	Institute of Astronomy, Univ. of Cambridge, UK
178	Wolter Anna	INAF, Italy
179	Wright Ned (*invited*)	UCLA, USA
180	Xavier Dupac	ESA-ESTEC
181	Yepes Gustavo	Universidad Autónoma de Madrid, Spain
182	Zanichelli Alessandra	Istituto di Radioastronomia & CNR, Italy
183	Zimer Marc	MPE, Germany
184	Zucca Elena	INAF - Oss.Astronomico di Bologna, Italy

Astrophysics and Space Science Library

Volume 299: ***Open Issues in Local Star Fomation,*** edited by Jacques Lépine, Jane Gregorio-Hetem
Hardbound, ISBN 1-4020-1755-3, December 2003

Volume 298: ***Stellar Astrophysics - A Tribute to Helmut A. Abt,*** edited by K.S. Cheng, Kam Ching Leung, T.P. Li
Hardbound, ISBN 1-4020-1683-2, November 2003

Volume 297: ***Radiation Hazard in Space,*** by Leonty I. Miroshnichenko
Hardbound, ISBN 1-4020-1538-0, September 2003

Volume 296: ***Organizations and Strategies in Astronomy, volume 4,*** edited by André Heck
Hardbound, ISBN 1-4020-1526-7, October 2003

Volume 295: ***Integrable Problems of Celestial Mechanics in Spaces of Constant Curvature,*** by T.G. Vozmischeva
Hardbound, ISBN 1-4020-1521-6, October 2003

Volume 294: ***An Introduction to Plasma Astrophysics and Magnetohydrodynamics,*** by Marcel Goossens
Hardbound, ISBN 1-4020-1429-5, August 2003
Paperback, ISBN 1-4020-1433-3, August 2003

Volume 293: ***Physics of the Solar System,*** by Bruno Bertotti, Paolo Farinella, David Vokrouhlický
Hardbound, ISBN 1-4020-1428-7, August 2003
Paperback, ISBN 1-4020-1509-7, August 2003

Volume 292: ***Whatever Shines Should Be Observed,*** by Susan M.P. McKenna-Lawlor
Hardbound, ISBN 1-4020-1424-4, September 2003

Volume 291: ***Dynamical Systems and Cosmology,*** by Alan Coley
Hardbound, ISBN 1-4020-1403-1, November 2003

Volume 290: ***Astronomy Communication,*** edited by André Heck, Claus Madsen
Hardbound, ISBN 1-4020-1345-0, July 2003

Volume 287/8/9: ***The Future of Small Telescopes in the New Millennium,*** edited by Terry D. Oswalt
Hardbound Set only of 3 volumes, ISBN 1-4020-0951-8, July 2003

Volume 286: ***Searching the Heavens and the Earth: The History of Jesuit Observatories,*** by Agustín Udías
Hardbound, ISBN 1-4020-1189-X, October 2003

Volume 285: ***Information Handling in Astronomy - Historical Vistas,*** edited by André Heck
Hardbound, ISBN 1-4020-1178-4, March 2003

Volume 284: ***Light Pollution: The Global View,*** edited by Hugo E. Schwarz
Hardbound, ISBN 1-4020-1174-1, April 2003

Volume 283: ***Mass-Losing Pulsating Stars and Their Circumstellar Matter,*** edited by Y. Nakada, M. Honma, M. Seki
Hardbound, ISBN 1-4020-1162-8, March 2003

Volume 282: ***Radio Recombination Lines,*** by M.A. Gordon, R.L. Sorochenko
Hardbound, ISBN 1-4020-1016-8, November 2002

Volume 281: ***The IGM/Galaxy Connection,*** edited by Jessica L. Rosenberg, Mary E. Putman
Hardbound, ISBN 1-4020-1289-6, April 2003

Volume 280: ***Organizations and Strategies in Astronomy III,*** edited by André Heck
Hardbound, ISBN 1-4020-0812-0, September 2002

Volume 279: ***Plasma Astrophysics , Second Edition,*** by Arnold O. Benz
Hardbound, ISBN 1-4020-0695-0, July 2002

Volume 278: ***Exploring the Secrets of the Aurora,*** by Syun-Ichi Akasofu
Hardbound, ISBN 1-4020-0685-3, August 2002

Volume 277: ***The Sun and Space Weather,*** by Arnold Hanslmeier
Hardbound, ISBN 1-4020-0684-5, July 2002

Volume 276: ***Modern Theoretical and Observational Cosmology,*** edited by Manolis Plionis, Spiros Cotsakis
Hardbound, ISBN 1-4020-0808-2, September 2002

Volume 275: ***History of Oriental Astronomy,*** edited by S.M. Razaullah Ansari
Hardbound, ISBN 1-4020-0657-8, December 2002

Volume 274: ***New Quests in Stellar Astrophysics: The Link Between Stars and Cosmology,*** edited by Miguel Chávez, Alessandro Bressan, Alberto Buzzoni,Divakara Mayya
Hardbound, ISBN 1-4020-0644-6, June 2002

Volume 273: ***Lunar Gravimetry,*** by Rune Floberghagen
Hardbound, ISBN 1-4020-0544-X, May 2002

Volume 272:***Merging Processes in Galaxy Clusters,*** edited by L. Feretti, I.M. Gioia, G. Giovannini
Hardbound, ISBN 1-4020-0531-8, May 2002

Volume 271: ***Astronomy-inspired Atomic and Molecular Physics,*** by A.R.P. Rau
Hardbound, ISBN 1-4020-0467-2, March 2002

Volume 270: ***Dayside and Polar Cap Aurora,*** by Per Even Sandholt, Herbert C. Carlson, Alv Egeland
Hardbound, ISBN 1-4020-0447-8, July 2002

Volume 269: ***Mechanics of Turbulence of Multicomponent Gases,*** by Mikhail Ya. Marov, Aleksander V. Kolesnichenko
Hardbound, ISBN 1-4020-0103-7, December 2001

Volume 268: ***Multielement System Design in Astronomy and Radio Science,*** by Lazarus E. Kopilovich, Leonid G. Sodin
Hardbound, ISBN 1-4020-0069-3, November 2001

Volume 267: ***The Nature of Unidentified Galactic High-Energy Gamma-Ray Sources,*** edited by Alberto Carramiñana, Olaf Reimer, David J. Thompson
Hardbound, ISBN 1-4020-0010-3, October 2001

Volume 266: ***Organizations and Strategies in Astronomy II,*** edited by André Heck
Hardbound, ISBN 0-7923-7172-0, October 2001

Volume 265: ***Post-AGB Objects as a Phase of Stellar Evolution,*** edited by R. Szczerba, S.K. Górny
Hardbound, ISBN 0-7923-7145-3, July 2001

Volume 264: ***The Influence of Binaries on Stellar Population Studies,*** edited by Dany Vanbeveren
Hardbound, ISBN 0-7923-7104-6, July 2001

Volume 262: ***Whistler Phenomena - Short Impulse Propagation,*** by Csaba Ferencz, Orsolya E. Ferencz, Dániel Hamar, János Lichtenberger
Hardbound, ISBN 0-7923-6995-5, June 2001

Volume 261: ***Collisional Processes in the Solar System,*** edited by Mikhail Ya. Marov, Hans Rickman
Hardbound, ISBN 0-7923-6946-7, May 2001

Volume 260: ***Solar Cosmic Rays,*** by Leonty I. Miroshnichenko
Hardbound, ISBN 0-7923-6928-9, May 2001

Volume 259: ***The Dynamic Sun,*** edited by Arnold Hanslmeier, Mauro Messerotti, Astrid Veronig
Hardbound, ISBN 0-7923-6915-7, May 2001

Volume 258: ***Electrohydrodynamics in Dusty and Dirty Plasmas- Gravito-Electrodynamics and EHD,*** by Hiroshi Kikuchi
Hardbound, ISBN 0-7923-6822-3, June 2001

Volume 257: ***Stellar Pulsation - Nonlinear Studies,*** edited by Mine Takeuti, Dimitar D. Sasselov
Hardbound, ISBN 0-7923-6818-5, March 2001

Volume 256: ***Organizations and Strategies in Astronomy,*** edited by André Heck
Hardbound, ISBN 0-7923-6671-9, November 2000

Volume 255: ***The Evolution of the Milky Way- Stars versus Clusters,*** edited by Francesca Matteucci, Franco Giovannelli
Hardbound, ISBN 0-7923-6679-4, January 2001

Volume 254: ***Stellar Astrophysics,*** edited by K.S. Cheng, Hoi Fung Chau, Kwing Lam Chan, Kam Ching Leung
Hardbound, ISBN 0-7923-6659-X, November 2000

Volume 253: ***The Chemical Evolution of the Galaxy,*** by Francesca Matteucci
Paperback, ISBN 1-4020-1652-2, October 2003
Hardbound, ISBN 0-7923-6552-6, June 2001

Volume 252: ***Optical Detectors for Astronomy II***, edited by Paola Amico, James W. Beletic
Hardbound, ISBN 0-7923-6536-4, December 2000

Volume 251: ***Cosmic Plasma Physics***, by Boris V. Somov
Hardbound, ISBN 0-7923-6512-7, September 2000

Volume 250: ***Information Handling in Astronomy***, edited by André Heck
Hardbound, ISBN 0-7923-6494-5, October 2000

Volume 249: ***The Neutral Upper Atmosphere***, by S.N. Ghosh
Hardbound, ISBN 0-7923-6434-1, July 2002

Volume 247: ***Large Scale Structure Formation***, edited by Reza Mansouri, Robert Brandenberger
Hardbound, ISBN 0-7923-6411-2, August 2000

Volume 246: ***The Legacy of J.C. Kapteyn***, edited by Piet C. van der Kruit, Klaas van Berkel
Paperback, ISBN 1-4020-0374-9, November 2001
Hardbound, ISBN 0-7923-6393-0, August 2000

Volume 245: ***Waves in Dusty Space Plasmas***, by Frank Verheest
Paperback, ISBN 1-4020-0373-0, November 2001
Hardbound, ISBN 0-7923-6232-2, April 2000

Volume 244: ***The Universe***, edited by Naresh Dadhich, Ajit Kembhavi
Hardbound, ISBN 0-7923-6210-1, August 2000

Volume 243: ***Solar Polarization***, edited by K.N. Nagendra, Jan Olof Stenflo
Hardbound, ISBN 0-7923-5814-7, July 1999

Volume 242: ***Cosmic Perspectives in Space Physics***, by Sukumar Biswas
Hardbound, ISBN 0-7923-5813-9, June 2000

Missing volume numbers have not yet been published.
For further information about this book series we refer you to the following web site:
http://www.wkap.nl/prod/s/ASSL

To contact the Publishing Editor for new book proposals:
Dr. Harry (J.J.) Blom: harry.blom@wkap.nl

DURHAM UNIVERSITY LIBRARY
3 0104 01254602 8